Sustainable Development Goals Series

The **Sustainable Development Goals Series** is Springer Nature's inaugural cross-imprint book series that addresses and supports the United Nations' seventeen Sustainable Development Goals. The series fosters comprehensive research focused on these global targets and endeavours to address some of society's greatest grand challenges. The SDGs are inherently multidisciplinary, and they bring people working across different fields together and working towards a common goal. In this spirit, the Sustainable Development Goals series is the first at Springer Nature to publish books under both the Springer and Palgrave Macmillan imprints, bringing the strengths of our imprints together.

The Sustainable Development Goals Series is organized into eighteen subseries: one subseries based around each of the seventeen respective Sustainable Development Goals, and an eighteenth subseries, "Connecting the Goals", which serves as a home for volumes addressing multiple goals or studying the SDGs as a whole. Each subseries is guided by an expert Subseries Advisor with years or decades of experience studying and addressing core components of their respective Goal.

The SDG Series has a remit as broad as the SDGs themselves, and contributions are welcome from scientists, academics, policymakers, and researchers working in fields related to any of the seventeen goals. If you are interested in contributing a monograph or curated volume to the series, please contact the Publishers: Zachary Romano [Springer; zachary.romano@springer.com] and Rachael Ballard [Palgrave Macmillan; rachael.ballard@palgrave.com].

Dale Dominey-Howes
Ashleigh Rushton
William Leonard
Marcilyn Cianfarani
Lisa Overton • Haorui Wu
Editors

Queering Disasters, Climate Change and Humanitarian Crises

palgrave
macmillan

Editors
Dale Dominey-Howes
The Sydney Environment Institute and School of Geosciences
The University of Sydney
Sydney, NSW, Australia

William Leonard
Monash University
Accident Research Centre
Clayton, VIC, Australia

Lisa Overton
Middlesex University
London, UK

Ashleigh Rushton
Disaster and Emergency Planner
Emergency Services
Cleveland, UK

Marcilyn Cianfarani
Disaster and Emergency Management Practitioner
Ontario, ON, Canada

Haorui Wu
School of Social Work
Faculty of Health
Dalhousie University
Halifax, NS, Canada

ISSN 2523-3084 ISSN 2523-3092 (electronic)
Sustainable Development Goals Series
ISBN 978-981-96-3856-7 ISBN 978-981-96-3857-4 (eBook)
https://doi.org/10.1007/978-981-96-3857-4

This Palgrave Macmillan imprint is published by the registered company Springer Nature Singapore Pte Ltd.
The registered company address is: 152 Beach Road, #21-01/04 Gateway East, Singapore 189721, Singapore

We dedicate this collection to gender and sexual diverse individuals and communities affected by disasters, climate change, and humanitarian crises, as well as to the gender and sexual diverse practitioners and professionals in emergency management and humanitarian fields, whose unwavering dedication and tireless efforts make a profound impact.

Foreword

We are in a period of global uncertainty where the frequency, intensity and impacts of disasters are increasing. There is a growing urgency among all those working in disaster risk reduction to address the underlying causes of disasters, climate change and humanitarian crises and to mitigate and reduce their impacts. We also know that disasters disproportionality affect marginalised and minority communities, placing members of these communities at even greater risk at a time of escalating climate change and humanitarian crises.

There is an established body of research and policy exploring the impacts of disasters on the vulnerabilities, needs and capacities of different marginalised and at-risk groups including, among others, women and girls, ethnic and faith-based minorities, Indigenous and First Nations people, persons with disabilities, and refugees and asylum seekers. However, less is known about the risks and impacts of disasters, climate change and humanitarian crises on gender and sexual diverse people and communities.

This edited collection, *Queering disasters, climate change and humanitarian crises*, draws together an emerging body of disaster, climate change, and humanitarian research and policy looking at the experiences and service needs of gender and sexual diverse people and communities. It includes contributions from across the globe, from culturally and geographically diverse communities, and from disaster researchers, policy makers, community advocates and workers. The book proposes a novel approach to addressing the risks and impacts of disasters on gender and sexual diverse communities, providing both detailed empirical studies and first-hand accounts of gender and sexual diverse people's experiences as researchers and workers in the disaster and emergency sectors.

UNDRR is committed to an approach where no one is left behind. This commitment to inclusion is also a commitment to equality and justice. As part of Palgrave's series on the Sustainable Development Goals (SDGs), this collection provides both a challenge and opportunity for all of us involved in disaster risk reduction. It provides an evidence base on which to advocate for the inclusion of gender and sexual diversity in national and international disaster risk reduction policies. Whilst the collection showcases the impacts of systemic discrimination on gender and sexual diverse people, it also provides recommendations to assist governments, non-government agencies, emergency management and humanitarian organisations be more inclusive of gender and sexual diversity.

I commend this book to students and practitioners of disaster risk reduction, climate change adaptation and humanitarian crises work everywhere.

Kamal Kishore

Special Representative of the United Nations Secretary-General for Disaster Risk Reduction, and Head of UNDRR

Geneva, Switzerland
December 2024

Kamal Kishore

Preface

Queering Disasters, Climate Change and Humanitarian Crises foregrounds the complex but untheorised relationships across social inequality and systemic discrimination, and disasters, climate change and humanitarian crises. It introduces disaster justice as a framework that addresses this complex nest of relationships. The framework provides new ways of considering and addressing the relationships between human-led environmental degradation and the forces at work in systemic discrimination and marginalisation of vulnerable and minority groups, including gender and sexual diverse communities.

This book also highlights the need to address the impacts of colonialism and western hegemony on gender and sexual diverse people and communities in the Global South and Global North, while not losing sight of what many of these communities have in common—their shared but culturally specific experiences of heterosexist discrimination. The book offers a critique of international disaster risk and resilience (DRR) frameworks that do not adequately address the disaster experiences and service needs of gender and sexual diverse people, including *The Sendai Framework for Disaster Risk Reduction 2015–2030* and the United Nations Sustainable Development Goals (SDGs).

This collection combines research and policy with personal vignettes (short chapters), detailing sexual and gender diverse people's experiences of disasters and as workers in the disaster and emergency sectors. It opens a critical space to document and discuss how the disaster-related experiences and needs of gender and sexual diverse people and communities manifest and differ across national, cultural and regional boundaries. At the same time, the collection identifies areas in which these differences can be mobilised to further the development of just and equitable DRR frameworks and polices that include gender and sexual diverse people and communities.

Sydney, NSW, Australia — Dale Dominey-Howes
Cleveland, UK — Ashleigh Rushton
Clayton, VIC, Australia — William Leonard
Ontario, ON, Canada — Marcilyn Cianfarani
London, UK — Lisa Overton
Halifax, NS, Canada — Haorui Wu

Acknowledgements

Thank you to all the gender and sexual diverse people and communities whose voices, struggles and triumphs in the face of disaster are the subject matter and heart of this edited collection.

Huge gratitude goes to all the contributors for their excellent submissions and for their support and commitment to the project. Thanks to our expert external reviewers whose constructive feedback deepened the quality of the book.

Thanks to Jessica at Palgrave Macmillan for her warmth, guidance, and support in seeing this project to completion and to the other series editors for finding a place for this book in their Sustainable Development Goals (SDGs) series.

A great thanks goes to Kamal Kishore, Special Representative of the United Nations Secretary-General for Disaster Risk Reduction, and Head of the United Nations Office for Disaster Risk Reduction for acknowledging the importance of this book in promoting the inclusion of gender and sexual diversity in the ongoing development of national and international DRR frameworks and policies.

Finally, many thanks to Dale, who, as lead editor, has faced many challenges over the course of this project and who has been instrumental in its genesis and completion.

Contents

Notes on Contributors

Supriya Akerkar is a reader, Centre for Development and Emergency Practice, Oxford Brookes University, Oxford, United Kingdom.

Edward J. Alessi is an associate professor and MSW Director, School of Social Work, Rutgers, the State University of New Jersey, New Brunswick, NJ, USA.

Naro Alonzo is a Wellness Activist, KERI: Caring for Activists, Philippines.

Melanie M. Bowers is an assistant professor, Department of Political Science, Rutgers University, Camden, New Jersey, USA.

Clare E. B. Cannon is an associate professor, Department of Human Ecology, University of California; and Davis and Research Fellow, Department of Social Work, University of the Free State, Bloemfontein, South Africa.

Marcilyn Cianfarani is a queer disaster and emergency management practitioner with over 20 years of experience. She has an MA in Disaster and Emergency Management and has worked as a paramedic, a disaster responder, an educator and an emergency management consultant for government. Her research on Two Spirit, Lesbian, Gay, Bisexual, Trans, Queer, and Intersex people was one of the first of its kind in Canada. She currently works in an Emergency Management role for Health Canada. She enjoys walking through forests, swimming, advocating for accessibility and mapping precipitation in her spare time. She now lives in Ontario.

Neil J. W. Crawford is Lecturer in International Politics and Public Policy, Division of History, Heritage and Politics, University of Stirling, FK9 4LA, Scotland, UK

Andrew Dalton is a Chair and CEO of ReportOUT, c/o Pride Action North, Unit D, Old Casino, Forth Lane, Newcastle-upon-Tyne, Tyne and Wear, United Kingdom. NE1 5HX

Dale Dominey-Howes is Professor of Hazard and Disaster Risk Sciences at the University of Sydney, Australia. Dale, a Geographer with globally recognised expertise in risk science, utilises the risk assessment and management framework to tackle complex, trans-disciplinary risk problems at multiple scales. His work spans the earth and environmental sciences and the human and policy sciences domains of environmental sustainability, climate change, impact and adaptation, disaster risk reduction, and risk policy. Specific expertise includes natural hazards, disaster risk reduction, climate change impacts and adaptation, and the associated public policy, decision-making, and public affairs. He has experience related to specific sectors, economic and governance institutions (e.g. the emergency management, insurance, and tourism sectors) and in working with marginalised groups including gender and sexual diverse individuals, First Nations people, and people with disability. He is globally recognised as one of the leading advocates of the emerging field of 'queering disasters.' Dale has worked for the United Nations, the World Bank, insurance and reinsurance companies, State and Federal disaster agencies. He is Chairman of the United Nations UNESCO-IOC Post-disaster Policy and Protocols Working Group. He has produced 250+ publications of various types. Dale has received multiple awards including the 2024 Australia Scott Johnson Memorial Prize for services to Queers in STEMM; the 2019 Vice Chancellor's Award for Outstanding Service to People with Disability; the 2014 and 2016 Resilient Australia State and Federal government prizes for services to disaster management; and the 2015 Disabilities Industries Innovation Award for Community Leadership.

Louise Elstow Fynbos Consulting Limited and University of Cambridge

Rikky Muchammad Fajar is the coordinator of Sanggar Seroja and a freelance video journalist in Jakarta, Indonesia. Riyah.

Andrew Gorman-Murray is Professor of Geography School of Social Sciences, Western Sydney University Locked Bag 1797, Penrith, NSW, 2751, Australia

Lowine Stella Hill is a Lead Organizer and Facilitator Decolonising Methodologies for Sustainability Research, School of Environment Resources and Sustainability, University of Waterloo, ON, Canada

Matthew Hogan is a Resilience Officer, London Unit, London Fire Brigade.

Sarilee Kahn is an associate professor, School of Social Work, McGill University, Montreal, QC, Canada

Yarrow Koning is a Co-Director of *Can't Stop Change.*

Katherine Kovach is a Research Fellow, Department of Social Work, University of the Free State, Bloemfontein, South Africa.

Brigid Larkin is a senior lecturer, School of Medicine and Health Sciences, Edith Cowan University.

Jon Martin Larsen is a PhD candidate and Assistant Professor of Journalism, Kristiania University College, Norway.

Yong Gun Lee is an assistant professor, Department of Social Work and Social Administration, the University of Hong Kong, Pokfulam, Hong Kong.

William Leonard is an Adjunct Research Fellow at Monash University, Disaster and Resilience Initiative. He was Director GLHV (now Rainbow Health Australia) at La Trobe University; Manager Policy and Training, Gender and Disaster Australia; and Senior Lecturer in Sociology and Gender and Sexuality Studies. His areas of expertise include queer theory, LGBTIQ+ health, LGBTIQ+-inclusive professional practice and social policy. Liam has led over 20 state and national (Australian) LGBTIQ+ research projects including *Private Lives 2*, one of the world's largest surveys of LGBTQ+ physical and mental health and the development of the world's first LGBTIQ+ service accreditation programme, *the Rainbow Tick*. He was instrumental in Australian LGBTIQ+ legislative reform including the expungement of gay men's prior homosexual convictions and banning LGBTIQ+ conversion practices. Liam has developed and delivered LGBTIQ+ disaster training and inclusive services to government, academic and international agencies. He co-led research on LGBTQ+ refugees, lesbian women's disaster experiences, and the impacts of Covid-19 on LGBTIQ+ people's mental health and family violence. Liam has secured over $20 million in research grants and government contracts and published over 100 articles, book chapters, government reports and research monographs.

Teater Manekin is a community-based theater group based in Kampung Duri, West Jakarta. After a long hiatus, it was reestablished on April 22, 2024. It was originally founded in 2012 by two trans women, Mama Atha and Madam Seroja.

Jess Martínez is a Co-Director of *Can't Stop Change.*

Scott McKinnon is an Honorary Research Fellow, Department of Archaeology and History, La Trobe University, Melbourne Vic, Australia, 3086.

Katie McQuaid is an associate professor and UKRI Future Leaders Fellow, School of Geography, University of Leeds

Lisa Overton holds a PhD from Middlesex University on young women's experiences of gender, sexuality and youth growing up in post-disaster New Orleans, and two MScs—Social Science Research Methods from Middlesex University and Gender from The London School of Economics. She is Senior Lecturer in Politics at Middlesex University, where she has worked since 2013, Co-Chair of Ethics, and teaches Sustainable Development and Human Rights. Lisa's areas of expertise include disaster sociologies; lived experience of disasters; gender, sexuality and youth; queer feminist theories and methodologies; creative methods; and writing.

Vanessa Raditz is a PhD student and queer biocultural geographer, University of Georgia, Athens, Georgia, USA; founding member of the Queer Ecojustice Project and Co-Director of *Can't Stop Change*

Kate Roy is an Australian emergency management professional.

Ashleigh Rushton holds a PhD in Emergency Management from the Joint Centre for Disaster Research (JCDR), Massey University, and an MSc in International Disaster Management from the University of Manchester. She is currently working in emergency management for a local authority in the UK. Previously, she has held a Post-Doctoral Fellowship at the University of the Fraser Valley in British Columbia, Canada, and Research Officer position at the JCDR, Massey University, Aotearoa New Zealand. Ashleigh's areas of expertise include disaster risk reduction, gender and sexuality, human geography, resilience, climate change, geographies of health, and feminist and qualitative methodologies.

Borja Velez is a Research Affiliate, Centre for Development and Emergency Practice, Oxford Brookes University, Oxford, United Kingdom.

Natalia Villarán-Quiñones is a Co-Director of *Can't Stop Change.*

Cameron T. Whitley is an assistant professor, Department of Sociology, Western Washington University, Bellingham, WA, USA.

Firdhan Aria Wijaya is a lecturer at Universitas Pattimura, Indonesia, and artist.

Haorui Wu is the Canada Research Chair in Resilience, the Building A Culture of Resilience (BACOR) lab Director, and an associate professor in the School of Social Work at Dalhousie University. With an interdisciplinary background (architecture, landscape architecture, regional and community planning, and social work), his community-based interdisciplinary research and emerging practice have nuancedly explored human-animal-environment interplay through the lens of environmental justice, social justice, and health justice in the global context of climate change, disaster, and wilful acts of violence. His 18-year practice/policy-oriented research strategy, operating across micro-, mezzo-, and macro-levels, has been primarily focusing on building resilience for vulnerable and structurally marginalised communities (particularly Indigenous, youth and older adults, African diaspora, 2SLGBTQIA+, immigrants and refugees, people with (dis)Abilities, and homeless populations). At the micro-level, he developed an innovative social-health protection approach that examines and addresses social and health disparities in vulnerability to global climate change and disaster. At the mezzo-level, he had nurtured partnerships with community-based service sectors, effectively bridging the gaps between micro-level grassroots engagement and macro-level policy improvement. His macro-level interventions improve the social and health impacts through climate change and public health policies at the regional, national, and global levels. He has secured over CA$10M in research grants, fostering collaborations across public, private, and not-for-profit sectors from over 30 countries and publishing over 300 publications of various types. He received the 2024 Dalhousie University President's Research Excellence Award and the Delegate for Science Meets Parliament in Canada in 2023.

List of Figures

List of Tables

CHAPTER 1

Queering Disasters, Climate Change and Humanitarian Crises: An Introduction

Lisa Overton, Marcilyn Cianfarani, Ashleigh Rushton, Haorui Wu, Dale Dominey-Howes, and William Leonard

Background

There has been a dramatic shift in our understanding and response to disasters, climate change and humanitarian crises. Until the mid-1970s, disasters were seen as natural events, 'acts of God' whose causes were beyond human control. Disaster planning, response and recovery centred on improvements to systems and infrastructure aimed at minimising the

L. Overton (✉)
Middlesex University, London, UK
e-mail: L.A.Overton@mdx.ac.uk

M. Cianfarani
Disaster and Emergency Management Practitioner, Ontario, ON, Canada

A. Rushton
Disaster and Emergency Planner, Emergency Services, Cleveland, UK

D. Dominey-Howes et al. (eds.), *Queering Disasters, Climate Change and Humanitarian Crises*, Sustainable Development Goals Series, https://doi.org/10.1007/978-981-96-3857-4_1

impacts of disasters rather than addressing their underlying causes. White (1973) was one of the first researchers to question this disciplinary orthodoxy. However, the most significant shift in thinking was led by Blakie and colleagues in the 1990s. Blakie, building on White's insights, argued that while disasters involve natural forces their impacts and causes are related to systemic inequalities and vary according to the unequal distribution and access to material resources and social capital (Blaikie et al., 1994).

This shift from a naturalistic account to one that highlights the role of human agency and social inequality in the causes and impacts of disasters has had a profound effect on research, policy and practice into the twenty-first century. It has led to an urgency among researchers and policy makers to tackle the common drivers of catastrophic events, including disasters, climate change and humanitarian crises (Soden & Owen, 2021). This includes research on the links among regional socio-political conflicts, disasters and humanitarian crises (Burkle et al., 2014; Archer & Dodman, 2017) and policy documenting the cumulative impacts of unregulated human activity on the environment (United Nations Intergovernmental Panel on Climate Change, 2019).

This paradigmatic shift has also fuelled interest in the differential impacts of disasters, climate change and humanitarian crises on vulnerable and marginalised populations. This has included disaster research and policy on 'at risk' groups including women and girls, ethnic and faith-based minorities, Indigenous populations, people with disabilities, refugees and migrants, and the homeless (Begum, 1993; Krishnaraj et al., 1997; Enarson et al., 2007; Marable, 2006, 2008; Lambert, 2014; Phibbs et al., 2014; Chandran et al., 2015; Drolet et al., 2015; Good et al., 2016; Gray, 2017; Hou & Wu, 2020; Wu & Karabanow, 2020; Kailes & Lollar, 2021; Calgaro et al., 2021).

H. Wu
School of Social Work, Faculty of Health, Dalhousie University, Halifax, NS, Canada

D. Dominey-Howes
The Sydney Environment Institute and School of Geosciences, The University of Sydney, Sydney, NSW, Australia

W. Leonard
Monash University, Accident Research Centre, Clayton, VIC, Australia

More recently, there has been a push to consider how the drivers of disasters, climate change and humanitarian crises may be linked to those that underpin discrimination and violence against vulnerable and marginalised populations. Ecofeminists, for example, argue that patriarchy underpins both the subordination of women and the exploitation of the natural environment (Gaard, 2017; Parkin-Albayrakoglu, 2022). Similarly, some proponents of the Anthropocene argue that capitalism's emphasis on profit and material wealth has relied on a disregard for the environment and the exploitation of minority and marginalised population's lands and labour (Bradshaw, 2013; Moore, 2016; Saldanha, 2020).

However, what is absent from these accounts is the role that heterosexist discrimination and inequality play in disasters, climate change and humanitarian crises and their impacts on gender and sexual diverse people's disaster experiences (Dominey-Howes et al., 2014, 2016; Gaillard et al., 2017b; Wisner et al., 2017; Rushton et al., 2020; Edge Effect, 2021). Gender and sexual diverse minorities are rarely included in data collection on the impacts of disasters (Larkin, 2019). Nor are they included in the lists of minority and marginalised populations that automatically appear in international inclusive disaster planning and management policies and programmes (Dominey-Howes et al., 2016; Gaillard et al., 2017a; Rushton et al., 2019; Carter, 2021; UN OHCHR, 2021).

Knowledge of the experiences, vulnerabilities and needs of gender and sexual diverse people and communities in disasters and how they vary within and between geographic locations and cultural contexts lags far behind that of women, minority and other marginalised populations. Active discrimination against gender and sexual diverse people combined with their absence from planning, response and recovery increases their disaster risk and impacts, and limits their resilience and capacity to recover from future disasters (Larkin, 2019).

Rationale and Aims

This collection addresses these gaps in our knowledge and practice. It includes contributions from the Global North and Global South, and from culturally diverse communities. It draws together people working in and across different sectors, including university and research institutions, government and non-government agencies, emergency management, and community-based and humanitarian organisations. It includes contributions from gender and sexual diverse people with lived experience of disasters and people who have been subject to heterosexist discrimination as service recipients, employees or volunteers.

This collection aims to make a significant contribution to the development of a new field of inquiry, ***queer disaster studies***. The work undertaken on gender and sexual diverse people and disasters, climate change and humanitarian crises is limited in scope and involves small-scale projects, dotted across multiple locations, sectors and disciplines. The published work lacks an overarching conceptual framework and shared terms. The current literature has relied, almost exclusively, on western, scientific taxonomies and concepts reflected in the dominance of acronyms such as LGBTIQA+ (lesbian, gay, bisexual, transgender, intersex, queer, asexual and other gender and sexual diverse identities). It has drawn from studies of the experiences of other vulnerable, minority and marginalised populations that may or may not be applicable to the varied situations of gender and sexual diverse people and communities. This collection aims to broaden the scope of queer disaster studies by considering how research methodologies, disaster risk reduction (DRR) policies and frameworks, and dominant 'disaster terminology' can be tailored to better meet the needs of all gender and sexual diverse communities.

To promote more inclusive DRR policies and frameworks, the collection contributes to what Dalton refers to as 'queering the SDGs' (this collection, Chap. 21). The collection is part of Palgrave Macmillan's series on the United Nation's Sustainable Development Goals (SDGs). The series aims to assist governments and key national and international agencies meet their targets under each of the 17 SDGs outlined in the Sendai Framework for Disaster Risk Reduction (2015–2030). This collection raises concerns about the degree to which the SDGs acknowledge the specific challenges facing gender and sexual diverse people and communities. While many of the SDGs address diversity and inclusion, there are no detailed references to gender and sexual diversity (Dalton, 2023). The United Nations Office for Disaster Risk Reduction's (UNDRR) midterm review of the Sendai Framework was critical of the lack of disaggregated data yet failed to include a detailed discussion of the inclusion of data on gender and sexual diverse people and communities (UNDRR, 2023). The midterm report only mentions 'sexual minorities' once as a 'group at higher risk' of disaster (p. 60).

The collection also raises concerns about the inclusion of gender and sexual diversity under the umbrella of gender. The Sendai Framework's use of gender is indicative of how gender is understood in many key national and international DRR policies. SDG-5 addresses gender directly and aims to 'achieve gender equality and empower all women and girls' (United Nations for Disaster Risk Reduction, 2023; United Nations SDGs, 2023). Yet this definition relies on a binary, scientific model that can exclude Indigenous, non-western gender and sexual diverse people, and non-heteronormative gender identities and forms of expression.

The exclusion of 'gender and sexual diversity' from inclusive international DRR frameworks or its disappearance in generic diversity and gender statements fails to address what is particular to gender and sexual diverse people's lives. The failure to acknowledge and address the needs of gender and sexual diverse people in these frameworks is symptomatic of broader, heterosexist discrimination and touches on many of the issues raised in this collection.

Outline of the Book

Organisation

This chapter combines research and policy chapters with shorter chapters documenting individuals' personal experiences of disasters and emergency services. The weaving together of research and personal accounts is central to the novelty, depth and impact of the collection. The research and policy chapters provide a broader context in which to locate and respond to the complexities and uniqueness of individual experience captured in the chapters that rely on first-person narrative or story telling.

Several chapters present the findings of local and national research on the impacts of disasters on gender and sexual diverse people and communities. Some of these rely on previously unpublished data while others report on the impacts of recent events, including COVID-19. The remaining chapters offer critical reviews of the absence of gender and sexual diversity in current national and international disaster and emergency management policies, programmes and services and how these absences can be addressed. What is clear is that gender and sexual diverse people's experiences have changed very little since the first studies conducted in the late 2000s nor have there been significant improvements in their access to inclusive and appropriate disaster and emergency relief services.

The first-person chapters include a diversity of modes of presentation, from storytelling (Chaps. 5 & 22), prose-poetry (Chap. 2) and interviews (Chap. 14), to autoethnographic accounts of the deeply traumatic experiences of queer individuals as disaster and emergency services' employees (Chaps. 7, 8, 13, 17 & 19). These more personal accounts and imaginative forms of engagement can be thought of as vignettes, defined by Merriam-Webster as a 'short, literary sketch'. Razack (1998, p. 36) argues that the vignette, as a 'mode of storytelling', stands in opposition to established knowledges. For many sexual and gender diverse people, how they want their disaster experiences told and documented are at odds with academic protocols and conventions. For some, these 'truth-telling' and evidentiary conventions are part of the same western, neo-colonial machinery that has silenced the voices of non-western and Indigenous communities. For others, these protocols marginalise or exclude a detailed telling of the complexities of the systemic discrimination they have experienced, often daily, as employees in the disaster and emergency sectors. This collection gestures to new ways of presenting the experiences of marginalised communities that rely on the blending of academic, creative and personal narrative, one that challenges an assumed opposition between storytelling and established knowledges.

Themes and Overview of Chapters and Vignettes

The book is divided into five discreet but overlapping sections, each dealing, in different ways, with the collection's overall themes. These are acknowledging and responding to gender and sexual diverse peoples' lived experiences of disasters, climate change and humanitarian crises; developing research, policy and practice that acknowledge how these experiences differ within and across different geographic locations and cultural contexts; and the development of culturally sensitive gender and sexual diverse inclusive disaster policies, programmes and services.

Part I, 'Queer(y)ing disaster studies', considers the need for the emerging field of queer disaster studies to develop its own conceptual framework and terminology. Each research and policy chapter and shorter, first-person vignette explores the impacts of systemic, heterosexist discrimination on gender and sexual diverse people and communities' experiences of disasters, climate change and humanitarian crises in different locations and cultural contexts. The part presents an argument for critical engagement with current terminology and concepts in the field of disaster research and policy, and the development of new theoretical frameworks better able to understand and address the drivers of gender and sexual diverse communities increased vulnerabilities to disasters and their impacts.

This part is bookended by two vignettes (Chaps. 2 & 5). The first is a prose poem by Naro Alonzo, documenting their observations as a volunteer in crisis relief centres in the Philippines between 2015 and 2018. The second is by Firdhan Aria Wijaya, Rikky Muchammad Faja and Teater Manekin and presents stories from members of the Sannggar Seroja collective detailing the impacts of systemic discrimination on waria/transpuan in Indonesia. Chapters 3 & 4 sit between the vignettes. Chapter 3 by William Leonard, Dale Dominey-Howes, Ashleigh Rushton, Marcilyn Cianfarani, Lisa Overton and Haorui Wu. It presents a brief history of the different strands of research, policy and practice that have informed the emergence of 'queer disaster studies' and argues that disaster justice provides a promising framework to guide the ongoing development of this emerging field. Chapter 4 by Cameron Whitely and Melanie Bowers looks at how the combination of social and political marginalisation makes LGBTQ+ people and communities particularly vulnerable to climate change in the United States of America (USA) and how queer political culture (QPC) is being used to advocate for change.

Part II, 'Disaster justice and gender and sexual diversity', takes up Chap. 3's challenge. It presents disaster justice as a promising framework for addressing how the intersection of heterosexist and other types of systemic discrimination increase the risks and impacts of disaster for gender and sexual diverse minorities. Disaster justice also allows us to consider how the processes underpinning social inequality and the marginalisation of minority populations are linked to those driving environmental damage and degradation globally.

This part opens with a chapter by Scott McKinnon, Dale Dominey-Howes and Andrew Gorman-Murray. It explores how mobility figured in people's recovery processes in Christchurch, New Zealand, following the 2011 earthquake and how disaster justice provides a framework for responding to the mobility barriers they faced. This chapter is followed by two vignettes, each describing how, in very different contexts, gender and sexual diverse people can be made to feel unwelcome and *unhomed*. Kate Roy's vignette (Chap. 7), takes up their personal experiences of working in emergency management in Australia while Jon Martin Larsen tells the stories of two gender and sexual diverse refugees fleeing homophobic and transphobic persecution in Iran (Chap. 8).

Chapter 9, by Edward Alessie, Yong Gun Lee and Sarilee Kahn, reflects on the findings of a 2017 study involving LGBTQ+ refugees from the Middle East, North Africa and South Asia who fled to Austria and the

Netherlands. The authors argue that disaster justice provides an effective framework for understanding and responding to the rapid increase in humanitarian crises and their impacts on LGBTQ+ refugees. This part's final chapter by Katie McQuaid and Neil Crawford examines how the everyday exclusions and discrimination experienced by gender and sexual diverse people intersect with urban environmental hazards in the context of global climate change.

Part III, 'Sins of omission: Procedural vulnerabilities and the (re)marginalisation of queer people in disaster response and recovery', looks at the indirect consequences of historical power imbalances and discrimination against gender and sexual diverse people. Disaster justice provides a way of understanding how the absence of any detailed consideration of the needs of gender and sexual diverse people constitutes a type of *indirect discrimination*, a 'procedural vulnerability' that reduces their capacity to plan for, respond to and recover from disaster.

Borja Velez and Supriya Akerkar's opening chapter presents an intersectional analysis of gender and sexual diverse individuals and communities experiences of the 2010 Chilean earthquake and tsunami. Clare Cannon and Katherine Kovach's chapter addresses the lack of research and data on how disasters impact rates of intimate partner violence (IPV) within sex and gender diverse minority (SGM) communities in the USA. This part concludes with a personal and unsettling vignette by Marcilyn Cianfarani. Cianfarani ties her past research on 2SLGBTQI peoples' experiences of Canadian disaster and emergency management to the violence she experienced from right-wing extremists and the lack of protection and support from law enforcement personnel when reporting incidents of harassment and abuse in her local community in Atlantic Canada.

Part IV, 'Doing it for themselves: Queering disaster scholarship, media representations and allyship', moves from a critique of the heterosexist conceits of current disaster research, policies and media reporting to positive examples of queer people 'doing it for themselves'.

In the opening vignette (Chap. 14), Vanessa Raditz discusses with their co-directors, Yarrow Koning, Natalia Villarán-Quiñones and Jess Martínez, the making of their documentary film *Can't Stop Change: Queer Climate Stories from the Florida Frontline*. They discuss how the film's production and content foregrounded queer climate change. In Chap. 15, Ashleigh Rushton critiques the heterosexist and cis-normative biases of current, dominant disaster research frameworks, methodologies and practices. She offers a guide on the ethical and methodological considerations driving the inclusion of gender and sexual diverse voices in all aspects of disaster research and practice. Vanessa Raditz' chapter critically engages with queer theory to broaden the dialogue within disaster justice on how LGBTQ+ people and communities are represented in media reports of disasters and their impacts. Dale Dominey-Howes' vignette (Chap. 17) focuses on the public abuse that queer researchers face in conducting research on queer people and disasters and the need for training and support for those undertaking this work. Louise Elstow and Matthew Hogan's concluding chapter provides one of the first reviews of how current UK emergency management policies and practices address gender and sexual diversity.

Part V, 'No longer an optional extra: Queer-inclusive disaster policies and services', extends the argument for the development and implementation of queer-led, inclusive disaster and emergency management policies outlined in Part IV.

This part opens with Brigid Larkin's vignette (Chap. 19) in which they reflect on their PhD journey and the strengths, limitations and costs to queer people of undertaking research that overlaps with their own lived experience. Lowine Hill's chapter explores if and how disaster risk governance processes in the Commonwealth of Dominica address the lived experiences of gender and sexual minorities (GSM). Andrew Dalton's concluding chapter offers a critical analysis of the heteronormative and cisnormative framing of the United Nations Sustainable Development Goals (SDGs) and identifies opportunities for gender and sexual diverse organisations to 'queer' the SDGs. The final vignette by Lisa Overton (Chap. 22) is a fictionalised account of how the pressures young queer people experience in their everyday lives are magnified under disaster conditions.

What Next? Queering the Way Forward

The contributions to *Queering Disasters, Climate Change and Humanitarian Crises* detail many of the issues facing gender and sexual diverse people. They range from how the interactions of heterosexist and other forms of systemic, identity-based discrimination increase gender and sexual diverse people's disaster risk and impacts, to how their exclusion from DRR policies reduces their capacity to withstand and recover from disasters, climate change and humanitarian crises. They also present examples of positive change, from grass roots queer disaster activism to governments and DRR agencies working with gender and sexual diverse communities to develop diverse, inclusive policies and services.

The collection includes submissions from all over the world. Nonetheless, it remains 'top heavy' with an overrepresentation of material from the Global North. There is an urgent need for more contributions from gender and sexual diverse people living and working in the Global South and from First Nations and Indigenous people and communities. The collection also lacks contributions from minorities that don't always 'fit' within the dominant categories that different gender and sexual diverse communities use to define themselves, including people with variations in sex characteristics. If the emerging field of queer disaster studies is to realise its commitment to diversity, inclusion and justice, then it needs to address the barriers that prevent different gender and sexual diverse people and communities from participating in and leading the development of research, policies and programmes. In the absence of such strategies and supports, it will remain top heavy, perpetuating many of the neo-colonial and heterosexist inequalities it critiques.

The collection also provides accounts of the pressures gender and sexual diverse people face in undertaking research on communities they identity with or being their authentic selves as employees in disaster and emergency management and services. More work is needed on how queer disaster research has been targeted in the escalation of culture wars and how this impacts the mental health and well-being of those who undertake this work. Similarly, we need to know more about the experiences of gender and sexual diverse people who work in disaster and emergency services. The contributions that tackle these issues highlight the hyper-masculinist and heterosexist cultures that still dominate this sector. They detail the personal and professional costs to gender and sexual diverse employees who advocate for cultural and organisational reform and those who mask their authentic selves at work for fear of discrimination and abuse.

It is no coincidence that the vignettes are home to these deeply personal accounts. This collection is unique in combining academic research and public policy with individuals' stories of their experiences as clients and employees in the disaster and emergency sectors and as researchers pushing against the constraints of dominant research methodologies and academic modes of 'self-presentation'. The play between academic research and the vignettes suggests new ways of doing DRR policy that gives voice to and addresses the complexities of queer people's lives.

The collection poses disaster justice as one possible framework for addressing both the causes and disproportionate impacts of disasters, climate change and humanitarian crises on gender and sexual diverse people and communities. More work is needed to see if disaster justice can provide a conceptual, ethical and pragmatic framework capable of drawing together past and current work and guiding the development of new collaborative queer disaster research, policy and programmes. This will involve broadening queer disasters disciplinary base and coalitions, fostering research and policy in law, political sciences, history, development studies and more.

Finally, the collection calls on the UN and other national and international agencies that are committed to diverse inclusive DRR policies and practices to honour that commitment. This involves future strategic visions and the successor to the Sendai Framework acknowledging and addressing, in detail, the causes and disproportionate impacts of disasters, climate change and humanitarian crises on gender and sexual diverse people and communities, globally.

As we approach 2030, we hope this collection provides frameworks, recommendations and tools to assist in the development of more inclusive, people-centred approaches to disaster research, policy and services where reality and rhetoric are no longer at odds, where 'no-one is left behind'.

Conflict of Interest None of the authors have any conflict of interests.

References

Archer, D., & Dodman, D. (2017). Editorial: The urbanization of humanitarian crises. *Environment and Urbanization, 29*(2), 339–348. https://doi.org/10.1177/0956247817722731

Begum, H. (1993). Family planning and social position of women. *Bioethics, 7*(2/3), 218–223.

Blaikie, P., Cannon, T., Davies, I., & Wisner, B. (1994). *At risk: Natural hazards, people's vulnerability, and disasters.* Routledge.

Bradshaw, D. (2013). The *Logoi* of beings in Greek patristic thought. In J. Chryssavgis & B. V. Foltz (Eds.), *Toward an ecology of transfiguration* (pp. 9–22). Fordham University Press.

Burkle, F. M., Greenough, P. G., & Martone, G. (2014). The changing face of humanitarian crises. *Brown Journal of World Affairs, 20*(2), 19–36.

Calgaro, E., et al. (2021). Silent no more: Identifying and breaking through the barriers that d/deaf people face in responding to hazards and disasters. *International Journal of Disaster Risk Reduction, 57*, 102156. https://doi.org/10.1016/j.ijdrr.2021.102156

Carter, B. (2021). *Impact of social inequalities and discrimination on vulnerability to crises. K4D Helpdesk Report 994.* Institute of Development Studies. https://doi.org/10.19088/K4D.2021.049

Chandran, D., et al. (2015). Psychosocial impact of the Uttarakhand flood disaster on elderly survivors. *Indian Journal of Gerontology, 29*(1), 62–76.

Dalton, D. (2023). Have we left behind the rainbow warriors? The climate emergency and its impact on global queer people and their communities. In D. Dalton & A. Smith (Eds.), *Gender, sexuality, and the UN'S SDGs* (pp. 17–40). Palgrave Macmillan.

Dominey-Howes, D., Gorman-Murray, A., & McKinnon, S. (2014). Queering disasters: On the need to account for LGBTI experiences in natural disaster contexts. *Gender, Place and Culture, 21*(7), 905–918. https://doi.org/10.1080/0966369x.2013.802673

Dominey-Howes, D., et al. (2016). Emergency management response and recovery plans in relation to gender and sexual minorities in New South Wales, Australia. *International Journal of Disaster Risk Reduction, 16*, 1–11.

Drolet, J., et al. (2015). Women rebuilding lives post-disaster: Innovative community practices for building resilience and promoting sustainable development. *Gender and Development, 23*(3), 433–448. https://doi.org/10.1080/13552074.2015.1096040

Edge Effect. (2021). *The only way is up: Monitoring and encouraging diverse SOGIESC inclusion in the humanitarian and DRR sectors.* Empower - Women for Climate-Resilient societies programme. Available at: https://www.edgeeffect.org/wp-content/uploads/2021/03/TheOnlyWayIsUp_Web.pdf

Enarson, E., Fothergill, A., & Peek, L. (2007). Gender and disaster: Foundations and directions. In H. Rodriquez, E. L. Quarantelli, & R. R. Dynes (Eds.), *Handbook of disaster research* (pp. 130–146). Springer.

Gaard, G. (2017). Feminism and environmental justice. In R. Holifield & J. Chakraborty (Eds.), *Handbook of environmental justice* (pp. 74–88). Routledge.

Gaillard, J. C., Gorman-Murray, A., & Fordham, M. (2017a). Gender and sexual minorities in disaster. *Gender, Place and Culture, 24*(1), 18–26. https://doi.org/10.1080/0966369X.2016.1263438

Gaillard, J. C., et al. (2017b). Beyond men and women: A critical perspective on gender and disaster. *Disasters, 41*(3), 429–447.

Good, G. A., Phibbs, S., & Williamson, K. (2016). Disoriented and immobile: The experiences of people with visual impairments during and after the Christchurch, New Zealand, 2010 and 2011 earthquakes. *Journal of Visual Impairment and Blindness, 110*(6), 425–435. https://doi.org/10.1177/0145482X1611000605

Gray, L. (2017). Social determinants of health, disaster vulnerability, severe and morbid obesity in adults: Triple jeopardy? *International Journal of*

Environmental Research and Public Health, 14(12), 1452. https://doi.org/10.3390/ijerph14121452

Hou, C., & Wu, H. (2020). Rescuer, decision-maker, and breadwinner: Women's predominant leadership across the post-Wenchuan earthquake efforts in rural areas, Sichuan, China. *Safety Science, 125*, 1–6. https://doi.org/10.1016/j.ssci.2020.104623.Human Rights Campaign (2023) 'LGBTQ+ Americans Under Attack: A Report and Reflection on the 2023 State Legislative Session', *Human Rights Campaign*, 6 August 2023. Available at: https://hrc-prod-requests.s3-us-west-2.amazonaws.com/Anti-LGBTQ-Legislation-Impact-Report.pdf

Kailes, J. I., & Lollar, D. J. (2021). Disasters and disability: Rhetoric and reality. In D. J. Lollar, W. Horner-Johnson, & K. Froehlich-Grobe (Eds.), *Public health perspectives on disability* (pp. 251–268). Springer. https://doi.org/10.1007/978-1-0716-0888-3_12

Krishnaraj, M. with Sudershan, R. M., & Shariff, A. (Eds.). (1997). *Gender, population and development*. Oxford University Press.

Lambert, S. (2014). Indigenous peoples and unban disaster: Maori responses to the 2010-12 Christchurch earthquake. *Australasian Journal of Disaster and Trauma Studies, 18*(1), 39–48.

Larkin, B. (2019). Pride and prejudice: LGBTIQ community responses to disaster events worldwide. *Australian Journal of Emergency Management, 34*(3), 60–66.

Marable, M. (2006). Katrina's unnatural disaster A tragedy of black suffering and White denial. *Souls, 8*(1), 1–8.

Marable, M. (2008). Introduction: Seeking higher ground: Race, public policy and the hurricane katrina crisis. In M. Marable & K. Clarke (Eds.), *Seeking higher ground: The hurricane katrina crisis, race and public policy reader* (pp. ix–xvi). Basingstoke.

Moore, J. W. (Ed.). (2016). *Anthropocene or Capitalocene? Nature, history and the crisis of capitalism*. PM Press.

Parkin-Albayrakoglu, E. (2022). Out and proud in the field: Eco-queers for climate adaptation. *Peace Review: A Journal of Social Justice, 34*(1), 51–63.

Phibbs, S., et al. (2014). What about us? Reported experiences of disabled people related to the Christchurch earthquakes. *Procedia Economics and Finance, 18*, 190–197. https://doi.org/10.1016/S2212-5671(14)00930-7

Razack, S. (1998). *Looking White people in the eye: Gender, race, and culture in courtrooms and classrooms*. University of Toronto Press.

Rushton, A., Gray, L., Canty, J., & Blanchard, K. (2019). Beyond binary: (re) defining "gender": For 21st century disaster risk reduction research, policy, and practice. *International Journal of Environmental Research and Public Health, 16*(20), 3984. https://doi.org/10.3390/ijerph16203984

Rushton, A., Phibbs, S., Kenney, C., & Anderson, C. (2020). The gendered body politic in disaster policy and practice. *International Journal of Disaster Risk Reduction, 47*(2), 101648. https://doi.org/10.1016/j.ijdrr.2020.101648

Saldanha, A. (2020). A date with destiny: Racial capitalism and the beginnings of the Anthropocene. *Environment and Planning D: Society and Space, 38*(1), 12–34. https://doi.org/10.1177/0263775819871964

Soden, R., & Owen, E. (2021). Dilemmas in mutual aid: Lessons for crisis informatics from an emergent community response to the pandemic. In *Proceedings of the ACM on Human-Computer Interaction* (CSCW2), pp. 1–19. https://doi.org/10.1145/3479862

United Nations for Disaster Risk Reduction. (2023). *Main findings and recommendations of the Midterm Review of the implementation of the Sendai Framework for Disaster Risk Reduction 2015–2030.* Available at: https://sendaiframework-mtr.undrr.org/publication/main-findings-and-recommendations-midterm-review-implementation-sendai-framework

United Nations Intergovernmental Panel on Climate Change. (2019). *UN annual report: Climate change 2018.* United Nations Framework Convention on Climate Change (UNFCCC). Available at: https://unfccc.int/sites/default/files/resource/UN-Climate-Change-Annual-Report-2018.pdf

United Nations Office of the High Commissioner for Human Rights. (2021). *Data collection and management as a means to create heightened awareness of violence and discrimination based on sexual orientation and gender identity.* Report of the Independent Expert on protection against violence and discrimination based on sexual orientation and gender identity (A/HRC/41/45). Available at: https://documents-dds-ny.un.org/doc/UNDOC/GEN/G19/138/27/PDF/G1913827.pdf?OpenElement

United Nations SDGs. (2023). *Homepage.* (Online) Available at: https://sgds.un.org/

White, G. F. (1973). Natural hazards research. In R. Chorley (Ed.), *Directions in geography* (pp. 193–216). Routledge.

Wisner, B., Berger, G., & Gaillard, J. C. (2017). We've seen the future, and it's very diverse: Beyond gender and disaster in West Hollywood, California. *Gender, Place and Culture, 24*(1), 27–36.

Wu, H., & Karabanow, J. (2020). *COVID-19* and beyond: Social work interventions for supporting homeless population. *International Social Work, 63*(6), 790–794. https://doi.org/10.1177/0020872820949625

PART I

Queer(y)ing Disaster Studies

CHAPTER 2

Notes from Evacuation Centres in Mindanao 2015–2018

Naro Alonzo

They blamed the rain. They blamed PAGASA.[1]
They blamed the mine, illegal loggers, activists.
They blamed politicians. Lack of faith, destiny.
They blamed *them*, believing this disaster was God's
Punishment for those Miss Gay pageants, for tolerance.
I listened, providing psychological first aid,
Guilt carving my throat into the longest apology.

* * *

He prophesized the flood, an elderly fisherman said.
There were omens. Before the warming temperature,
There were invasive species. Fish he'd never seen even

[1] Philippine Atmospheric, Geophysical and Astronomical Services Administration.

N. Alonzo (✉)
Graduate Student of Psychology, The University of the Philippines,
Quezon City, Philippines
e-mail: mlalonzo@up.edu.ph

D. Dominey-Howes et al. (eds.), *Queering Disasters, Climate Change and Humanitarian Crises*, Sustainable Development Goals Series, https://doi.org/10.1007/978-981-96-3857-4_2

When his father was alive. New, foreign, dangerous,
"Parang yung mga—". Just like *them*. Then he laughed
To cope with the loss of his family, his home.
This joke, a bridge to hope. I laughed too, to cope.

* * *

In this basketball court were a hundred stories
Of loss. By the 3-point line, a mound of donated clothing.
Given generously by those far from disaster. Volunteers
Did not know what to do with this pile. Underwear with
Broken garters and holes, girl scout uniforms, election
Campaign shirts. In this court of makeshift tents, *they* made
A runway. Like a model, one slipped on a loose bridesmaid dress,
Strutting, hands over hips, ignoring the red warning of her
Father's rubber slipper, gaze far beyond this court.

* * *

He jokes that his gaydar was his way of coping. How a flashing
Screen from his phone and the buzz of Grindr were
"relief goods". That he may not know what grief looks like,
But with one look, can tell the nationality of the aid worker
He'll be mourning with tonight. Accepting that,
Like all other times, they would leave forgetting
To add him as a friend on Facebook. He forgets them
With each flood that enters his phone, but dreams again.
Resilience is to bury a broken phone in a sack of rice.

* * *

In their most honest confessions, *they* told me:
How horrible it was, wanting an abusive relative to never
Be found. Praying they won't be called by search
And rescue to identify a body. That she threw away the meds
To help her breasts make milk, or how with the excuse
Of finding internet connection, he and someone's husband
From the same shelter would kiss
behind a classroom that had the sign "prayer only".

* * *

What they couldn't tell the nurse: their fatigue
was because all energy was spent holding in their urine.
Afraid of what could happen in those portalets. Shortness of breath
from a sour bandage wound too tight around their chest.
Migraines from having to be called a name
Type-written on a birth certificate, lost in a landslide.
Pain from having privacy only in our 10-minute chat.
From censoring every movement of their body
Since there was no world to distract their family's eyes.

* * *

Here comes the time when the news have moved on.
Interviews were conducted, needs assessed.
Donations stop coming, manila papers of names, folded.
Done with debriefing, responders drop by a fast-food chain
for pagpag.[2] Let go, forget. There's another disaster
To face. While survivors are left to find
A way home, wearing clothes they will soon own
Inside-out, I go back to my small, safe space.

* * *

Description of *Notes from Evacuation Centres in Mindanao 2015–2018*

This poem is woven from small notes I took while being deployed as a volunteer disaster relief crisis responder. The Philippines is a Southeast Asian country that is vulnerable to the effects of climate change and natural hazards. Disaster exposure, issues in disaster risk reduction management, and disaster capitalism (Porteria, 2015; Yee, 2018) have negatively impacted the well-being of communities (Edwards et al., 2021), particularly Filipino gender and sexual minorities (McSherry et al., 2015;

[2] Literally, to shake off dirt. In Filipino superstitious belief, pagpag is a practice where Filipinos drop by another place, believing that one should not go straight home after attending a wake. Some mental health practitioners use the same word but in the context of disasters.

Alibudbud, 2023). Environmental and climate defenders, some of which also self-identify as part of gender and sexual minorities, have had to struggle with human rights violations and political repression (Global Witness, 2023; Tran, 2023) to seek justice and accountability for environmental degradation.

Each stanza of this poem is a moment captured from narratives of families, mothers, children, indigenous peoples, and staff while facilitating psychological first aid as a volunteer for different human rights and psychology/mental health organisations in Mindanao. In a way, these form a collective memory of all those times spent looking, listening, and linking alongside survivors of typhoons, earthquakes, and floods. The evacuation centre, I found, is a temporally unique space—where life appears to be suspended, arrested—where boundaries, such as of bodies and privacies, are also fluid.

This poem also navigates feelings and dialectics of guilt, self-censorship, and rapport-building, when supporting communities that prioritise cis-heteronormative attitudes. A colleague once commented, "They lost their home, farms, and are still looking for family members…why would homophobia matter now? How on earth could *you* be *hurt*?"

In the face of repetitive disasters, I've seen how narratives of support seekers and support providers could be reduced to documentation and statistics. This poem aspires to reflect the intimacies of gender and sexuality in disasters. More personally, this is my way of preserving memory, so easily displaced and equally vulnerable to forgetting, coming from a context where another tragedy could only be a day away.

Conflict of Interest The author has no conflict of interest.

References

Alibudbud, R. (2023). Gender in climate change: Safeguarding LGBTQ+ Mental Health in the Philippine climate change response from a minority stress perspective. *Journal of Preventive Medicine and Public Health, 56*(2), 196–199. https://doi.org/10.3961/jpmph.22.501

Edwards, B., Gray, M., & Borja, J. (2021). The influence of natural disasters on violence, mental health, food insecurity, and stunting in the Philippines: Findings from a nationally representative cohort. *SSM - Population Health, 15*, 100825. https://doi.org/10.1016/j.ssmph.2021.100825

Global Witness. (2023). *Standing firm: Land and environmental defenders on the frontlines of the climate crisis.* Taken from https://www.globalwitness.org/en/campaigns/environmental-activists/standing-firm/

McSherry, A., Manalastas, E. J., Gaillard, J. C., & Dalisay, S. N. (2015). From deviant to Bakla, strong to stronger: Mainstreaming gender and sexual minorities into disaster risk reduction in the Philippines. *Forum for Development Studies, 42*(1), 27–40. https://doi.org/10.1080/08039410.2014.952330

Porteria, A. (2015). Making money out of people's misery: Has disaster capitalism taken over post-Haiyan Philippines? *Philippine Sociological Review*, *63*, 179–206. http://www.jstor.org/stable/24717192

Tran, D. (2023). Gendered violence martyring Filipina environmental defenders. *The Extractive Industries and Society*, 13. https://doi.org/10.1016/j.exis.2023.101211

Yee, D. (2018). Violence and disaster capitalism in post-Haiyan Philippines. *Peace Review, 30*(2), 160–167. https://doi.org/10.1080/10402659.2018.1458943

CHAPTER 3

Something Borrowed, Something New: Navigating the Emerging Field of Queer Disaster Studies

William Leonard, Dale Dominey-Howes, Ashleigh Rushton, Marcilyn Cianfarani, Lisa Overton, and Haorui Wu

Introduction

This chapter provides a guide to the emerging field of gender and sexual diversity in disasters, climate change and humanitarian crises, or 'queer disaster studies.' It begins with a history of queer disaster studies,

W. Leonard (✉)
Monash University, Accident Research Centre, Clayton, VIC, Australia

D. Dominey-Howes
The Sydney Environment Institute and School of Geosciences, The University of Sydney, Sydney, NSW, Australia

A. Rushton
Disaster and Emergency Planner, Emergency Services, Cleveland, UK

D. Dominey-Howes et al. (eds.), *Queering Disasters, Climate Change and Humanitarian Crises*, Sustainable Development Goals Series, https://doi.org/10.1007/978-981-96-3857-4_3

including its debt to feminist and queer theory; queer community responses to the HIV/AIDS pandemic; and the 'queer turn' in human geography. Whilst not an exhaustive account, it reviews some of the key disciplines, research and policy involved in the emergence and development of queer disaster studies.

The chapter introduces disaster justice as a framework for better understanding and responding to the drivers and impacts of disasters, climate change and humanitarian crises on gender and sexual diverse people. The framework addresses the relationships between human-led environmental degradation and the forces at work in systemic discrimination and marginalisation of minority and vulnerable groups, including gender and sexual diverse communities. Its focus on social and environmental justice also highlights the need to address the impacts of colonialism and western hegemony on gender and sexual diverse Indigenous and First Nations peoples and communities in the Global South and Global North without losing sight of what many of these communities have in common—their shared, but culturally specific experiences of heterosexist discrimination.

The chapter concludes with an introduction to some of the terminological debates within the emerging field of queer disaster studies. These include how best to describe, while acknowledging, the cultural diversity of gender and sexual diverse people and communities and how to adapt the current, dominant terminology in disaster studies to better meet their needs. It suggests that addressing these issues is crucial to the ongoing development and growth of this important field of enquiry.

Legacy

The growing interest and consolidation of research, policy and practice on gender and sexual diverse people's experiences of disasters, climate change and humanitarian crises has a complex history. It is impossible to do justice

M. Cianfarani
Disaster and Emergency Management Practitioner, Ontario, Canada

L. Overton
Middlesex University, London, UK

H. Wu
School of Social Work, Faculty of Health, Dalhousie University, Halifax, NS, Canada

to the complexity of that history here. However, the development of queer disaster studies has relied on the mutually supportive efforts of gender and sexual diverse advocates, researchers, policy makers and practitioners from different disciplines, sectors and countries.

Feminist and Queer Intersections

The work of feminist, and later queer advocates and scholars, has been instrumental in the development of a critique of the heteronormative conceits and biases of disaster-related research and studies. In the 1980s, feminist scholars began to explore the impacts of patriarchy and systemic gender inequality on women's experience of disasters, climate change and humanitarian crises. Initially, this work focused on the lack of a gendered analysis of the differential impacts of disasters on women and girls (Anderson, 1994; Fothergill, 1999; Shreve et al., 2017). Feminist research demonstrated that in many locations and social contexts, women were at increased risk of disaster impacts and experienced worse outcomes than men, including increased unemployment, domestic labour and family violence (Ikeda, 1995; Fothergill, 1996; Enarson & Morrow, 1998).

Simultaneously, queer theorists had begun questioning the limits of current understandings of gender. Butler and others, working at the intersection of feminist and queer studies, argued that sex, gender and sexuality were not separate but mutually constitutive categories (Rubin, 1984; Butler, 1990). According to Butler, heteropatriarchal power and privilege relied on the generation and policing of a rigid, binary model of sex, gender and sexuality. Butler (1990) called this the heterosexual matrix and argued that the regulation of gender necessarily involved the devaluation and subordination of both women and gender and sexual diverse minorities.

However, over the past decade, several feminist and queer scholars have criticised the continued dominance of a binary model of gender in disaster-related research, policy and practice. Whilst early feminist research was critical of the impacts of a binary model of gender on women, it nonetheless relied on this model to highlight the differences between women and men's disaster experiences. Gaillard and others have argued that the continued use of a binary model in disaster studies excludes trans and gender diverse women and gender non-binary people (Dolan, 2014; Dominey-Howes et al., 2014; Gaillard et al., 2017b; Rushton et al., 2019). What is needed, they suggest, is a more flexible and culturally sensitive understanding of gender identities and expressions (Gaillard et al., 2017a).

The HIV/AIDS Pandemic

Queer and feminist critiques of heteronormativity provided the oxygen for interdisciplinary research and policy on gender and sexual diverse individuals', families' and communities' experiences of disasters. However, it was the HIV/AIDS pandemic emerging in the 1980s that led, for the first time, to the inclusion of gender and sexual diverse people in national public health policies (Brown, 1997; Leonard, 2012).

In many countries, gay men, sex workers, drug users and other stigmatised minorities were blamed for the onset and spread of HIV and AIDS (Grover, 1988; Overs & Druce, 1994; Leonard, 2014). The COVID-19 pandemic and the monkey pox virus are reminders that gender and sexual diverse people continue to be blamed for disasters by conservative governments and religious extremists (Kane, 2020; Quinn et al., 2021; Rushton & Scarlett, 2023). The United Nations has noted the increase of homophobic, biphobic and transphobic hate speech, globally, in the wake of COVID-19 and the use of public health measures by some national governments as a pretence for winding back gender and sexual diverse people's rights and funding for their community organisations (ILGA Europe, 2020; LGBT Foundation, 2020; United Nations, 2020).

Whilst HIV was a public health crisis, it was also a pandemic. National and international responses to HIV/AIDS are, perhaps, the first disaster prevention and risk reduction strategies to include gender and sexual diverse people. Simultaneously, the response to HIV relied on the capacity and resilience of gender and sexual diverse people and communities to 'take care of their own' (Omoto & Crain, 1995). In the absence of government support, and in the face of public hostility, gender and sexual diverse people and their allies established volunteer-led organisations (Rayside & Lindquist, 1992; Haworth et al., 2022). This self-reliance has become a feature of how gender and sexual diverse people have responded to subsequent disasters (Gorman-Murray et al., 2018; Larkin, 2019; Haworth, 2021).

Queer(y)ing Geography

Haworth et al. (2022), reviewing research on disaster and gender and sexual diverse minorities, argue that 'geography [has been] a leading discipline in this research' (p. 3). Human geography explores how place and space both shape and are shaped by the underlying socio-political and cultural processes that create and reproduce systemic inequalities.

A focus within human geography on how social inequality is implicated in both the causes and the impacts of disasters and other catastrophic events has led to two discreet but interrelated streams of research and policy. One examines how *place* is implicated in the risks and impacts of disasters, climate change and humanitarian crises on individuals and groups (Choudhury & Wu, 2023). This has involved exploring differences between the Global North and Global South and how the impacts of colonisation, environmental exploitation, forced displacement and access to wealth and resources have led to significant regional and national variations in the risk and impacts of disasters, climate change and humanitarian crises (Drolet et al., 2015). It has also involved exploring variations in the risks and disaster impacts within countries including variations within and between urban and rural areas (Birkmann et al., 2015).

The second has focused on how *identity*-based discrimination increases the risks and disaster impacts for vulnerable and marginalised communities. There has been growing interest in how forms of systemic discrimination shape the disaster experiences of marginalised and minority populations, including patriarchy, racism, ethnocentrism, ageism and ableism (Krishnaraj, 1997; Marable, 2006; Enarson et al., 2007; Chandran et al., 2015; Drolet et al., 2015; Lambert & Mark-Shadbolt, 2021; Phibbs et al., 2015; Wu & Karabanow, 2020; Calgaro et al., 2021; Kailes & Lollar, 2021).

Queer geography has relied heavily on the work of human(ist) geographers and models that highlight the central role social inequality plays in the causes, impacts and response to disasters, climate change and humanitarian crises (Alexander, 2004). It has combined this model with feminist and queer critiques of heteronormativity to understand and name the socio-spatial determinants of gender and sexual diverse peoples' and communities' increased risk of disasters, climate change and humanitarian crises and their negative outcomes.

Two recent reviews of global research on gender and sexual diverse people and disasters document the geographic diversity and different types of disaster studied (Acciari et al., 2021; Haworth et al., 2022). Pincha and Krishna's (2008) study of the impacts of the December 2004 Indian Ocean tsunami on the Aravanis in India is credited as the first to focus on gender and sexual diverse communities. D'Ooge (2008) then documented the impacts of Hurricane Katrina on lesbian, gay and transgender communities in New Orleans. Subsequent studies have addressed the impacts of earthquakes, fires, COVID-19 and other catastrophic events on gender and sexual diverse communities. These include studies in Nepal (Knight & Sollom, 2012), Haiti (International Gay and Lesbian Human Rights Commission and SEROVie, 2011), Japan (Ozawa, 2012), the Philippines (Gaillard, 2011), Indonesia (Balgos et al., 2012), Samoa (Smith, 2013), Canada (Cianfarani, 2013), the USA (Eads, 2002; Fothergill, 2004; D'Ooge, 2008; Fontanez, 2019; Goldsmith et al., 2022) and Australia and New Zealand (McKinnon et al., 2016; Gorman-Murray et al., 2017, 2018). They also include several studies with a global rather than a single country focus (e.g., Larkin, 2019).

Current State of Affairs

The development within human geography of a subfield exploring the impacts of disasters on gender and sexual diverse people and communities has sparked, and in some cases reignited, interest within other disciplines. These include the social sciences, cultural studies, history, law, urban design and planning, queer studies and many more. All bring their own research histories, interests and methods. Collectively, these disciplines represent the development of a new, multidisciplinary field—queer disaster studies—that focuses on the impacts of disasters, climate change and humanitarian crises on gender and sexual diverse people and communities.

Goldsmith et al. (2022) and Larkin (2019) in their reviews of this emerging field have identified issues and themes common to gender and sexual diverse peoples' experiences of disasters, climate change and humanitarian crises across geographic and cultural boundaries (see also Dominey-Howes et al., 2022; Blanchard et al., 2023). These include increased risk of homophobic, biphobic and transphobic violence, harassment and discrimination; homelessness; physical and mental health issues; and further reductions in gender and sexual diverse people's safety and access to health and other essential services. Additionally, researchers identify loss of safe domestic and community spaces; the failure of local and national media to report the impacts of disaster on gender and sexual diverse people and communities; the delivery of disaster and emergency relief services by faith-based organisations; and the often-underestimated resilience and capacity of gender and sexual diverse people and communities (Dominey-Howes et al., 2022; Goldsmith et al., 2021).

Frameworks and Concepts

Few existing disaster frameworks address the needs of gender and sexual diverse people and communities. Even fewer attempt to address the links between interrelated catastrophic events and discrimination and violence against vulnerable and minority groups. The need for new, more expansive frameworks that explore how power, privilege and position are implicated in both disasters and systemic social inequality is critical given the global escalation of disasters, climate change and humanitarian crises and violence against gender and sexual diverse people.

Disaster Justice

Disaster justice is a relatively new and evolving framework that applies human rights principles to the field of disaster studies and management (Lukasiewicz, 2020). Lukasiewicz contends that 'the term, disaster justice … recognises that disasters can expose, magnify, and deepen existing injustices in society' (p. 9). Disaster justice focuses on the fair and equitable treatment of people and communities in disasters and provides a useful way of framing and dealing with many of the issues specific to gender and sexual diverse communities.

Disaster justice also addresses the devastating, cumulative impacts of human activity on the environment. It advocates a new form of environmental justice that recognises humans are part of and dependent on nature, and legislation limiting exploitative and harmful human development. As Lukasiewicz (2020) writes, disaster justice is located at 'the confluence of environmental, climate and social justice' (p. 9). It is this unique and timely confluence that makes disaster justice a powerful framework for addressing the relationships among systemic social inequality and injustice, the dramatic increase in the incidence of interrelated catastrophic events, and gender and sexually diverse people's increased risk of disasters.

Douglass and Miller (2018, p. 271) characterise disaster justice as 'a moral claim on governance, which arises from anthropogenic interventions in nature that incubate environmental crises and magnify their socially and spatially uneven impacts.' They highlight two key operational elements: procedure and governance. Both relate to the exclusion of minority and marginalised voices in decision-making and the unequitable distribution of disaster relief and recovery goods, resources and services. These procedural vulnerabilities not only reflect but also reproduce and magnify existing social inequalities. The absence of any consideration of the needs of gender and sexual diversity in national and international disaster planning, such as the UN's SDGs and the Sendai Framework (United Nations Office for Disaster Risk Reduction (2015), reduces gender and sexual diverse people's access to appropriate disaster risk reduction (DRR) services and their capacity to deal with and recover from current and future disasters.

Lukasiewicz (2020) offers a complex typology of the different approaches, principles and processes that underpin disaster justice and its emergence as 'a new research focus in disaster management' (p. 8). Despite or perhaps because of its newness and complexity, Lukasiewicz provides neither definitions nor descriptions of the discreet but intersecting types of justice that constitute 'disaster justice.' Table 3.1 addresses this gap. It focuses on eight of the ten types of justice that Lukasiewicz associates with procedure and governance in her complex modelling of disaster justice. It provides a short definition of each and how this definition can be applied to gender and sexual diverse people and communities. Table 3.1 also provides examples of how each of these eight types of disaster justice can improve our understanding and response to gender and sexual diverse peoples' and communities' experiences of hazards, disasters, climate change and humanitarian crises.

Intersectionality

Closely tied to disaster justice is an understanding of how intersecting forms of structural and identity-based discrimination impact the disaster experiences and needs of people who belong to more than one minority population. Crenshaw (1991) developed the notion of intersectionality to understand how interactions among race, gender and socio-economic status shape black women's experiences of disaster in the USA, including their access to essential services (Crenshaw, 1991; Katz, 2008; Hopkins, 2019). Intersectional feminist scholarship paved the way for similar work on the impacts of multiple forms of discrimination on other minority

Table 3.1 Queer(y)ing elements of Lukasiewicz's (2020) typology of disaster justice

Type of justice	*Definition[a]/Gender and sexual diverse description*	*Gender and sexual diverse examples*
Intergenerational	The fair and equitable intergenerational distribution of benefits and burdens. *Ensuring that present actions do not jeopardise the well-being, resources and access to services of future generations of gender and sexual diverse people when dealing with disasters, climate change and humanitarian crises.*	• The potential lack of connection between older and younger cohorts of gender and sexual diverse people • The isolation of older cohorts of gender and sexual diverse people from both mainstream and gender and sexual diverse communities, organisations and services • The paucity of positive role models for younger gender and sexual diverse people • How younger gender and sexual diverse people may be subject to discrimination, abuse and rejection from members of their 'natal families' and have reduced access to resources, benefits and supports compared with their non-gender and non-sexual diverse siblings

(*continued*)

Table 3.1 (continued)

Type of justice	*Definition[a]/Gender and sexual diverse description*	*Gender and sexual diverse examples*
Temporal	The fair and equitable distribution of benefits and burdens across different time periods. *Ensuring that the timing of interventions, investments and policies minimises the impacts of disasters, climate change and humanitarian crises on gender and sexual diverse people and communities.*	• How jurisdictions can exclude or delay addressing systemic, historical discrimination against gender and sexual diverse people • How jurisdictions can exclude or delay ensuring that gender and sexual diverse people have the same rights and responsibilities (including access to appropriate and effective services and resources) as cisgender and heterosexual people
Spatial	The fair and equitable distribution of benefits and burdens across different geographical areas. *Ensuring that policies and interventions do not disproportionately affect certain regions or communities, including gender and sexual diverse minorities, and that resources and assistance are equitably distributed during and after disasters, climate change and humanitarian crises.*	• Gender and sexual diverse people and communities' loss of safe spaces and places (e.g., homes, clubs, bars, meeting places) • How the cisgender and heteronormative constitution of spaces shape, govern and limit gender and sexual diverse people's everyday lives • How systemic heterosexist discrimination can lead to gender and sexual diverse communities living in areas of increased disaster risk

(*continued*)

Table 3.1 (continued)

Type of justice	*Definition[a]/Gender and sexual diverse description*	*Gender and sexual diverse examples*
Historical/ restorative	Addressing and providing legal compensation or redress for historical injustices and harms. *Acknowledging and addressing the historical factors that contribute to gender and sexual diverse people and communities' increased vulnerability and unequal access to resources and protections in the face of disasters, climate change and humanitarian crises.*	• Identifying inequality under the law and a lack of legal protections and rights for gender and sexual diverse people and communities • How lack of equality and legal protections lea to the marginalisation and increased vulnerability of gender and sexual diverse people and communities to disasters, climate change and humanitarian crises and their negative impacts
Corrective	The rectification of injustices caused by wrongful actions or events. *Legislative and other measures that put an end to actions that discriminate against and harm gender and sexual diverse people and communities and that increase their vulnerability to disasters, climate change and humanitarian crises.*	• Providing gender and sexual diverse people full equality under the law • Ensuring gender and sexual diverse people's equal treatment and access to services, resources and support needs

(*continued*)

Table 3.1 (continued)

Type of justice	*Definition[a]/Gender and sexual diverse description*	*Gender and sexual diverse examples*
Interactional	The fair and respectful treatment of individuals during interactions with others (individuals and groups) often in relation to decision-making processes. *Ensuring that representatives of affected communities, including representatives of gender and sexual diverse communities, are included in decision-making processes related to disaster preparedness, response and recovery.*	• Addressing the absence, marginalisation or exclusion of gender and sexual diverse people and communities (their voices and experiences) in all aspects of disaster planning, response and recovery • Ensuring all aspects of disaster planning, response and recovery are diverse and inclusive and include representatives of gender and sexual diverse communities
Informational	Fair and equitable distribution of, and access to, information. *Ensuring that all communities, including gender and sexual diverse communities, have access to accurate and timely information about disasters, climate change and humanitarian crises. It also involves ensuring that gender and sexual diverse people and communities have access to appropriate and effective risk-reduction measures, as well as the opportunity to contribute their knowledge and perspectives.*	• Understanding and addressing how informational (in) justice occurs because the specific needs of gender and sexual diverse people and communities are not recognised or ignored • Understanding and addressing how informational (in) justice occurs when the voices, knowledges and capacities of gender and sexual diverse people and communities are overlooked, ignored or actively excluded

(*continued*)

Table 3.1 (continued)

Type of justice	*Definition[a]/Gender and sexual diverse description*	*Gender and sexual diverse examples*
Relational	Recognising and addressing the importance of relationships and social networks in the distribution of benefits and burdens. *Recognising and supporting the role of community networks, social capital and local institutions in building resilience and responding to disasters, climate change and humanitarian crises. This includes measures aimed at increasing the capacity and resilience of gender and sexual diverse communities and organisations to respond to disasters, climate change and humanitarian crises.*	• Utilising, supporting and resourcing well-established gender and sexual diverse social networks and community organisations • Authorising well-established gender and sexual diverse social networks and community organisation to make decisions and oversee the distribution of resources within the communities they represent

[a]The definitions provided in this table were developed during informal discussions with disaster scholars including suggestions provided by Justin Sanir, to whom we are most grateful (editor's note)

groups (Habib, 1995; Fordham, 1999; Perera-Mubarak, 2013; Jacobs, 2019). Recent disaster studies have used intersectionality to understand the differential impacts of gender identity, sexuality, race, ethnicity, age, socio-economic means and culture on the vulnerabilities and service access of gender and sexual diverse minorities (Gaillard, 2010; Brown, 2011; Haworth et al., 2022).

While intersectionality has been used to highlight the negative impacts of multiple forms of discrimination, researchers in different disciplines have also begun to consider its positive impacts (see Fredriksen-Goldsen et al., 2014). In disaster research, growing interest in resilience and *resistance* has led to work on how the capacities of different minorities *within* gender and sexual diverse communities can be mobilised to mitigate the risks and impacts of disasters (Dominey-Howes et al., 2018; Gorman-Murray et al., 2018).

Vulnerability, Marginality, Resilience and Capacity

Vulnerability

Vulnerability is one of the key concepts for understanding the differential impacts of disasters (Wisner, 2009). It gained traction in disaster studies in the mid-1970s as the focus shifted from viewing disasters as consequences of natural hazards to seeing them as linked to systemic inequalities that increased certain populations' susceptibility and vulnerability (Gaillard, 2010; Steckley & Doberstein, 2011). While the meaning of vulnerability continues to be contested, it is understood as the structural conditions—physical, social, cultural, economic and political— that render people and communities susceptible to the impacts of hazards (Wisner, 2009; Thomas et al., 2013).

Ton and Adamson (2020), in a discussion of disaster justice for people with disabilities, make a distinction between two discrete but interrelated aspects of vulnerability. One refers to place-based factors, 'the characteristics of communities and the built environment, such as level of urbanisation, growth rates and economic vitality, that contribute to the social vulnerability of places' (Cutter et al., 2003 quoted in Ton & Adamson, 2020, n.p.). The second refers to social factors 'that influence or shape the susceptibility of various groups to harm and that also govern their ability to respond' (n.p.; See also Thomas et al., 2013). Similarly, Gaillard (2010) argues that vulnerability is an effect of structural adversity that compromises the ability of people and communities to withstand the impacts of hazards and disasters. Like Ton and Adamson, Gaillard draws a distinction between two aspects of vulnerability: one related to reduced access to material and economic resources, and the other to structural and identity-based inequality and discrimination (see also Phillips & Fordham, 2009).

This understanding of vulnerability demonstrates how material, structural and identity-based inequality against gender and sexual diverse people and communities interact and increase their susceptibility to disasters. It also highlights how heterosexist discrimination leads to the marginalisation of gender and sexual diverse people within mainstream social institutions and, with that, their omission from disaster relief and recovery policies and programmes and reduced access to essential services (Gilbert & Barkun, 1981; Urbatsch, 2016; Cowan, 2022; Leonard et al., 2022). This is evident in gender and sexual diverse people's absence in high-level disaster relief and recovery policies such as the Sendai Framework despite a compelling body of evidence documenting their increased vulnerability (Dominey-Howes et al., 2016; Atkin et al., 2024).

Marginality

Marginality has much in common with vulnerability and some researchers use the terms interchangeably (Wisner, 1998; Gurung & Kollmair, 2007; Gaillard, 2010). However, marginality is more overtly concerned with power and privilege, and how they operate to overvalue certain types of people and groups at the expense of others (Douglass & Miller, 2018; Cullen & Pretes, 2000). The negative impacts of marginality are linked to the transformation of social differences into social gradients or hierarchies. These include socio-economic status, culture, race and ethnicity, religion, disability, gender and sexual diversity, geographic location and age. For example, heteropatriarchy can be understood as both cause and effect of a sexed and gendered hierarchy that values and rewards individuals who are exclusively cisgender and heterosexual at the expense of those who are gender and sexually diverse (Butler, 1990; Victorian Government Department of Human Services, 2003; Robinson, 2016).

Cullen and Pretes (2000) found that social scientists' (including human geographers) use of the terms 'marginal' and 'marginalised' relies heavily on the spatial metaphor of centre and periphery. According to this metaphor, power and privilege are located at the centre and drop away as we move outward towards the periphery or margins. Those who are located at the centre have access to power and privilege while those who are relegated to the margins have little or none. This has implications for how we understand and react to gender and sexual diverse communities' increased risk of disasters, climate change and humanitarian crises and their poorer medium- to longer-term response and recovery outcomes. For many gender and sexual diverse communities, the socio-symbolic processes that seek to minimise their individual and collective worth are implicated in their physical separation and distancing from the more powerful and privileged. Consequently, gender and sexual diverse people are more likely to live and work in areas of increased disaster risk and reduced access to essential mainstream response and recovery services (Calgaro et al., 2014a, 2014b).

Marginality can be viewed solely through the lens of social inequality and oppression. But, like intersectionality, it also provides opportunities for minority groups to generate their own values, norms and ways of belonging. hooks (1990, p. 206) argues that while marginality never comes with the guarantee of safety it can function 'as a space of radical possibility [and] of resistance ... for the production of a counter-hegemonic discourse that is not just found in words but in habits of being and the way one lives.'

Many marginalised groups develop strategies and ways of being that allow them to not only cope with everyday abuse and systemic discrimination but also reclaim and value what makes them different from the 'mainstream' (Cody & Welch, 1997; Webster, 2021). Focusing solely on the negative impacts of marginality and vulnerability ignores gender and sexual diverse minorities' capacity to draw on these everyday strategies to mitigate the impacts of disasters, climate change and humanitarian crises (Balgos et al., 2012; Kelly & Smith, 2012; Gorman-Murray et al., 2017). This more expansive understanding of marginality straddles the divide between risk, vulnerability and oppression on one side and resistance, resilience and affirmation on the other.

Resilience and Capacity

Over the past two decades disaster scholars and practitioners have shifted their energies from disaster response to understanding vulnerabilities and DRR (e.g., UNDRR, 2005, 2015). The reimagined goal of many disaster management policies is to build resilient communities as a way of mitigating the risks and impacts of disaster (UNDRR, 2015). Resilience, like vulnerability and marginality, is a contested concept (Aldrich, 2012; Cretney, 2014). However, at its core, disaster resilience describes an adaptive process in which individuals and communities build on their response to previous disasters. This increases their capacity to deal with disasters while reducing their susceptibility to future catastrophic events (Bonanno et al., 2007; Zhou et al., 2010; Brown & Westaway, 2011; Haworth et al., 2018).

Resilience is often framed as the positive flipside to vulnerability (Kendra et al., 2018). According to King (2022), '[W]hile vulnerability recognises the susceptibility to risk in individuals and groups, resilience builds on the strengths of people and communities' (p. 2). However, Calgaro et al. (2014a, b) argue that vulnerability and resilience are better conceptualised as opposite ends of a spectrum. They suggest that an individual or community may be, simultaneously, vulnerable and resilient, and that the ratio of the one to the other fluctuates in response to ongoing changes in their social context.

Regardless of these different understandings, vulnerability highlights the external structural conditions that expose different social groups to disasters' adverse effects. In contrast, resilience focuses on the capacities internal to social groups or communities including existing knowledge, resources, skills and solidarity networks that can be mobilised as coping strategies during crises (Gaillard, 2010; Brown & Westaway, 2011). Including resilience and capacity-building in disaster relief and recovery planning and policy does not mean that marginalised and minority communities should be left to their own devices. As Brown and Westaway (2011) argue, the capacity of marginal communities is limited by structural inequality and can be enabled and enhanced by the provision of resources by governments, authorities and other organisations.

What is of greater concern, however, is how a singular focus on resilience and capacity reflects a neoliberal tendency to 'off-load' responsibility for DRR and mitigation from government to minority groups and communities. Providing additional resources may increase the capacity and resilience of gender and sexual diverse communities and lessen the negative impacts of disasters, climate change and humanitarian crises. However, if resilience and capacity-building are not located within a broader disaster justice framework, they risk leaving unchallenged the structural inequalities that lead to the marginalisation of gender and sexual diverse communities in the first place (Dominey-Howes, 2018).

Terminology

Identity politics and terminology have become minefields in current debates about gender and sexual diversity. At one level, this reflects a conservative, global backlash against progressive and inclusive legislative and social reforms (Butler, 2024). At another, it reflects vigorous debates within gender and sexual diverse communities, globally, about who is included in the widening orbit of gender and sexual diversity (Hines, 2019; McLean, 2021; Overton & Hepple, 2022).

Taxonomies

The scientific taxonomies of the Global North have dominated popular and professional understandings of gender and sexual identity. These taxonomies have relied on three discreet but interrelated categories: sex, gender and sexual orientation. Sex is understood as the division of humans into two separate classes—female and male—based primarily on differing reproductive capacities. In this binary model, gender describes the social roles available to females and males, determined by their biological differences. Sexual orientation is the reciprocal attraction between sexed and gendered opposites, that is between women and men, and guarantees both species and social reproduction (Leonard, 2005).

Queer and feminist scholars have argued that this binary model of social life has little to do with nature or biology but everything with power and inequality (Rubin, 1984; Butler, 1990; Stryker, 2008; Robinson, 2016). It works to generate and maintain heterosexual male privilege and power at the expense of women and gender and sexual diverse people. According to queer critiques, gender and sexual diverse people challenge the *hetero*-gendered binary logic and order of this system (Lay et al., 2018, p. 31). In so doing they threaten established hierarchies of power and privilege and are subject to differing levels of violence, discrimination and abuse (Rubin, 1984; Jagose, 1996; Leonard, 2005).

Identities

In the Global North, the acronym LGBTIQA+—lesbian, gay, bisexual, trans, intersex, queer/questioning, asexual, other—and its many variations, has become the dominant way of reclaiming and drawing together these different gender and sexual diverse minorities. Queer has also been reclaimed as an umbrella term that includes both people who identify as LGBTIQA+ and new and emerging gender and sexual diverse identities and expressions (Jones, 2023).

In disaster studies, these debates have centred on the imposition of terminology from the Global North on communities in the Global South. Gaillard et al. (2017b), Baumann et al. (2022) and others have been highly critical of the use of western identity categories, including LGBTIQA+ and queer, by international aid and disaster relief and recovery agencies. They argue that the use of these terms reflects western hegemony and neo-colonial oppression and does not represent indigenous knowledges or the lives of gender and sexual diverse people from non-western cultures.

Drawing on feminist and queer critiques they have also criticised the heterogendered binary that informs these models. They argue that in many non-western countries gender and sexual diversity are not framed within and limited by 'the binary.' The Asia Pacific Transgender Network

(APTN) describes some of the indigenous terms for transgender and gender diverse peoples many of whom also describe themselves as 'third gender.' They include 'terms for people assigned male at birth who identify as female or transgender ... [and] for people assigned female at birth and who identify as male' (APTN quoted in Amnesty International, 2022, n.p.).

Postcolonial critics argue that the imposition of a western binary model has the potential to render these and other indigenous gender and sexual diverse people and communities invisible. It also pressures members of these communities to conform to western expectations and categories when seeking assistance and support from international aid and recovery agencies following disaster (Gaillard et al., 2017b; Rushton et al., 2019; Fox & Wu, 2023).

Categories Versus Identities

What is not highlighted in these debates is the critical distinction between categories and identities. Categories represent a set of common characteristics which are expressed differently by different individuals and groups. It is commonly assumed that everyone has a sexual orientation but how this is expressed varies according to sexual identity and sexual practices. 'Gender and sexual diversity' and the categories included in the acronym SOGIESC—sexual orientation, gender identity and expression, and sex characteristics—are attempts to offer looser, less-constrained definitions. These are categories less enmeshed in western hierarchies of sex and gender and better able to accommodate a more geographically and culturally diverse collection of identities and expressions (Rushton et al., 2020; Edge Effect, 2021). For example, the Canadian Government has adopted the acronym 2SLGBTQIA+ to include both Indigenous and non-Indigenous gender and sexual diverse identities (Pruden & Edmo, 2016; Brotman, 2002; Women and Gender Equality Canada, 2022). Two Spirit (2S) is 'a term that incorporates Indigenous views of gender and sexual diversity and encompasses sexual, gender and spiritual identity' (Native Women's Association of Canada, n.d.). In this case, an Indigenous identity is added to the list of western identities leaving the overarching categories under which they sit unchanged.

This distinction between categories and identities has been vital to the development of legal protections that address discrimination against gender and sexual diverse minorities. The categories name protected attributes that are common to the different identities that sit under them.

Although heavily indebted to western terminology and jurisprudence, these categories have nonetheless been used in non-western contexts to provide varying levels of legal protections for gender and sexual diverse minorities (Brown, 2011; The Yogyakarta Principles, 2017; Cowan, 2022).

While several non-western cultures have a place for non-binary gender and sexual diverse identities and expressions, in many cases that place remains highly circumscribed within heteronormative systems of social control and regulation (Khan et al., 2009; Ong, 2017; Simmonds et al., 2022). Studies have documented the limited social roles available to gender and sexual diverse people and communities in several non-western countries including the Netuwas and Metic in Nepal (Knight & Welton-Mitchell, 2013) and the Aravani and Hijra in India and Bangladesh (Roy et al., 2021; Choudhury & Harini, 2023). Recent studies on the impacts of COVID-19 restrictions on gender and sexual diverse minorities in India, Bangladesh, Nepal and several Pacific Islands have documented the escalation of existing discrimination including severe restrictions on traditional forms of employment, reductions in income and access to financial and material support, and increased prejudice-motivated violence and abuse (Banerjee & Sathyanarayana Rao, 2021; Dasgupta et al., 2021; Al-Mamun et al., 2022; Amnesty International, 2022; Pandya & Redcay, 2022).

What Next

There are several challenges and opportunities facing the emerging field of gender and sexual diverse studies in disasters, climate change and humanitarian crises. One is how to develop conceptual frameworks that affirm difference and diversity while acknowledging the negative impacts of neo-colonial disaster relief and recovery programmes on non-western gender and sexual diverse people and communities. Another is how to ensure that a renewed focus on diversity does not lose sight of what is common to the lives of many gender and sexual diverse minorities across geographic and cultural boundaries—their shared, but culturally specific experiences of heteronormative marginalisation and discrimination.

Perhaps disaster justice provides a framework for reconciling these different, and at times contrary, imperatives: affirming the principles of justice, equity and diversity in disaster relief and recovery; acknowledging the ongoing impacts of neo-colonialism in the delivery of disaster services to non-western gender and sexual diverse people and communities; and providing opportunities to work across cultural, geographic and other borders to reduce the causes and impacts of systemic discrimination on all gender and sexual diverse people.

Conflict of Interest None of the authors of this chapter have any conflicts of interest.

References

Acciari, L., et al. (2021). *Gender and disaster bibliography and reference guide—Volume 1* (2nd ed., pp. 76–80). IRDR Centre for Gender and Disaster, University College London.

Aldrich, D. P. (2012). *Building resilience: Social capital in post-disaster recovery*. University of Chicago Press.

Alexander, D. (2004). Natural hazards on an unquiet Earth. In J. Matthews & D. Herbert (Eds.), *Unifying geography* (pp. 266–282). Routledge.

Al-Mamun, M., et al. (2022). Discrimination and social exclusion of third-gender population (Hijra) in Bangladesh: A brief review. *Heliyon, 8*(10), e 10840. https://doi.org/10.1016/j.heliyon.2022.e10840

Amnesty International. (2022). *'Pandemic or not, we have the right to live': The urgent need to address structural barriers undermining transgender people's rights across Asia and the Pacific Islands*. Amnesty International. https://creativecommons.org/licenses/by-nc-nd/4.0/legalcode

Anderson, M. (1994). Understanding the disaster-development continuum: Gender analysis is the essential tool. *Focus on Gender, 2*(1), 7–10.

Atkin, S., et al. (2024). Queer vulnerability and disaster situations. *AIMS Geosciences, 10*(1), 196–207. https://doi.org/10.3934/geosci.2024011

Balgos, B., Gaillard, J. C., & Sanz, K. (2012). The *warias* of Indonesia in disaster risk reduction: The case of the 2010 Mt Merapi eruption in Indonesia. *Gender and Development, 20*(2), 337–348. https://doi.org/10.1080/13552074.2012.687218

Banerjee, D., & Sathyanarayana Rao, T. S. (2021). "The graying minority": Lived experiences and psychological challenges of older transgender adults during *COVID--19* pandemic in India—A qualitative exploration. *Frontiers in Psychology, 11*, 604472. https://doi.org/10.3389/fpsyt.2020.604472

Baumann, L., Sharan, A., & Gaillard, J. C. (2022). Queering "gender and disaster" for inclusive disaster risk reduction. In *Oxford research encyclopaedia of natural hazard science*. https://doi.org/10.1093/acrefore/9780199389407.013.412

Birkmann, J., et al. (2015). Scenarios for vulnerability: Opportunities and constraints in the context of climate change and disaster risk. *Climatic Change, 133*, 53–68. https://doi.org/10.1007/s10584-013-0913-2

Blanchard, K., et al. (2023). *Gender and sexual minorities in disaster risk reduction: A reference guide*. GRRIPP Programme. IRDR Centre for Gender and Disaster, University College London.

Bonanno, G. A., Galea, S., Bucciarelli, A., & Vlahov, D. (2007). What predicts psychological resilience after disaster? The role of demographics, resources, and life stress. *Journal of Consulting and Clinical Psychology, 75*(5), 671–682. https://doi.org/10.1037/0022-006X.75.5.671

Brotman, S. (2002). Reclaiming space-regaining health: The health care experiences of two-spirit people in Canada. *Journal of Gay and Lesbian Social Services, 14*(1), 67–87.

Brown, M. (1997). *Replacing citizenship: AIDS activism and radical democracy*. The Guilford Press.

Brown, M. (2011). Gender and sexuality I: Intersectional anxieties. *Progress in Human Geography, 36*(4), 541–550.

Brown, K., & Westaway, E. (2011). Agency, capacity, and resilience to environmental change: Lessons from human development, well-being, and disasters. *Annual Review of Environment and Resources, 36*, 321–342. Available at: https://ssrn.com/abstract=1955076

Butler, J. (1990). *Gender trouble: Feminism and the subversion of identity*. Routledge.

Butler, J. (2024). *Who's afraid of gender*. Farrar.

Calgaro, E., Lloyd, K., & Dominey-Howes, D. (2014a). Application of the *destination sustainability framework (DFS)* to explore the drivers of vulnerability in Thailand following the 2004 Indian Ocean tsunami. *Journal of Sustainable Tourism, 22*(3), 361–383.

Calgaro, E., Lloyd, K., & Dominey-Howes, D. (2014b). From vulnerability to transformation: A framework for assessing the resilience and vulnerability of tourism destinations in a world of uncertainty. *Journal of Sustainable Tourism, 22*(3), 341–360.

Calgaro, E., Craig, N., Craig, L., Dominey-Howes, D., & Allen, J. (2021). Silent no more: Identifying and breaking through the barriers that d/deaf people face in responding to hazards and disasters. *International Journal of Disaster Risk Reduction, 57*, 102156. https://doi.org/10.1016/j.ijdrr.2021.102156

Chandran, D., et al. (2015). Psychosocial impact of the Uttarakhand flood disaster on elderly survivors. *Indian Journal of Gerontology, 29*(1), 62–76.

Choudhury, N. R., & Harini, C. (2023). Aravani as citizen: The forging of a sexual identity. *Advanced Journal of Social Science, 12*(1), 30–38. https://doi.org/10.21467/ajss.12.1.30-38

Choudhury, M., & Wu, H. (2023). Improving the feedback loop between community- and policy-level learning: Building resilience of coastal communities in Bangladesh. *Sustainable Development, 32*, 1–17. https://doi.org/10.1002/sd.2686

Cianfarani, M. (2013). *Supporting the resilience of LGBTQI2S people and households in Canadian disaster and emergency management*. MA thesis. Royal Roads University.

Cody, P. J., & Welch, P. L. (1997). Rural gay men in northern New England: Life experiences and coping styles. *Journal of Homosexuality, 33*(1), 51–67. https://doi.org/10.1300/J082v33n01_04

Cowan, C. (2022). *LGBTQ+ people and disasters: A queer human rights-based critique of vulnerability*. MA thesis. Lund University. Available at: https://lup.lub.lu.se/luur/download?func=downloadFileandrecordOId=9093394andfileOId=9093399

Crenshaw, K. (1991). Mapping the margins: Intersectionality, identity politics, and violence against women of colour. *Stanford Law Review, 43*(6), 1241–1299. https://doi.org/10.2307/1229093

Cretney, R. (2014). Resilience for whom? Emerging critical geographies of socio-ecological resilience. *Geography Compass, 8*(9), 627–640. https://doi.org/10.1111/gec3.12154

Cullen, B. T., & Pretes, M. (2000). The meaning of marginality: Interpretations and perceptions in social science. *The Social Science Journal, 37*(2), 215–229. https://doi.org/10.1016/S0362-3319(00)00056-2

Cutter, S. L., Boruff, B. J., & Shirley, W. L. (2003). Social vulnerability to environmental hazards. *Social Science Quarterly, 84*(2), 242–261.

D'Ooge, C. (2008). Queer Katrina: Gender and sexual orientation matters in the aftermath of the disaster. In B. Willinger (Ed.), *Katrina and the women of New Orleans* (pp. 22–24). Newcomb College Centre for Research on Women—Tulane University.

Dasgupta, S., Sinha, S., & Roy, R. (2021). "We are helpless, hopeless and living in despair": Impact of *COVID-19* on the overall health and well-being and participation of the transgender community in India. *The International Journal of Community and Social Development, 3*(4), 372–389. https://doi.org/10.1177/25166026211050743

Dolan, C. (2014). Letting go of the gender binary: Charting new patterns of humanitarian interventions on gender-based violence. *International Review of the Red Cross, 96*(894), 485–501.

Dominey-Howes, D. (2018). Natural hazards and disasters in the Anthropocene: Some critical reflections for the future. *Geoscience Letters, 5*(7). https://doi.org/10.1186/s40562-018-0107-x

Dominey-Howes, D., Gorman-Murray, A., & McKinnon, S. (2014). Queering disasters: On the need to account for LGBTI experiences in natural disaster contexts. *Gender, Place and Culture, 21*(7), 905–918. https://doi.org/10.1080/0966369x.2013.802673

Dominey-Howes, D., et al. (2016). Emergency management response and recovery plans in relation to gender and sexual minorities in New South Wales, Australia. *International Journal of Disaster Risk Reduction, 16*, 1–11.

Dominey-Howes, D., Gorman-Murray, A., & McKinnon, S. (2018). On the disaster experiences of sexual and gender (LGBTI) minorities: Insights to support inclusive disaster risk reduction policy and practice. *Australian Journal of Emergency Management*, 60–68.

Dominey-Howes, D., McKinnon, S., Gorman-Murray, A., & Eriksen, C. (2022). *Gender and sexual minorities in disaster*. Encyclopedia Entry ETH Zurich. https://doi.org/10.4324/9780367854584-19

Douglass, M., & Miller, M. A. (2018). Disaster justice in Asia's urbanising anthropocene. *Environment and Planning E: Nature and Space, 1*(3), 271–287. https://doi.org/10.1177/2514848618797333

Drolet, D., et al. (2015). Women rebuilding lives post-disaster: Innovative community practices for building resilience and promoting sustainable development. *Gender and Development, 23*(3), 433–448. https://doi.org/10.1080/13552074.2015.1096040

Eads, M. (2002). *Marginalized groups in times of crisis: Identity, needs, and response*. Quick Response Report #152. Natural Hazards Canter.

Edge Effect. (2021). *The only way is up: Monitoring and encouraging diverse SOGIESC inclusion in the humanitarian and DRR sectors*. Empower—Women for Climate-Resilient societies programme. Available at: https://www.edgeeffect.org/wp-content/uploads/2021/03/TheOnlyWayIsUp_Web.pdf

Enarson, E., & Morrow, B. H. (1998). Conclusion: Toward gendered disaster policy, practice, and research. In E. Enarson & B. H. Morrow (Eds.), *The gendered terrain of disaster: Through women's eyes* (pp. 225–249). Praeger/Greenwood.

Enarson, E., Fothergill, A., & Peek, L. (2007). Gender and disaster: Foundations and directions. In H. Rodriguez, E. L. Quarantelli, & R. R. Dynes (Eds.), *Handbook of disaster research. Handbooks of sociology and social research* (pp. 130–146). Springer. https://doi.org/10.1007/978-0-387-32353-4_8

Fontanez, J. A. (2019). *Transgender an at-risk population during and following emergencies and disasters*. PhD dissertation, Walden University. https://scholarworks.waldenu.edu/cgi/viewcontent.cgi?article=8567andcontext=dissertations

Fordham, M. H. (1999). The intersection of gender and social class in disaster: Balancing resilience and vulnerability. *International Journal of Mass Emergencies and Disaster, 17*(1), 15–36.

Fothergill, A. (1996). Gender, risk, and disaster. *International Journal of Mass Emergencies and Disasters, 14*(1), 33–56. https://doi.org/10.1177/028072709601400103

Fothergill, A. (2004). *Heads above water: Gender, class, and family in the Grand Forks flood.* State University of New York Press.

Fox, M., & Wu, H. (2023). Terminology and language used in indigenous-specific gender and sexuality diversity studies: A systematic review. *Social Sciences, 12*(3), 145. https://doi.org/10.3390/socsci12030145

Fredriksen-Goldsen, K. I., et al. (2014). The health equity promotion model: Reconceptualisation of lesbian, gay, bisexual and transgender (LGBT) health disparities. *The American Journal of Orthopsychiatry, 84*, 653–663.

Gaillard, J. C. (2010). Vulnerability, capacity and resilience: Perspectives for climate and development policy. *Journal of International Development, 22*(2), 218–232. https://doi.org/10.1002/jid.v22:2

Gaillard, J. C. (2011). *People's response to disaster: Vulnerability, capacities, and resilience in The Philippines.* Centre for Kapampangan Studies.

Gaillard, J. C., Gorman-Murray, A., & Fordham, M. (2017a). Gender and sexual minorities in disaster. *Gender, Place and Culture, 24*(1), 18–26. https://doi.org/10.1080/0966369X.2016.1263438

Gaillard, J. C., et al. (2017b). Beyond men and women: A critical perspective on gender and disaster. *Disasters, 41*(3), 429–447.

Gilbert, A., & Barkun, M. (1981). Disaster and sexuality. *Journal of Sex Research, 17*(3), 288–299. Available at: http://www.jstor.org/stable/3812564

Goldsmith, L., Raditz, V., & Méndez, M. (2021). *Policy brief: The need for equitable disaster response for LGBTQ+ communities.* University of California, Irvine.

Goldsmith, L., Raditz, V., & Mendez, M. (2022). Queer and present danger: Understanding the disparate impacts of disasters on LGBTQ+ communities. *Disasters, 46*(4), 946–973.

Gorman-Murray, A., et al. (2017). Problems and possibilities on the margins: LGBT experiences in the 2011 Queensland floods. *Gender, Place and Culture, 24*(1), 37–51.

Gorman-Murray, A., et al. (2018). Listening and learning: Giving voice to trans experiences of disasters. *Gender, Place and Culture, 25*(2), 166–187. https://doi.org/10.1080/0966369x.2017.1334632

Grover, J. Z. (1988). AIDS: Key words. In D. Crimp (Ed.), *AIDS: Cultural analysis cultural activism* (pp. 17–30). The MIT Press.

Gurung, G. S., & Kollmair, M. (2007). *Marginality: Concepts and their limitations* (2nd ed.). NCCR North-South Dialogue.

Habib, L. A. (1995). "Women and disability don't mix!": Double discrimination and disabled women's rights. *Gender and Development, 3*(2), 49–53.

Haworth, B. T. (2021). Learning from LGBTIQ+ experiences of *COVID-19* in the UK for future crises. *HCRI Policy Brief Series, 1*(21), 1–4. Available at: https://documents.manchester.ac.uk/display.aspx?DocID=53042

Haworth, B. T., Bruce, E., Whittaker, J., & Read, R. (2018). The good, the bad, and the uncertain: Contributions of volunteered geographic information to community disaster resilience. *Frontiers of Earth Science 6*, Article 183, 1–15. https://doi.org/10.3389/feart.2018.00183

Haworth, B. T., McKinnon, S., & Eriksen, C. (2022). Advancing disaster geographies: From marginalisation to inclusion of gender and sexual minorities. *Geography Compass, 16*(11), e12664. https://doi.org/10.1111/gec3.12664

Hines, S. (2019). The feminist frontier: On trans and feminism. *Journal of Gender Studies, 28*(2), 145–157.

hooks, b. (1990). *Yearning: Race, gender and cultural politics.* South End Press.

Hopkins, P. (2019). Social geography I: Intersectionality. *Progress in Human Geography, 43*(5), 937–947. https://doi.org/10.1177/0309132517743677

Ikeda, K. (1995). Gender difference in human loss and vulnerability in natural disasters: A case study from Bangladesh. *Indian Journal of Gender Studies, 2*(2), 171–193.

ILGA Europe. (2020). *COVID--19 impacts on LGBTI communities in Europe and Central Asia: A rapid assessment report.* Available at: https://www.ilga-europe.org/sites/default/files/covid19-lgbti-assessment-2020.pdf (Accessed 2 July 2020).

International Gay and Lesbian Human Rights Commission and SEROVie. (2011). *The impact of the earthquake, and relief and recovery programs on Haitian LGBT people.* The International Gay and Lesbian Human Rights.

Jacobs, F. (2019). Black feminism and radical planning: New directions for disaster planning research. *Planning Theory, 18*(1), 24–39.

Jagose, A. (1996). *Queer theory.* Melbourne University Press.

Jones, T. W. (2023). Reviled, reclaimed, and respected: The history of the word queer. *The Conversation*, 19 January. Available at: https://theconversation.com/reviled-reclaimed-and-respected-the-history-of-the-word-queer-197533

Kailes, J. I., & Lollar, D. J. (2021). Disasters and disability: Rhetoric and reality. In D. J. Lollar, W. Horner-Johnson, & K. Froehlich-Grobe (Eds.), *Public health perspectives on disability* (pp. 251–268). Springer. https://doi.org/10.1007/978-1-0716-0888-3_12

Kane, P. L. (2020). 'Unresolved grief: Coronavirus presents eerie parallels for many aids advocates. *The Guardian*, 22 March. Available at: https://www.theguardian.com/world/2020/mar/22/coronavirus-aids-epidemic-san-francisco

Katz, C. (2008). Bad elements: Katrina and the scoured landscape of social reproduction. *Gender, Place and Culture, 15*(1), 15–29.

Kendra, M. J., Clay, L. A. & Gill, K. B. (2018). 'Resistence and disasters'. In H. Rodriguez, W. Donner & J. E. Trainor (Eds.), *Handbook of Disaster Research, Second edition* (pp. 87–107). Springer.

Kelly, S., & Smith, W. (2012). Marginality, adaptation and farming in the New Zealand high country. *Journal of Alpine Research, 100*(1) http://rga.revues.org/1711

Khan, S. I., et al. (2009). Living on the extreme margin: Social exclusion of the transgender population (Hijra) in Bangladesh. *Journal of Health, Population, and Nutrition, 27*(4), 441–451. https://doi.org/10.3329/jhpn.v27i4.3388

King, D. (2022). Hearing minority voices: Institutional discrimination towards LGBTQ in disaster and recovery. *Journal of Extreme Events, 8*(4), 2241005-1-20. https://doi.org/10.1142/S2345737622410056

Knight, K., & Sollom, R. (2012). Making disaster risk reduction and relief programmes LGBTI inclusive: Examples from Nepal. *Humanitarian Exchange, 55*, 36–39.

Knight, K., & Welton-Mitchell, C. (2013). Gender identity and disaster response in Nepal. *Forced Migration Review, 42*, 57–58. https://www.fmreview.org/sogi/knight-weltonmitchell

Krishnaraj, M. (1997). Gender issues in disaster management: The Latur earthquake. *Gender, Technology and Development, 1*(3), 395–411. https://doi.org/10.1177/097185249700100304

Lambert, S., & Mark-Shadbolt, M. (2021). Indigenous knowledges of forest and biodiversity management: How the watchfulness of Māori complements and contributes to disaster risk reduction. *AlterNative: An International Journal of Indigenous Peoples, 17*(3), 368–377. https://doi.org/10.1177/11771801211038760

Larkin, B. (2019). Pride and prejudice: LGBTIQ community responses to disaster events worldwide. *Australian Journal of Emergency Management, 34*(3), 60–66.

Lay, Y., Leonard, W., Horsely, P., Carmen, M., & Parsons, M. (2018). *Primary prevention of family violence against people from LGBTI communities: An analysis of existing research*. OurWatch.

Leonard, W. (2005). Queer occupations: Development of Victoria's gay, lesbian, bisexual, transgender and intersex health and wellbeing action plan. *Gay and Lesbian Issues and Psychology Review, 1*(3), 92–97.

Leonard, W. (2012). Safe sex and the aesthetics of gay men's HIV/AIDS prevention in Australia: From Rubba me in 1984 to F**k me in 2009. *Sexualities, 15*(7), 834–849.

Leonard, W. (2014). From disease prevention to health promotion. In M. Temple-Smith (Ed.), *Sexual health: A multidisciplinary approach* (pp. 335–350). IP Communications.

Leonard, W., et al. (2022). Under pressure: Developing lesbian, gay, bisexual, transgender and intersex (LGBTI) inclusive emergency services. *Australian Journal of Emergency Management, 37*(1), 52–58.

LGBT Foundation. (2020). *The essential briefing on the impact of COVID--19 on LGBT communities in the UK*. Available at: www.lgbt.foundation

Lukasiewicz, A. (2020). The emerging imperative of disaster justice. In A. Lukasiewicz & C. Baldwin (Eds.), *Natural hazards and disaster justice: Challenges for Australian and its neighbours* (pp. 3–24). Palgrave Macmillan.

Marable, M. (2006). Race, class and the Katrina crisis. *The Journal of Labor and Society, 9*(2), 155–160.

McKinnon, S., Gorman-Murray, A., & Dominey-Howes, D. (2016). The greatest loss was a loss of our history: Natural disasters, marginalised identities and sites of memory. *Social and Cultural Geography, 17*(8), 1120–1139.

McLean, C. (2021). The growth of the anti-transgender movement in the United Kingdom. The silent radicalization of the British electorate. *International Journal of Sociology, 51*(6), 473–482.

Native Women's Association of Canada. (n.d.). *Sex and gender inclusive terminology.* Available at: https://nwac.ca/assets-knowledge-centre/NWAC-2SLGBTQ-Factsheet-EN.pdf

Omoto, A. M., & Crain, A. L. (1995). AIDS volunteerism: Lesbian and gay community-based responses to HIV. In G. M. Herek & A. L. Crain (Eds.), *AIDS, identity, and community: The HIV epidemic and lesbians and gay men* (Vol. 2, pp. 187–209). Sage.

Ong, J. C. (2017). Queer cosmopolitanism in the disaster zone: "My Grindr became the United Nations". *International Communication Gazette, 48*, 656. https://doi.org/10.1177/1748048517727177

Overs, C., & Druce, N. (1994). Sex work, HIV and the state: An interview with Nel Druce. *Feminist Review, 48*, 114–121.

Overton, L., & Hepple, J. (2022). Negotiating pleasure and danger: Intersections of LGBTQ+ social spaces using gender analysis and the social model. In J. Healy & B. Colliver (Eds.), *Contemporary intersectional criminology in the UK: Examining the boundaries of intersectionality and crime* (pp. 204–219). Bristol University Press. https://doi.org/10.1332/policypress/9781529215946.003.0013

Ozawa, K. (2012). Relief activities for LGBTI people in the affected areas. *Voices from Japan, 26*, 21–22.

Pandya, A., & Redcay, A. (2022). Impact of *COVID--19* on Transgender Women and Hijra: Insights from Gujarat, India. *Journal of Human Rights and Social Work, 7*(2), 148–157.

Perera-Mubarak, K. N. (2013). Positive responses, uneven experiences: Intersections of gender, ethnicity, and location in post-tsunami Sri Lanka. *Gender, Place and Culture, 20*(5), 664–685.

Phibbs, S., Good, G., Severinsen, C., Woodbury, E., & Williamson, K. (2015). Emergency preparedness and perceptions of vulnerability among disabled people following the Christchurch earthquakes: Applying lessons learnt to the Hyogo framework for action. *Australasian Journal of Disaster and Trauma Studies, 19*(Special issue), 37–46.

Phillips, B., & Fordham, M. (2009). Introduction. In *Social vulnerability to disasters*. Taylor and Francis.

Pincha, C., & Krishna, H. (2008). Aravanis: Voiceless victims of the tsunami. *Humanitarian Exchange, 41*, 41–43.

Pruden, H., & Edmo, S. (2016). Two-spirit people: Sex, gender and sexuality in historic and contemporary Native America. *National Congress of American Indians Policy Research Centre*. Available at: https://www.ncai.org/policy-researchcentre/initiatives/Pruden-Edmo_TwoSpiritPeople.pdf

Quinn, K. G., Walsh, J. L., John, S. A., & Nyitray, A. G. (2021). I feel almost as though I've lived this before': Insights from sexual and gender minority men on coping with *COVID--19. AIDS and Behavior, 25*, 1–8.

Rayside, D. M., & Lindquist, E. A. (1992). AIDS activism and the state in Canada. *Studies in Political Economy, 39*, 37–76.

Robinson, B. A. (2016). Heteronormativity and homonormativity. In *The Wiley Blackwell encyclopedia of gender and sexuality studies*. Wiley. https://doi.org/10.1002/9781118663219

Roy, S. K., Sarker, B. K., & Chowdhury, S. H. (2021). Socioeconomic status of Hijra (third gender) Community at Dhaka City in Bangladesh: A descriptive study. *The American Journal of Humanities and Social Science Research, 4*(1), 84–95.

Rubin, G. (1984). Thinking sex: Notes for a radical theory of the politics of sexuality. In C. Vance (Ed.), *Pleasure and danger: Exploring female sexuality* (pp. 267–319). Routledge and Kegan Paul.

Rushton, A., & Scarlett, J. (2023). The blame game: Disaster, queerness and prejudice. *Disaster Prevention and Management, 32*(2), 352–363. https://doi.org/10.1108/DPM-11-2022-0222

Rushton, A., Gray, L., Canty, J., & Blanchard, K. (2019). Beyond binary: (re) defining "gender" for 21st century disaster risk reduction research, policy, and practice. *International Journal of Environmental Research and Public Health, 16*(20), 3984. https://doi.org/10.3390/ijerph16203984

Rushton, A., Phibbs, S., Kenney, C., & Anderson, C. (2020). The gendered body politic in disaster policy and practice. *International Journal of Disaster Risk Reduction, 47*(2), 101648. https://doi.org/10.1016/j.ijdrr.2020.101648

Shreve, C., Davis, B., & Fordham, M. (2017). Gender and disaster resilience. In D. Paton & D. Johnston (Eds.), *Disaster resilience: An integrated approach* (2nd ed., p. 182). Charles C. Thomas.

Simmonds, K., et al. (2022). Health impacts of climate changes of gender diverse populations: A scoping review. *Journal of Nursing Scholarship, 54*(81), 81–91.

Smith, F. (2013). *Gender minorities and disaster: A case study of Fa'afafine in the 2012 cyclone in Samoa*. BA Hons thesis, The University of Auckland.

Steckley, M., & Doberstein, B. (2011). "Tsunami survivors" perspectives on vulnerability and vulnerability reduction: Evidence from Koh Phi Phi Don and Khao Lak, Thailand. *Disasters, 35*(3), 465–487. https://doi.org/10.1111/j.1467-7717.2010.01221.x

Stryker, S. (2008). Transgender history, homonormativity, and disciplinarity. *Radical History Review, 100*, 145–157. https://doi.org/10.1215/01636545-2007-026

The Yogyakarta Principles. (2017). *Principles on the application of international human rights law in relation to sexual orientation and gender identity*, and The Yogyakarta principles plus 10.

Thomas, D. S. K., Phillips, D., Lovekamp, W. E., & Fothergill, A. (Eds.). (2013). *Social vulnerability to disasters* (2nd ed.). CRC Press, Taylor and Francis Group.

Ton, K. T., & Adamson, C. E. (2020). Disaster justice for people with disabilities. *Disaster Prevention and Management, 30*, 1–12. https://doi.org/10.1108/DPM-08-2020-0248

United Nations. (2020). *Covid-19 and the human rights of LGBTI People. COVID-19* response. Office of the High Commissioner. Available at: https://www.ohchr.org/sites/default/files/Documents/Issues/LGBT/LGBTIpeople.pdf

United Nations Office for Disaster Risk Reduction. (2005). *Hyogo framework for action 2005–2015*. UNDRR. Available at: https://www.unisdr.org/we/coordinate/hfa

United Nations Office for Disaster Risk Reduction. (2015). *Sendai framework for disaster risk reduction 2015–2030*. UNDRR. Available at: http://www.preventionweb.net/files/43291_sendaiframeworkfordrren.pdf

Urbatsch, R. (2016). Judgement days: Moral attitudes in the wake of local disasters. *Disasters, 40*(1), 26–44.

Victorian Government Department of Human Services. (2003). *Health and sexual diversity: A health and wellbeing action plan for gay, lesbian, bisexual, transgender and intersex (GLBTI) Victorians*. Prepared by William Leonard for the Ministerial Advisory Committee on Gay and Lesbian Health. Victorian Government Department of Human Services.

Webster, A. K. (2021). 'Embedded agency: Lessons from marginalized identities. In D. David Shoemaker (Ed.), *Oxford studies in agency and responsibility 7* (Oxford online ed., pp. 104–129). https://doi.org/10.1093/oso/9780192844644.003.0006

Wisner, B. (1998). Marginality and vulnerability: Why the homeless of Tokyo don't "count" in disaster preparations. *Applied Geography, 18*(1), 25–33. https://doi.org/10.1016/S0143-6228(97)00043-X

Wisner, B. (2009). Vulnerability. In I. Hay & R. Kitchin (Eds.), *International encyclopaedia of human geography* (pp. 176–182). Oxford University Press.

Women and Gender Equality Canada. (2022). *2SLGBTQI+ terminology—Glossary and common acronyms*. Government of Canada. Available at: https://women-gender-equality.canada.ca/en/free-to-be-me/2slgbtqi-plus-glossary.html

Wu, H., & Karabanow, J. (2020). *COVID-19* and beyond: Social work interventions for supporting homeless population. *International Social Work, 63*(6), 790–794. https://doi.org/10.1177/0020872820949625

Zhou, H., Wan, J., & Ji, H. (2010). Resilience to natural hazards: A geographic perspective. *Natural Hazards, 531*(1), 21–41. https://doi.org/10.1007/s11069-009-9407-y

CHAPTER 4

Queer Vulnerability and Political Power in the Face of Climate Changes and Associated Disasters

Cameron T. Whitley and Melanie M. Bowers

Introduction

LGBTQ+ individuals make up a relatively small group within the USA,[1] accounting for about 7.1 percent of the population (Jones, 2022). Nonetheless, they face social and economic marginalisation linked to higher rates of homelessness (Ecker et al., 2019; Fraser et al., 2019); mental health distress and suicidal ideation (Clark et al., 2020; Green et al., 2022); substance abuse (Moazen-Zadeh et al., 2019); social and family

[1] This chapter focuses on the USA. For examples of current research on LGBTQ+ climate change issues in other countries, see Dalton (2023), Kilpatrick et al. (2023).

C. T. Whitley (✉)
Department of Sociology, Western Washington University,
Bellingham, WA, USA
e-mail: whitlec@wwu.eu

M. M. Bowers
Department of Political Science, Rutgers University, Camden, NJ, USA

D. Dominey-Howes et al. (eds.), *Queering Disasters, Climate Change and Humanitarian Crises*, Sustainable Development Goals Series, https://doi.org/10.1007/978-981-96-3857-4_4

violence (Newcomb et al., 2019; Adamson et al., 2022; Turner & Hammersjö, 2023); and poverty (Badgett et al., 2019). At the same time, LGBTQ+ people are also subject to political marginalisation, as evidenced by the hundreds of US state-level policies that have been introduced and passed to limit their visibility, rights and autonomy (Kilander, 2023). Among LGBTQ+ people, transgender people have been subjected to the greatest levels of abuse and discrimination, including bias-motivated incidents and anti-transgender legislation (Barr et al., 2021; Barbee et al., 2022; Cunningham et al., 2022; Drabish & Theeke, 2022).

Research on how climate change affects the LGBTQ+ community in the USA suggests it will exacerbate the causes and impacts of discrimination against them, with the most socially, economically and politically marginalised experiencing the worst outcomes (Thomas et al., 2019). The impacts of climate change on LGBTQ+ people include reducing accessible housing, exacerbating mental health challenges, creating fragmented communities, reducing access to LGBTQ+ affirming healthcare and potentially limiting access to emergency services (Richards, 2010; Dominey-Howes et al., 2014; Vinyeta et al., 2016; Gaillard et al., 2017; Technical Advisory Group, 2017; Goldsmith et al., 2022; Goldsmith & Bell, 2022; Whitley & Bowers, 2023).

In this chapter, we discuss the relationship between climate change and the LGBTQ+ population in the USA. The chapter focuses on how the current US political climate is likely to limit LGBTQ+ people's access to key resources, what LGBTQ+ people are thinking about climate change and how they are mobilising to address it. The chapter also considers what future research is needed to better understand the intersections between sexuality and gender identity and climate change impacts, attitudes and advocacy.

The Impact of Climate Change on LGBTQ+ People in the USA

The Fourth National Climate Assessment, produced by the U.S. Global Change Research Program, details how climate change is exacerbating extreme events such as hurricanes, heatwaves, wildfires, flooding and disease outbreaks in the USA. The Assessment suggests that these events will become more frequent and intense over time (U.S. Global Change Research Program, 2018). Although it is understood that underrepresented and marginalised communities are most at risk from climate change (Thomas et al., 2019; Dietz et al., 2020), the impacts of climate change on LGBTQ+ communities have received limited attention (Cappelli et al., 2021; Dietz et al., 2020). Despite the lack of attention, LGBTQ+ communities face social, economic and political marginalisation, which are likely to present distinct climate change related challenges.

One of the biggest factors influencing both a group's vulnerability and ability to respond to climate-related disasters in the USA is socioeconomic status, which influences everything from one's ability to evacuate to one's ability to resettle and recover after disaster. LGBTQ+ populations have higher proportions of individuals with extremely low socioeconomic status compared to the general population. The USA Census Bureau estimates that approximately 11 percent (37.2 million people) of the total USA population lives in poverty[2] (US Census Bureau, 2022), though racial and structural inequalities mean that this figure varies by demographics. For example, just 8 percent of white individuals live in poverty compared to 20 percent of Black individuals (US Census Bureau, 2022), and women and gender variant individuals experience higher poverty rates than cisgender men (Badgett et al., 2019; Smith & Mazure, 2021). LGBTQ+ communities also face higher poverty rates, with nearly a quarter (22 percent) of LGBTQ+ people and one third of transgender people in the USA living in poverty (Badgett et al., 2019). These figures are worse for those with intersectional identities. For example, LGBTQ+ people of colour experience poverty rates ranging from 22.9 to 42.1 percent (Badgett et al., 2019).

[2] The Census provides annual geographic and population-based poverty estimates. Poverty rates are based on family size and income. If a family's total cash income before taxes (and excluding non-cash payments like housing benefits) falls below a time-dependent threshold, they are considered in poverty. In 2024, the income threshold for a family of four was $31,200.

In addition to having direct implications for one's ability to evacuate and rebuild or resettle after disaster, poverty can also be an indicative characteristic that is likely to either amplify or be amplified by the impacts of climate change. For example, LGBTQ+ populations have high levels of homelessness (Glynn et al., 2021). This means that more people are susceptible to extreme temperatures, water shortages, natural disasters and food instability, while living on the streets or in shelters. Those in poverty and/or who are homeless are also more likely to experience health issues that can be exacerbated by things like wildfire smoke and high temperatures, and are more likely to have limited access to healthcare before, during and after disaster (Clarke & Erreygers, 2020).

Beyond poverty, a lack of recognition of LGBTQ+ families, bias in federal disaster programmes, and the prevalence and reliance on faith-based disaster relief programmes all put LGBTQ+ people at further risk in climate-related disasters (Goldsmith et al., 2022). Limited research assessing the impacts of disasters on LGBTQ+ individuals and communities has demonstrated that accessing shelter and healthcare can be particularly challenging for members of this group. As part of their research on how LGBTQ+ peoples' vulnerability in major climate-related disasters varies across the globe, Dominey-Howes et al. (2014) examine how the LGBTQ+ community was affected by Hurricane Katrina (Katrina).

Katrina made landfall as a Category 5 storm on August 29, 2005, over southeastern Louisiana and Mississippi. It quickly became one of the deadliest and most expensive disasters in US history (Dolfman et al., 2007). Between 1200 and 1836 people were killed and property damage topped $200 billion (Beven et al., 2008). Katrina was one of the deadliest hurricanes in US history largely because levees protecting New Orleans failed, flooding 80 percent of the city and damaging or destroying over 800,000 homes. LGBTQ+ people were among those with the worst outcomes from the storm because they had reduced resiliency, were significantly affected by loss of community, faced discriminatory government policies and had to confront organisational bias (Dominey-Howes et al., 2014).

Many shelter facilities use government sex assigned at birth to allocate space to individuals. This can put transgender and non-binary people who do not have gender-conforming documentation at risk of being placed in inappropriate locations. During Katrina there were also cases of LGBTQ+ folks who were excluded from shelters and other facilities based on perceived or documented gender and sexual orientation. Their stories often emerged long after the disaster and the documentation of and response to those stories has been inadequate (Richards, 2010; Cianfarani, 2012; Dominey-Howes et al., 2014). The journalistic accounts and academic studies that have emerged in the years following Katrina have demonstrated that LGBTQ+ people, and specifically transgender and intersex people, faced difficulties in accessing restrooms, evacuation shelters and affirming healthcare (D'Ooge, 2008).

In addition, they faced community backlash. Rather than focusing on the environmental and social conditions that exacerbated the impacts of the hurricane, many conservative Christian religious leaders discussed the hurricane as 'God's punishment' for the sin of man and particularly the depravity of the LGBTQ+ community. This type of response is indicative of a broader trend in the USA and in other parts of the world where LGBTQ+ people are blamed by religious leaders for environmental disasters, placing them at risk for verbal and physical violence, and exclusion from resources (Tierney, 2019; Greenhalgh, 2020; Goldsmith & Bell, 2022; King, 2022; Rushton & Scarlett, 2023). These challenges reflect failures at the federal, state and local levels, as well as in the private and non-profit sectors.

Katrina led to extensive discussions about how to better prepare for disasters and how to protect vulnerable populations (United States Executive Office of the President et al., 2006). These discussions focused on older adults, the disabled community and people of colour, while LGBTQ+ people were largely excluded. Since that time, cities and states across the USA have engaged in climate mitigation and adaptation efforts, but few have plans to specifically consider the needs of, and impacts on, LGBTQ+ people (Goldsmith & Bell, 2022; Whitley & Bowers, 2023).

Over a decade ago, the Human Rights Campaign (HRC), which is one of the largest LGBTQ+ organisations in the country, detailed how to best support LGBTQ+ people in disasters (HRC, 2012). The document focused on cultural competency in understanding different family structures and the types of support they might need. However, the majority of disaster plans have failed to incorporate these recommendations (Goldsmith & Bell, 2022). The state of California, which passed SB 160 in 2019 requiring the integration of cultural competency into emergency planning (Open States, 2023), is a notable exception. While the exclusion of LGBTQ+ people from climate change policy is likely an oversight in many places, it speaks to the cis-normative bias that exists in policy processes that place LGBTQ+ people at increased risk (Goldsmith et al., 2022).

Risk-related policies that do not address LGBTQ+ concerns do not just have a simple omission but also represent a serious exclusion, and can create opportunities for significant harm and discrimination (Caldwell, 2006).

The LGBTQ+ Political and Religious Landscape in the USA

The political and religious landscape in the USA heightens the vulnerability of LGBTQ+ people, particularly in climate-related disasters. In the USA, LGBTQ+ people have limited legal protections and face ongoing social stigma. Political and policy attacks on LGBTQ+ people are hardly new, but the increase in recent years reflects several things. These include amplified feelings of threat on the political and religious right due to growing LGBTQ+ rights and societal acceptance, the shift in the US Supreme Court to a conservative majority, the political rise of the far right, and an increasing focus on and vilification of LGBTQ+ populations and liberal social values amongst conservative elites (e.g. Lempinen, 2022).

In the first half of 2023, more than 500 anti-LGBTQ+ bills were introduced in state legislatures in the USA (Kilander, 2023) and the Supreme Court ruled that creative businesses can discriminate against LGBTQ+ people (*303 Creative LLC* et al. *V. Elenis* et al.). Political and social attacks on the LGBTQ+ community have taken the form of drag bans, limits on transgender affirming care, and increasingly open social discrimination. This discrimination ranges from book bans to hate crimes, judicial rulings that legalise discrimination in services and other issues, all of which highlight the challenges LGBTQ+ people face under ordinary conditions. The trend has continued into 2025 and will likely continue well into the future.

Although the direct attacks have largely occurred in Republican-led US states, they influence the entire nation by doing things like re-normalising open homo-, bi- and trans- phobia and facilitating national boycotts of companies that support LGBTQ+ people and rights. They also encourage or pressure LGBTQ+ people to exit red states and/or increasingly use resources in more LGBTQ+ supportive states. New laws allow doctors to refuse care to LGBTQ+ people based on religious or moral beliefs, prohibit public schools from using students' preferred pronouns and ban children from interacting with drag queens, though many of these laws are facing legal challenges (Rozsa, 2023). This new legislation compounds existing laws that restrict transgender youth from accessing transgender affirming care and make it increasingly difficult for transgender adults to continue receiving care (Rozsa, 2023). For instance, when Florida began passing anti-LGBTQ+ legislation in late 2022, LGBTQ+ people began to flee Florida for more supportive areas (Kilander, 2023). The stories of LGBTQ+ people leaving Florida are numerous. Among them is Floridian Sage Chelf, a transgender woman living in Orlando. After Florida's SB254 was signed into law by Governor Ron DeSantis, she was no longer able to receive hormone medication to continue her transition, causing her to leave the state (Kilander, 2023). Other transgender people have noted that the bathroom bill, HB1521, bans them from using a public restroom based on their presented gender, as the bill requires people to use the bathroom associated with the sex they were assigned at birth (Kilander, 2023).

The merging of Republicanism and evangelical Christian ideals has also increased social and political attacks on the LGBTQ+ community (Karol, 2023). When we begin to think about how these laws and social patterns would be amplified in climate-related disasters, the picture becomes more concerning. Many of the projected climate change impacts on the USA, such as an increase in the frequency and severity of hurricanes, wildfires, drought and heatwaves, are expected to occur in states that are actively restricting LGBTQ+ rights and facilitating discrimination. For example, states in the Southeast—Texas, Louisiana, Georgia, Alabama, Florida, South Carolina, North Carolina, and Mississippi—face the greatest risk from hurricanes (FEMA, 2024). Based on the US Census' Household Pulse Survey, in 2023, 7.3 percent of Louisiana's population was displaced by hurricanes, flooding and other natural disasters (Davis, 2023). These 8 states introduced 136 pieces of anti-LGBTQ+ legislation in 2023, representing about 27 percent of all anti-LGBT legislation in the nation (ACLU, 2023).

The combination of high disaster risk and anti-LGBTQ+ legislation could have devastating consequences for the LGBTQ+ community. Florida continues to seek some of the most restrictive policies, such as SB1580, which allow healthcare providers to refuse to provide any medical treatment that violates their religious or moral beliefs. This means that emergency medical technicians (EMTs), paramedics and doctors could refuse lifesaving care to transgender people during or after a disaster. These examples highlight why the scholarly and policy communities need to include this group in discussions about the impacts of climate change and consider how anti-LGBTQ+ policies and anti-LGBTQ+ sentiment interact with climate mitigation and disaster response policies. Without securing protections for LGBTQ+ people, we will repeat the problems that arose during hurricane Katrina and more recent disasters, where the fragmentation of the LGBTQ+ community left many to fend for themselves as they tried to find appropriate shelter and healthcare.

Queer Political Culture (QPC) and Climate Change Attitudes

Although LGBTQ+ communities face many challenges from climate change and climate-related disasters, they also possess strengths that make them uniquely positioned to pursue advocacy, policy change and climate resiliency. An expanding body of literature suggests that LGBTQ+ communities have a distinct political culture—queer political culture or QPC—that encourages higher levels of political participation and activism than is seen in the general public (Swank & Fahs, 2013, 2017, 2019; Bowers & Whitley, 2020; Strode & Flores, 2021). Although the concept lacks a single definition, the basic argument behind QPC is that the combination of shared grievances that require political recourse (such as discrimination and lack of civil rights), a tight-knit and historically isolated community, and common socialisation results in increased political and social awareness that activates political engagement. Researchers argue that QPC increases voter registration and voting in ways that help overcome barriers to political participation (Swank & Fahs, 2013; Bowers & Whitley, 2020; Strode & Flores, 2021), as well as encourages participation in protests (Swank & Fahs, 2017, 2019).

While earlier work (Egan, 2012, 2020; Swank, 2018a, b) suggested that increased political engagement among LGBTQ+ people may simply reflect demographic trends among those who *openly* identify as LGBTQ (e.g. historically out individuals tended to be more liberal and better educated), changing social and demographic trends challenge this explanation (Gates, 2017). As social acceptance of LGBTQ+ people has increased, a more diverse group of individuals has begun openly identifying as LGBTQ+ (Jones, 2022), suggesting that higher levels of politicisation reflect more than mere demographics. Scholars have identified two primary drivers of QPC that reflect a combination of socialisation and shared trauma or oppression.

LGBTQ+ communities offer unique spaces of social acceptance for many and provide individuals with extensive social ties with other members of the group, creating in-group networks (Bowers & Whitley, 2020; Swank, 2018a, b; Whitley & Bowers, 2023). Networks and social connections occur within all types of communities and identities, raising the question of what facilitates greater political responsiveness in LGBTQ+ communities. Research suggests that the answer lies in shared oppression, with both historic and contemporary roots, and shared socialisation (Swank, 2018a, b; Bowers & Whitley, 2020; Whitley & Bowers, 2023).

LGBTQ+ people often experience social isolation, harassment and restricted rights, even in the most accepting locations. It is only in the last decade that marriage equality, protections against workplace discrimination and other rights have become US law, but in many states and localities these rights remain contested. Negative experiences and shared oppression help unify communities and encourage a collective identity that manifests in identity-specific spaces such as LGBTQ+ community centres and bars, the creation of LGBTQ+ advocacy organisations, and higher levels of social pressure to engage in political initiatives (James, 2018; Bowers & Whitley, 2020; Cravens, 2020; Whitley & Bowers, 2023). Even without contemporary attacks on its rights, LGBTQ+ populations remain grounded in their history of political and social organising and resistance. In the USA, Stonewall, which was the 1969 riot credited with starting the LGBTQ+ rights movement, the AIDS crisis and the fight for civil rights inform contemporary understandings of what it means to be part of the LGBTQ+ community (e.g. Bernstein, 2002; Weststrate & McLean, 2010).

This understanding of organising, advocacy and activism has moved beyond LGBTQ+ specific issues. Although not often recognised, LGBTQ+ people have long been key contributors to mobilisation and activism around environmental issues, which now includes climate change (Sbicca, 2012; Dietz et al., 2020). Whitley and Bowers (2023) have documented distinct views on climate change, finding that LGBTQ+ individuals express higher agreement with climate change beliefs, identify climate change as a greater threat and worry more about climate change compared with their cis-heterosexual counterparts. These findings highlight the LGBTQ+ population's distinctive views on the issue, which often translate to action (Harris, 2010; Swank, 2018a, b).

Mobilisation and Activism on Environmental Issues

Non-binary environmental educator Eve Ettinger (2021) argues that environmentalism and climate concern is embedded in US LGBTQ+ culture and that the community will be one of the most profoundly impacted. While the US government and academics have historically paid little attention to this group, there are numerous examples of the LGBTQ+ community in the USA being at the forefront of environmental advocacy. American Marine Biologist Rachel Carson, a lesbian, brought attention to the unmitigated use of synthetic pesticides. Her 1962 book *Silent Spring* (Carson, 2002), led to the banning of dichloro-diphenyl-trichloroethane (DDT), spurred grassroots environmental organising and assisted in the creation of the Environmental Protection Agency (EPA).

Since this early action, dozens of LGBTQ+ environmental organisations have emerged, with many focusing on preserving land, addressing biodiversity loss, and supporting climate change mitigation and adaptation. Many of these organisations have roots in previous LGBTQ+ focused organisations like ACT Up (Whitley & Bowers, 2023) that emphasised a community-driven response to crisis. ACT Up and the LGBTQ+ community's response to the AIDS epidemic highlights its experience responding to disasters in the face of social backlash. In the 1980s, the LGBTQ+ community faced one of the most devastating epidemic of our time. The AIDS crisis was initially identified as a gay disease (Elliott, 2012; Stulberg, 2018), and because of this, the US government and medical community largely ignored the issue. Instead, the LGBTQ+ community had to help itself. Through resource mobilisation and community activism, the community revolutionised how disasters could be handled, demonstrating that when a government is not acting, local communities and individuals can provide resources, services and care.

In the case of ACT UP and AIDS, this response was to help the thousands of gay men who contracted and died from the disease, but the organising and activism developed through this crisis are embedded in the community and are being applied to climate change. LGBTQ+ climate and environmental organisations like Out for Sustainability, Queers x Climate Change, Our Climate Voices, Queer Nature, Queer Ecojustice Project and The Venture Out Project utilise hyper-local, intersectional activism to respond to climate change and help the community develop climate resiliency (Whitley & Bowers, 2023; Bowers & Whitley, 2024). In the USA, LGBTQ+ environmental activists have also pushed for increased inclusion in, and recognition by, traditional environmental organisations. Their work has increased acknowledgement and concern for the LGBTQ+ population among national organisations like the Sierra Club, which routinely profiles LGBTQ+ environmental organisations during June (considered Pride Month in the USA) (Richard-Craven, 2022). The LGBTQ+ community in the USA has been and will likely continue to be a force of change in climate activism.

Future Needs

LGBTQ+ individuals are powerful social and environmental justice organisers. Better inclusion of this group in addressing climate change mitigation and adaptation will be essential in the climate crisis. This inclusion must start with the recognition that LGBTQ+ people face greater risks from the climate crisis than their cisgender and heterosexual counterparts. Next, US states and localities need to take a critical look at how LGBTQ+ individuals and communities are impacted by the climate crisis and include the group in their climate plans. This must involve risk and needs assessments that reflect the realities of LGBTQ+ life and the social and political constraints this population faces. On this front, there is hope for change. In 2023, the Federal Emergency Management Agency (FEMA) and the US Department of Homeland Security worked with US-based LGBTQ+ environmental organisation Out for Sustainability to host webinars with over 250 federal, state and local representatives about how to incorporate the needs of LGBTQ+ populations in their disaster planning. The brainstorming and recommendations are now available in a report that is being circulated with the hopes that more communities will engage in this important work (Goldsmith, 2023).

There also needs to be greater connection between the LGBTQ+ community and environmental organisations in the USA. Specifically, environmental organisations need to speak out about anti-LGBTQ+ policies and how these policies are directly connected to environmental protections. In 2023, the HRC declared a state of emergency in the USA for LGBTQ+ people for the first time ever (HRC, 2023). In doing so, the organisation released a national warning and guidebook. The guidebook highlights the reality that LGBTQ+ people face heightened discrimination, inequitable treatment and a patchwork of resource access across the country. Climate change and specifically climate-induced disasters further threaten the safety of LGBTQ+ people. More needs to be done, not just by LGBTQ+

organisations but by environmental groups to address these discriminatory laws and practices. In return, LGBTQ+ organisations must treat the climate crisis as an LGBTQ+ issue if they are not already doing so and should use their extraordinary ability to inform and mobilise the community to focus LGBTQ+ individuals on environmental activism.

Finally, research shows that LGBTQ+ people care about climate change and are increasingly concerned (Whitley & Bowers, 2023). This research needs to be expanded to assess the degree to which concern translates into action. Additional research should also investigate LGBTQ+ people's environmental concerns and needs and evaluate how these compare to cisgender and heterosexual people. Given the current political landscape, there are states in the USA where being LGBTQ+ and specifically transgender is becoming impossible under normal circumstances. As a result, determining ways that environmental and disaster organisations can support LGBTQ+ individuals in these states is particularly important. Although beyond the scope of this chapter, greater attention needs to be placed on how LGBTQ+ people may be impacted by climate change induced disasters, and how they are responding, not just in the USA but across the globe.

Conflict of Interest The authors have no conflict of interests.

References

ACLU. (2023). *Mapping attacks on LGBTQ rights in US state legislatures*. Accessed June 20, 2024, from https://www.aclu.org/legislative-attacks-on-lgbtq-rights-2023

Adamson, T., Lett, E., Glick, J., et al. (2022). Experiences of violence and discrimination among LGBTQ+ individuals during the *COVID--19* pandemic: A global cross-sectional analysis. *BMJ Global Health, 7*(9), e009400.

Badgett, M., Choi, S. K., & Wilson, B. (2019). LGBT poverty in the United States. *Los Angeles, CA: The Williams Institute and American Foundation for Suicide*. Accessed July 17, 2024, from https://williamsinstitute.law.ucla.edu/publications/lgbt-poverty-us/

Barbee, H., Deal, C., & Gonzales, G. (2022). Anti-transgender legislation—A public health concern for transgender youth. *JAMA Pediatrics, 176*(2), 125–126.

Barr, S. M., Snyder, K. E., Adelson, J. L., et al. (2021). Posttraumatic stress in the trans community: The roles of anti-transgender bias, non-affirmation, and internalized transphobia. *Psychology of Sexual Orientation and Gender Diversity, 9*(4), 410–421.

Bernstein, M. (2002). Identities and politics: Toward a historical understanding of the lesbian and gay movement. *Social Science History, 26*(3), 531–581.

Beven, J. L., Avila, L. A., Blake, E. S., et al. (2008). Atlantic hurricane season of 2005. *Monthly Weather Review, 136*(3), 1109–1173.

Bowers, M. M., & Whitley, C. T. (2020). Assessing voter registration among transgender and gender non-conforming individuals. *Political Behavior, 42*(1), 143–164.

Bowers, M. M., & Whitley, C. T. (2024). Queer political culture in the face of climate change. In D. Sarathchandra & K. Haltinner (Eds.), *Resolving the climate crisis: US social scientists speak out.* Routledge Press.

Caldwell, J. (2006). This is Home. *The Advocate*, 32–40.

Cappelli, F., Costantini, V., & Consoli, D. (2021). The trap of climate change-induced "natural" disasters and inequality. *Global Environmental Change, 70*, 102329.

Carson, R. (2002). *Silent spring 40th Anniversary Edition.* Houghton Mifflin Harcourt.

Cianfarani, M. (2012). Integrating diversity into disaster risk reduction (DRR): A literature review. *International Association of Emergency Managers Bulletin, 29*, 26–27.

Clark, K. A., Cochran, S. D., Maiolatesi, A. J., et al. (2020). Prevalence of bullying among youth classified as LGBTQ who died by suicide as reported in the National Violent Death Reporting System, 2003-2017. *JAMA Pediatrics, 174*(12), 1211–1213.

Clarke, P., & Erreygers, G. (2020). Defining and measuring health poverty. *Social Science and Medicine, 244*, 112633.

Cravens, R. G. I. (2020). Political behavior of gender and sexual minorities. In D. Hauder-Markel (Ed.), *Oxford research encyclopedia of politics.* Oxford University Press. Accessed April 1, 2022, from https://doi.org/10.1093/acrefore/9780190228637.001.0001acrefore-9780190228637-e-1252

Cunningham, G. B., Watanabe, N. M., & Buzuvis, E. (2022). Anti-transgender rights legislation and internet searches pertaining to depression and suicide. *PLoS One, 17*(12), e0279420.

Kilpatrick, C., Higgins, K., Atkin, S., & Dahl, S. (2023). A rapid review of the impacts of climate change on the queer community. *Environmental Justice.* Accessed June 1, 2024, from https://doi.org/10.1089/env.2023.0010

Dalton, D. (2023). Have we left behind the rainbow warriors? The climate emergency and its impact on global queer people and their communities. In D. Dalton & A. Smith (Eds.), *Gender, sexuality and the UN'S SDGs. Sustainable development goals series.* Palgrave Macmillan. https://doi.org/10.1007/978-3-031-31046-1_2

Davis, M. (2023). 3-m -plus Americans displaced by natural disasters in past year, with Louisiana far ahead of other states. *Value Penguin.* Accessed July 1, 2024, from https://www.valuepenguin.com/natural-disasters-displacement-study

Dietz, T., Shwom, R. L., & Whitley, C. T. (2020). Climate change and society. *Annual Review of Sociology, 46*, 135–158.

Dolfman, M. L., Wasser, S. F., & Bergman, B. (2007). The effects of hurricane Katrina on the New Orleans economy. *Monthly Lab. Rev, 130*, 3.

Dominey-Howes, D., Gorman-Murray, A., & McKinnon, S. (2014). Queering disasters: On the need to account for LGBTI experiences in natural disaster contexts. *Gender, Place and Culture, 21*(7), 905–918.

D'Ooge, C. (2008). Queer Katrina: Gender and sexual orientation matters in the aftermath of the disaster. In B. Willinger (Ed.), *Katrina and the women of New Orleans* (pp. 22–24). Tulane University.

Drabish, K., & Theeke, L. A. (2022). Health impact of stigma, discrimination, prejudice, and bias experienced by transgender people: A systematic review of quantitative studies. *Issues in Mental Health Nursing, 43*(2), 111–118.

Ecker, J., Aubry, T., & Sylvestre, J. (2019). A review of the literature on LGBTQ adults who experience homelessness. *Journal of Homosexuality, 66*(3), 297–323.

Egan, P. J. (2012). Group cohesion without group mobilization: The case of lesbians, gays and bisexuals. *British Journal of Political Science, 42*(3), 597–616.

Egan, P. J. (2020). Identity as dependent variable: How Americans shift their identities to align with their politics. *American Journal of Political Science, 64*(3), 699–716.

Elliott, J. (2012). *An introduction to sustainable development*. Routledge.

Ettinger, E. (2021). *The next generation of climate activists is queer*. Accessed April 1, 2022, from https://www.them.us/story/next-generation-climate-activists-are-queer

FEMA. (2024). *National risk index: Hurricanes*. Accessed June 1, 2024, from https://hazards.fema.gov/nri/hurricane

Fraser, B., Pierse, N., Chisholm, E., et al. (2019). LGBTIQ+ homelessness: A review of the literature. *International Journal of Environmental Research and Public Health, 16*(15), 2677.

Gaillard, J. C., Gorman-Murray, A., & Fordham, M. (2017). Gender and sexual minorities in disaster. *Gender, Place and Culture, 24*(1), 18–26.

Gates, G. J. (2017). In US, more adults identifying as LGBT. *Gallup News*. Accessed June 15, 2024, from http://news.gallup.com/poll/201731/lgbt-identification-rises.aspx

Glynn, C., Byrne, T. H., & Culhane, D. P. (2021). Inflection points in community-level homeless rates. *The Annals of Applied Statistics, 15*(2), 1037–1053.

Goldsmith, L. (2023). *Inclusive and Equitable Emergency Management*. Out for Sustainability. Accessed April 29, 2025 from https://img1.wsimg.com/blobby/go/cd599488-9b66-4c7c-bfc2-68da8aef9363/Report%20Inclusive%20and%20Equitable%20Emergency%20Manag.pdf

Goldsmith, L., & Bell, M. L. (2022). Queering environmental justice: Unequal environmental health burden on the LGBTQ+ community. *American Journal of Public Health, 112*(1), 79–87.

Goldsmith, L., Raditz, V., & Méndez, M. (2022). Queer and present danger: Understanding the disparate impacts of disasters on LGBTQ+ communities. *Disasters, 46*(4), 946–973.

Green, A. E., Price, M. N., & Dorison, S. H. (2022). Cumulative minority stress and suicide risk among LGBTQ youth. *American Journal of Community Psychology, 69*(1–2), 157–168.

Greenhalgh, H. (2020). Religious figures blame lgbt+ people for coronavirus. *Reuters website*, 10 March. Available at: https://www.reuters.com/article/world/religious-figures-blame-lgbt-people-for-coronavirus-idUSKBN20W2HK/

Harris Interactive. (2010). Environmental advocacy grows stronger for LGBT Americans. *PR Newswire*, 15 December. Accessed November 5, 2023, from https://www.prnewswire.com/news-releases/environmental-advocacy-grows-strongerfor-lgbt-americans-111941139.html

HRC. (2012). *Working with the lesbian, gay, bisexual, and transgender community: A cultural competence guide for emergency responders and volunteers.* Accessed June 29, 2023, from https://assets2.hrc.org/files/assets/resources/EmergencyResponders_-_LGBT_Competency.pdf

HRC. (2023). *For the first time ever, Human Rights Campaign officially declares 'State of Emergency' for LGBTQ+ Americans; Issues national warning and guidebook to ensure safety for LGBTQ+ residents and travelers.* Accessed June 26, 2023, from https://www.hrc.org/press-releases/for-the-first-time-ever-human-rights-campaign-officially-declares-state-of-emergency-for-lgbtq-americans-issues-national-warning-and-guidebook-to-ensure-safety-for-lgbtq-residents-and-travelers

James, S. (2018). *From marginalized identities to maximized impact: The role of identity in the political attitudes and behaviors of transgender adults.* Georgetown University.

Jones, J. M. (2022). LGBT identification in U.S. ticks up to 7.1%. Accessed April 1, 2022, from https://news.gallup.com/poll/389792/lgbt-identification-ticks-up.aspx

Karol, D. (2023). How does party position change happen? The case of LGBT rights in the US. *Political Research Quarterly, 76*(4), 1736. ISSN: 10659129231176212.

Kilander, G. (2023). LGBT+ people and fleeing Florida in 'mass migration' over 'don't say gay' law. *Independent*, 1 June. Accessed June 20, 2023, from https://www.independent.co.uk/news/world/americas/us-politics/ron-desantis-florida-lgbt-law-b2348259.html

King, D. (2022). Hearing minority voices: Institutional discrimination towards LGBTQ in disaster and recovery. *Journal of Extreme Events, 8*(4), 2241005.

Lempinen, E. (2022). Attack on LGBTQ+ rights: The politics and psychology of a backlash. *Berkeley News*. Accessed June 20, 2023, from https://news.berkeley.edu/2022/05/02/attack-on-lgbtq-rights-thepolitics-and-psychology-of-a-backlash/

Moazen-Zadeh, E., Karamouzian, M., Kia, H., et al. (2019). A call for action on overdose among LGBTQ people in North America. *The Lancet Psychiatry, 6*(9), 725–726.

Newcomb, M. E., LaSala, M. C., Bouris, A., et al. (2019). The influence of families on LGBTQ youth health: A call to action for innovation in research and intervention development. *Lgbt Health, 6*(4), 139–145.

Open States. (2023). *SB 160*. Accessed June 29, 2023, from https://openstates.org/ca/bills/20192020/SB160/

Richard-Craven, M. (2022). *Check out these 4 LGBTQI environmental organizations*. Accessed June 22, 2023, from https://www.sierraclub.org/sierra/check-out-these-4-lgbtqi-environmental-organizations

Richards, G. (2010). Queering Katrina: Gay discourses of the disaster in New Orleans. *Journal of American Studies, 44*(3), 519–534.

Rozsa, L. (2023). Florida anti-LGBTQ laws prompt families who feel unsafe to Flee. *Washington Post*, 10 June. Available at: https://www.washingtonpost.com › nation › 2023/06/10.

Rushton, A., & Scarlett, J. (2023). The blame game: Disaster, queerness and prejudice. *Disaster Prevention and Management, 32*(2), 352–363.

Sbicca, J. (2012). Eco-queer movement(s). *European Journal of Ecopsychology, 3*, 33–52.

Smith, M. V., & Mazure, C. M. (2021). Mental health and wealth: Depression, gender, poverty, and parenting. *Annual Review of Clinical Psychology, 17*, 181–205.

Strode, D., & Flores, A. R. (2021). Voter registration rates and traits by sexual orientation and gender expression. *Public Opinion Quarterly, 85*(3), 913–928.

Stulberg, L. M. (2018). *LGBTQ social movements*. Wiley.

Swank, E. (2018a). Sexual identities and participation in liberal and conservative social movements. *Social Science Research, 74*, 176–186.

Swank, E. (2018b). Who voted for Hillary Clinton? Sexual identities, gender, and family influences. *Journal of GLBT Family Studies, 14*(1–2), 21–42.

Swank, E., & Fahs, B. (2013). Predicting electoral activism among gays and lesbians in the United States. *Journal of Applied Social Psychology, 43*(7), 1382–1393.

Swank, E., & Fahs, B. (2017). College students, sexualities identities, and participation in political marches. *Sexuality Research and Social Policy, 14*(2), 122–132.

Swank, E., & Fahs, B. (2019). Explaining the sexuality gap in protest participation. *Journal of Homosexuality, 66*(3), 324–348.

Technical Advisory Group. (2017). *Sacramento, California: California State Government*. Accessed March 2, 2022, from https://opr.ca.gov/docs/20180312-Vulnerable_Communities_Descriptions.pdf

Thomas, K., Hardy, R. D., Lazrus, H., et al. (2019). Explaining differential vulnerability to climate change: A social science review. *Wiley Interdisciplinary Reviews: Climate Change, 10*(2), e565.

Tierney, K. (2019). *Disasters: A sociological approach.* Wiley.

Turner, R., & Hammersjö, A. (2023). Navigating survivorhood? Lived experiences of social support-seeking among LGBTQ survivors of intimate partner violence. *Qualitative Social Work, 23*(2), 242–260.

United States Executive Office of the President, United States Assistant to the President for Homeland Security and Counterterrorism. (2006). *The federal response to hurricane Katrina: Lessons learned.* White House: Washington D.C. https://biotech.law.lsu.edu/katrina/govdocs/katrina-lessons-learned.pdf

US Census Bureau. (2022). *National Poverty in America Awareness Month: January 2022.* Accessed April 1, 2022, from https://www.census.gov/newsroom/stories/poverty-awareness-month.html

US Global Change Research Program. (2018). *Fourth National Climate Assessment: Impacts, risks, and adaptation in the United States.* Summary Findings. US Global Change Research Program.

Vinyeta, K., Whyte, K., & Lynn, K. (2016). *Climate change through an intersectional lens: Gendered vulnerability and resilience in indigenous communities in the United States.* PNW-GTR-923. United States Department of Agriculture. Accessed March 1, 2022, from https://www.fs.fed.us/pnw/pubs/pnw_gtr923.pdf

Weststrate, N. M., & McLean, K. C. (2010). The rise and fall of gay: A cultural-historical approach to gay identity development. *Memory, 18*(2), 225–240.

Whitley, C. T., & Bowers, M. M. (2023). Queering climate change: Exploring the influence of LGBTQ+ identity on climate change belief and risk perceptions. *Sociological Inquiry, 93*(3), 413–439.

CHAPTER 5

'Waria/Transpuan Will Drown First!' Rereading Climate Justice

Firdhan Aria Wijaya, Rikky Muchammad Fajar, and Teater Manekin

We are trapped
in flooding cities,
kampungs
prone to flood.
"Aren't there shelters
at sports arenas and mosques?"
"Well, can we be
placed in
the shelters?
Instead,
we'll be the centre of attention
of heteronormative mobs"

"Climate Change in Kampung Duri" Monolog
Madam Seroja, 2021

F. A. Wijaya (✉) • R. M. Fajar • T. Manekin
Universitas Pattimura, Ambon, Indonesia
e-mail: firdhan.wijaya@lecturer.unpatti.ac.id

D. Dominey-Howes et al. (eds.), *Queering Disasters, Climate Change and Humanitarian Crises*, Sustainable Development Goals Series, https://doi.org/10.1007/978-981-96-3857-4_5

Disclaimer

In 2012, Sanggar Seroja was first introduced as Teater Manekin. In October 2016, however, they renamed themselves Sanggar Seroja. They decided to go back to using the name Teater Manekin as of 2024. Since this study is centred on the 2023 event, Sanggar Seroja is still used.

A Space to Meet: Waria/Transpuan and Environmental Issues

Despite facing oppression and social exclusion (UNDP and USAID, 2014; Sanggar Swara, 2020), waria/transpuan[1] in Indonesia show resilience and strength. They create workspaces where they can be themselves, such as salons, creative businesses, and frenetic nightlife areas (Boellstorff, 2007; Toomitsu, 2019). Additionally, the waria/transpuan communities in Indonesia advocate for state recognition and acceptance, which is crucial for gaining basic rights as citizens. For instance, the community has aided individuals in obtaining national identity cards, creating access to national health insurance (Hegarty, 2022; Kjaran & Naeimi, 2022). Some individuals also dedicate themselves to religious spaces to fulfil their spiritual needs, despite these spaces having historically discriminated against waria/transpuan (Rodriguez, 2020).

Waria/transpuan communities advocate not only for acceptance and rights but also address environmental issues. During the Covid-19 pandemic, an organisation supporting diverse faiths and sexualities recognised the need for alternative food sources for vulnerable groups, as many waria/transpuan lost their informal sector livelihoods as buskers, sex workers, and salon workers. Virtual workshops taught participants to grow seasonal vegetables (Wijaya et al., 2021). Additionally, an Indonesian think tank[2] and other international institutions provided training on climate change impacts on gender and sexual minorities, inspiring waria/transpuan communities to prioritise environmental issues and engage in related activities.

The environmental experiences of waria/transpuan, like facing floods during monsoon season, are often overlooked in grassroots and policy

[1] We use both 'waria' and 'transpuan' to show their historical relationship that can never be separated from the trans women movement in Indonesia. The use of both phrases also refers to how the Sanggar Seroja members feel comfortable in identifying themselves. For more detailed information, please see the publication of Sanggar Seroja (2022) and Hegarty (2022).

[2] All the names must remain confidential.

discussions. Their experiences highlight policy gaps impacting marginalised groups, underscoring the need for inclusive climate policies. To address this, a queer perspective must disrupt the traditional heteronormative framework of sustainable living (Sandilands, 1997, 2002). While Indonesian climate justice efforts consider some marginalised groups, including women with disabilities (Paramitha & Wongkar, 2022), indigenous people (Manaysay, 2020), and smallholder farmers (Merten et al., 2021), waria/transpuan remain largely unrecognised.

This vignette focuses on Sanggar Seroja,[3] a waria/transpuan's collective that uses art performances to highlight environmental issues such as climate change. For almost 40 years, they have used various forms of art—dance, monologue, *lenong*,[4] poetry, drag performances, and lip-syncing—performed at various festivals. Recognising that climate impacts disproportionately affect waria/transpuan people, who often work in climate-vulnerable industries, Sanggar Seroja frequently explores topics like climate change and waste management through their creative work.

Kampung Duri, where this vignette focuses, is a densely populated urban village in West Jakarta (Kusno, 2020), and waria/transpuan residing in the area comprise both long-term residents and temporary immigrants. Some work as buskers and sex workers or as make-up artists and hairdressers. Waria/transpuan and the residents are forced to live in slum conditions and cope with a multitude of environmental problems such as poor air quality (Anies, 2023).

Through the voices and actions of Sanggar Seroja, the authors and Sanggar Seroja members reflect in this vignette on the often undiscussed understanding of the intersectionality of climate change and the unique issues that waria/transpuan face.

Methods and Context

In 2023, Sanggar Seroja reflected together on the work the organisation had done related to environmental issues. Please note that as a collective, Sanggar Seroja's membership base is not binding, but each person has

[3] Sanggar Seroja and its members are permitted to use their name in this chapter.

[4] *Lenong* is a traditional performance art that combines the elements of drama, comedy, music, and folk tradition originating from the Betawi culture of Jakarta.

responsibilities and roles when activities take place. We collected their narratives separately.

We conducted interviews with seven founders, five volunteers, collaborators, and representatives from two organisations to gather narratives on their environmental and climate initiatives. Additionally, we reviewed Sanggar Seroja's archives, including performance videos, climate campaign documentation, and fashion parades in Kampung Duri. The data reveals Sanggar Seroja's journey and the role of waria/transpuan in Kampung Duri in stimulating ecological awareness and strengthening queer voices. The stories presented highlight members' efforts to care for their environment.

Narrative Fragments

This section is a compilation of various story fragments, capturing Sanggar Seroja member's perspectives on the climate crisis and impacts to them. Listening, rewriting, and understanding their stories encourage hope for ecological awareness and a new perspective that the climate crisis not only affects cis-gendered women and men but also impacts individuals of diverse genders, including waria/transpuan.

Impacts on Physical and Mental Health

Jakarta experiences increasing temperatures and changing rainfall patterns, resulting in frequent flooding (Maheng et al., 2023). Climate change also generates high air pollution levels, especially fine particulate matter (PM2.5) and ground-level Ozone (O_3) (Handhayani, 2023). The heat also binds these pollutants to stay longer in the air, resulting in respiratory problems among Jakarta residents (Haryanto, 2017).

Flooding and temperature rises impact the health and wellness of warias/transpuan. Nuke, one member of Sanggar Seroja, commented that when the weather is hot, she cannot continue busking. She chooses to stop and look for shade in kiosks or roadside stalls, as she stated: 'I get headaches a lot. Sometimes my head spins around and around. Especially when I force myself to walk.' Overly hot weather hinders waria/transpuan who work outside, limiting their source of income. Additionally, most warias carry out their activities on the street and several members of Sanggar Seroja have complained about how their exposure to respiratory diseases has become more frequent.

Other problems for some Sanggar Seroja members are dermatological issues. Mak Lina shared that she often experiences itching, especially when the weather is hot. The room that Mak Lina rents has plywood (triplex) walls, so the room gets damp and stuffy. Mak Lina cannot move out as she has limited mobility and cannot afford other accommodations. Increases in temperature are also linked to accelerating a person's emotion, such as feelings of anger, which may be accompanied by acts of violence (Mora et al., 2018). The anger caused by the hot weather is also experienced directly by waria/transpuan in Kampung Duri, as expressed by three members of Sanggar Seroja,

> Well, this is our job (busking). So, yes, when it gets really hot, then your emotions will definitely be at their peak. Especially when we go home with no money, it feels some type of way, like we get really annoyed. It's like when you're under really hot weather, you'll obviously feel annoyed. Like you want to just explode. (Rika)

> Plus there's climate change now, maybe that's how you're affected. Not even that, me personally, the heat makes me so emotional when I hear stupid nonsense, I get angry even if we don't hit each other or something, we just nag and nag. Why? Because getting angry and the hot weather have gotta be related. (Mak Lina)

> Of course, climate change and violence are linked. Why do I say so? Because there's the uncertain climate change, sometimes it rains, sometimes it's really, really hot. And when it's hot, people are usually easily annoyed. Why? Maybe because it's so hot. Personally, I experience that too. Since it's too hot, I can get annoyed with my niece, even my own sibling. (Mama Atha)

Those experiences sometimes trigger anger in some members of Sanggar Seroja, leading to violence that they struggle to avoid and control. Choosing to restrain themselves may seem like a temporary solution. Many prefer to stay home instead of working to manage their emotions and prevent violence. However, this inevitably leads to losses in income.

Anxiety and Uncertainty

The turning point for Sanggar Seroja's focus on climate change was initiated by an art competition with the theme 'Climate and Gender (In) Justice in the City' by the University of Leeds in the GENERATE (Gender,

Generation, and Climate Change) research project in 2021. Some of Sanggar Seroja's members sent a monologue video about the lives of trans women and climate change entitled 'Climate Change in Kampung Duri', which was narrated by the founder and committee member of Sanggar Seroja, Madam Seroja. This monologue was crowned as the winner.

In the monologue, Madam Seroja spoke about her anxiety regarding flooding, which was increasingly uncertain due to extreme rain. The anxiety is shown in Madam Seroja's monologue:

> The media reports that the Jakarta floods are a disaster for residents. But for us, the flood is a mixed doubles disaster. Like slipping from the second floor. Falling down and struck by the ladders. Torn pants and a broken sandal. (Madam Seroja, 2021)

Madam Seroja's monologue reflects the complex vulnerability experienced by waria/transpuan when erratic rain occurs, leading to floods. When it rains continuously, waria/transpuan in Kampung Duri cannot work. They depend on their income from the street as buskers and sex workers. If they are forced to work, they risk their health. They make ends meet with their income each day; therefore, a lost day of work significantly impacts their ability to support themselves.

Sanggar Seroja members highlighted the difficulties waria/transpuan face during flood evacuations, as temporary shelters often create anxiety and fear. Shelters frequently use mosques with strict gender divisions, making restroom use challenging and exposing them to prejudice and ridicule. This discomfort and trauma often leads waria/transpuan to avoid shelters altogether, choosing instead to stay in their flooded homes (Mama Atha, pers. comm., 2023).[5]

Balgos et al. (2012) noted that very few disaster risk reduction policies address the needs of waria/transpuan in Southeast Asia. Often excluded from government efforts, waria/transpuan lack information on disasters, leading to social exclusion and heightened emotional trauma, increasing climate anxiety (Taylor, 2020). Furthermore, the discomfort and trauma experienced by waria/transpuan are not often adequately addressed, potentially leading them to choose to stay in disaster-affected areas without receiving resources or support. Unfortunately, this has been the

[5] This conversation stems from personal communication with Mama Atha after the data collection was finished. Therefore, it does not appear as a direct quote.

experience of some of Sanggar Seroja's members. Therefore, it is important to include the risks and needs of waria/transpuan in disasters, not only to protect their rights during the disaster but also to take preventive actions (before the disaster) and rehabilitative actions related to psychology (after the disaster).

Ending in Death

Mama Atha, another Sanggar Seroja founder and committee member, received high praise for her monologue work, 'Eka and the Rain'. This monologue highlights the concerns of Sanggar Seroja members regarding climate change and provides other very urgent and concerning nuances. The monologue was based on a true story of a waria/trans woman living in Kampung Duri named Eka. Eka worked as a busker and spent most of her time on the streets in all weather. In her monologue, Mama Atha read,

> Aw, it's raining again. What to do? I don't even have any *duta* (money) yet?! [*So complained Eka, a busking waria*]. Well, the weather hasn't been predictable for the past days … Aaw, what if it rains again later? *Alemong* (OMG)! We won't get any *duta* (money). (Mama Atha., 2021)

Weather uncertainty caused by climate change made Eka's income uncertain. While she was working, it could suddenly rain, forcing her to take shelter. Unfortunately, Jakarta has the most flood-prone areas, and Eka sometimes had difficulty getting home. As a result, when it frequently rained, Eka decided to stay home, losing income. Unfortunately, Eka had fallen ill and could not afford treatment. She was evicted from her home due to not being able to pay her rent. However, amid this, Eka still forced herself to work. Ultimately, Mama Atha shared that Eka tragically died because she could not meet her own needs despite all the efforts she had made.

Unpredictable weather and frequent rain might lead to economic instability and health problems, and even fatalities. In the interview, a member of Sanggar Seroja joked, 'If there is a flood disaster, transpuan will drown first!' Despite the irony, Sanggar Seroja remains motivated to act. They take a variety of steps to express their vulnerabilities and their participation in the fight against the climate crisis. Interpreting fragments of Eka's story is important to reflect on, as the relationship between waria/trans woman's life and climate crisis is intertwined.

Sanggar Seroja's Environmental Campaigns

Sanggar Seroja aligns itself to climate action. Starting in March 2022, the organisation created creative campaigns titled: 'WARIA FACING THE DOUBLE BURDEN OF MIXED HEAVY CLASS MARATHON' and 'WE ARE WORRIED ABOUT THE CLIMATE CRISIS' and created posters to show the existence as waria/transpuan who are fighting not only for their rights but also voicing the increasingly urgent climate crisis. Through #GlobalClimateStrike[6] and the commemoration of World Environment Day, Sanggar Seroja took action by taking to the streets wearing costumes to attract public attention and convey their messages (see Figs. 5.1 and 5.2). This resistance reflects the history of queer activism and the use of creative performances to demand political and/or public action (Jenzen & Lewin, 2022).

Using costumes, Sanggar Seroja was able to establish relationships with other organisations that fight environmental issues. Dewi and Sekar saw this as an opportunity to build partnerships with other actors, and united for the cause. These partnerships showcase how at-risk populations and environmental degradation share the same oppression by a dominant

Fig. 5.1 (left) With the theme 'Fight Against Oppressive Authorities, Uphold Climate Justice', Sanggar Seroja members joined climate action in Jakarta, March 2023 (Images credit: Teater Manekin)

[6] The Global Climate Strike is a worldwide movement aimed at raising awareness about the climate crisis. Its objectives include raising public awareness about climate change issues, urging world leaders to take more serious actions to reduce greenhouse gas emissions and promoting sustainable policies. In Indonesia, similar actions have also been carried out several times. These actions have taken place in several cities including Jakarta, Bandung, Yogyakarta, Bali, Medan, Makassar, and Jayapura.

Fig. 5.2 (right) Presenting a theatrical performance by Sanggar Seroja members amid climate action, March 2022 (Images credit: Teater Manekin)

power system (i.e. governmental institutions and political elites) (Thomas, 2022).

In 2022, Sanggar Seroja organised the 'Trans Climate Change Super Heroes Carnival' in Kampung Duri. They created four waria/transpuan superhero costumes produced from out-of-use materials such as gallon lids, bubble wraps, snack wrappers, used plastic, broken hoses, broken dusters, and unused clothes. Each costume had different strengths or meanings but had the same goal: to fight for equality and inclusiveness. As shown in Fig. 5.3, the waria/trans woman superheroes consisted of Vei Lan (the cleaner), Asih (the healer), Dana (the fundraiser), and Tara (the fighter for justice).

Through a creative activism approach, Sanggar Seroja opens emotional and direct connections within the community, especially the residents of Kampung Duri, to invite them to take action and strive to protect the environment. In this effort, Sanggar Seroja plays a significant role in spreading ecological awareness and encouraging enthusiasm for forming coalitions that may not have been previously anticipated.

Of Hope and the Future

Although the climate crisis has created a sense of uncertainty, Waria/transpuan will always 'move', 'cross', and 'intermingle' (Stryker et al., 2008). Sanggar Seroja's footsteps, of course, will not stop. They said, 'We will continue to carry out this action to set a reminder that we as waria/transpuan exist and that the issue of climate change is our issue too.'

The collective actions of Sanggar Seroja reflect the realisation of hope. With this, hope for a better future will not fade creating even stronger resilience within the community in the face of uncertainty. This also needs

Fig. 5.3 Members of Sanggar Seroja transformed into trans superheroes (Image credit: Teater Manekin)

to be supported by expanding networks/partners so that an intersectional knowledge exchange process occurs. For example, residents around Kampung Duri and representatives of environmental organisations also expressed their hopes to collaborate with Sanggar Seroja in climate action or work related to other environmental issues.

Another hope expressed by most members of Sanggar Seroja is to have a shelter house. Mama Atha said,

> As for my hope for the future, I [hope] Sanggar Seroja will have a shelter. That's my hope. My biggest hope for the longest time. So that our elderly friends who are immigrants, and who are no longer able to pay the rent, can stay at Sanggar Seroja's shelter house. (Mama Atha, interview, September 2023)

The shelter house, a long-held hope for Sanggar Seroja, would serve as a 'multi-purpose house' for activities and a safe evacuation space for waria/transpuan if needed. It aims to be free and safe for all waria/transpuan.

Sanggar Seroja reflects the noble spirit and determination to face the uncertainty and challenges of climate change in Kampung Duri. Environmental justice or climate justice will never be actualised without the process of listening, reading, and understanding the context of vulnerable groups, especially waria/transpuan. As Madam Seroja concluded in her monologue, 'We want to help those around us. We care, too, so involve us and care for us!'

Conflict of Interest The authors have no conflict of interests.

References

Anies. (2023, August 26) *Polusi Udara Kota*. Kompas.id. Retrieved October 21, 2023, from https://www.kompas.id/baca/opini/2023/08/25/polusi-udara-kota

Balgos, B., Gaillard, J. C., & Sanz, K. (2012). The warias of Indonesia in disaster risk reduction: The case of the 2010 Mt Merapi eruption in Indonesia. *Gender and Development, 20*(2), 337–348.

Boellstorff, T. (2007). *A coincidence of desires: Anthropology, queer studies*. Duke University Press.

Handhayani, T. (2023). An integrated analysis of air pollution and meteorological conditions in Jakarta. *Scientific Reports, 13*(1), 5798.

Haryanto, B. (2017). Climate change and urban air pollution health impacts in Indonesia. In R. Akhtar & C. Palagiano (Eds.), *Climate change and air pollution: The impact on human health in developed and developing countries*. Springer International Publishing.

Hegarty, B. (2022). An inter-Asia history of transpuan in Indonesia. In S. Tang & H. Y. Wijaya (Eds.), *Queer Southeast Asia* (pp. 15–32). Routledge.

Jenzen, O., & Lewin, T. (2022). Global queer and feminist activism: An introduction. *Journal of Cultural Analysis and Social Change, 7*(2), 09, 1–17. https://doi.org/10.20897/jcasc/12750

Kjaran, J. I., & Naeimi, M. (2022). Politics of (un)intelligibility: Trans* activism in Indonesia. In *Queer social movements and activism in Indonesia and Malaysia*. Springer International Publishing.

Kusno, A. (2020). Middling urbanism: The megacity and the kampung. *Urban Geography, 41*(7), 954–970.

Madam Seroja. (2021). *Perubahan Iklim di Kampung Duri*. Retrieved October 17, 2023, from https://www.instagram.com/p/CWGNzp4A_5N/

Maheng, D., Bhattacharya, B., Zevenbergen, C., & Pathirana, A. (2023). Changing urban temperature and rainfall patterns in Jakarta: A comprehensive historical analysis. *Sustainability, 16*(1), 350.

Mama Atha. (2021). *Eka dan Hujan*. Retrieved October 17, 2023, from https://www.youtube.com/watch?v=G36rBMbPefkandt=465s

Manaysay, F. V. (2020). Norms from above, movements from below: Climate change and global-local dynamics of indigenous resistance in The Philippines and Indonesia. *JSEAHR, 4*, 226–252.

Merten, J., Nielsen, J. Ø., & Faust, H. (2021). Climate change mitigation on tropical peatlands: A triple burden for smallholder farmers in Indonesia. *Global Environmental Change, 71*, 102388.

Mora, C., Spirandelli, D., Franklin, E. C., Lynham, J., Kantar, M. B., Miles, W., Smith, C. Z., Freel, K., Moy, J., Louis, L. V., Barba, E. W., Bettinger, K., Frazier, A. G., Colburn, J. F., Hanasaki, N., Hawkins, E., Hirabayashi, Y., Knorr, W., Little, C. M., Hunter, C. L., et al. (2018). Broad threat to humanity from cumulative climate hazards intensified by greenhouse gas emissions. *Nature Climate Change, 8*, 1062–1071.

Paramitha, Y., & Wongkar, E. E. L. T. (2022). Membangun Resiliensi dari Bawah: Perempuan dengan Disabilitas dan Keadilan Iklim di Indonesia. *Jurnal Perempuan, 27*(3), 6–28.

Rodriguez, D. G. (2020). Queer religious geographies? Qu(e)erying Indonesian Muslim selves. *Gender, Place and Culture, 27*(9), 1326–1347. https://doi.org/10.1080/0966369X.2019.1693343

Sandilands, C. (1997). Mother Earth, the cyborg, and the queer: Ecofeminism and (more) questions of identity. *Women, Ecology, and the Environment, 9*(3), 18–40.

Sandilands, C. (2002). Lesbian separatist communities and the experience of nature: Toward a queer ecology. *Organisation and Environment, 15*(2), 131.

Seroja, S. (2022). *Seroja Mekar di Tengah Krisis*. Sanggar Seroja.

Sanggar Swara. (2020). *A long journey to acceptance: Situational analysis of trans-women in Indonesia*. Equality Konsultan.

Stryker, S., Currah, P., & Moore, L. J. (2008). Introduction: Trans-, trans, or transgender? *Women's Studies Quarterly, 36*(3/4), 11–22.

Taylor, S. (2020). Anxiety disorders, climate change, and the challenges ahead: Introduction to the special issue. *Journal of Anxiety Disorders, 76*, 102313.

Thomas, L. (2022). *The intersectional environmentalist: How to dismantle systems of oppression to protect people + planet*. Souvenir Press Limited.

Toomitsu, T. (2019). Between abjection and world-making: Spatial dynamics in the lives of Indonesian Waria. *Journal of Ethnology and Folkloristics, XIII*(2), 90–107.

UNDP and USAID. (2014). *Being LGBT in Asia: Indonesia country report*. UNDP.

Wijaya, F. A., Sanubari, T. P. E., Rayanti, R. E., & Paolini, L. (2021). *A context analysis on the state of the LGBTI movement in Indonesia*. COC Nederland.

PART II

Disaster Justice and Gender and Sexual Diversity

CHAPTER 6

Moving on: Exploring Mobilities of LGBT People in Christchurch After the 2011 Earthquake

Scott McKinnon, Dale Dominey-Howes, and Andrew Gorman-Murray

Introduction

In February 2011, Christchurch experienced a catastrophic earthquake. It resulted in complete destruction and change of 'spaces' across Christchurch. The government 'red zoned' the central business district (CBD) and

S. McKinnon
Department of Archaeology and History, La Trobe University, Melbourne, VIC, Australia

D. Dominey-Howes (✉)
The Sydney Environment Institute and School of Geosciences, The University of Sydney, Sydney, NSW, Australia
e-mail: dale.dominey-howes@sydney.edu.au

A. Gorman-Murray
School of Social Sciences, Western Sydney University, Penrith, NSW, Australia

D. Dominey-Howes et al. (eds.), *Queering Disasters, Climate Change and Humanitarian Crises*, Sustainable Development Goals Series, https://doi.org/10.1007/978-981-96-3857-4_6

8000+ residential homes were eventually destroyed (Canterbury Earthquake Recovery Authority, 2016; Truebridge, 2017). Red zones were spaces deemed unsafe to enter and were monitored until June 2013 by armed services personnel who prevented unauthorised access.

This chapter examines the impacts of the earthquake on Christchurch's modest gender and sexual diverse community (Myrdahl, 2016). We explore how gender and sexual diverse people and communities[1]' mobility was impacted during post-disaster recovery—a little researched area (Cretney, 2017; Cretney & Bond, 2017). Our interviews with residents show that for some, mobility was linked to a desire to return to normal, to the status quo. For others, the status quo was itself implicated in the causes of the disaster and mobility was a desire for change, to 'build back better' (Fernandez & Ahmed, 2019).

The chapter concludes by considering mobility in the broader context of disaster justice. Disaster justice provides a way of understanding both the importance of mobility to queer people's safety and individual and collective identities post-disaster and how impediments and barriers to mobility increase their vulnerability, reduce their resilience and deny them full disaster justice (Cresswell, 2008; Dominey-Howes et al., 2014; Nash & Gorman-Murray, 2014, 2015; Cook & Butz, 2016; Doan, 2016; Yamashita et al., 2017; Gorman-Murray et al., 2018).

Rethinking Mobility in Disasters

One under-researched aspect of the recovery process is how to enhance and increase mobility and the kinds of mobility available within disaster zones (Cresswell, 2008; Cook & Butz, 2015, 2016). Cook and Butz (2015) argue that after disaster:

> short-term relief relies on re-establishing accustomed mobility platforms or creating new ones, and longer-term recovery strategies are guided by recognition that mobility is a capability fundamental to social organisation and human thriving. (p. 389)

Mobility—the *ability to move about*—is crucial to recovery and interlinked to connections of identity and space. Easthope (2009, p. 75) states,

[1] Our respondents self-identified as lesbian, gay, bisexual and transgender. None identified as intersex or as having a variation in sex characteristics. When talking about our respondents and their experiences we use the acronym LGBT. When talking about other studies and gender and sexual diverse people more broadly we use the term people with diverse 'SOGIESC.'

'People can and do talk in terms of the impact of both mobility and place attachment on their identities simultaneously.' Mobility and motility are determined by hierarchies of power and privilege (Cresswell, 2008). Cook and Butz (2016, p. 401) argue, '[D]isasters and their associated effects can impede motility and mobile lives by disrupting access to dominant mobility systems in ways that reshape or exacerbate inequalities.'

Vulnerability to disaster is increased by a lack of access to transport facilities to evacuate to safe spaces (Cresswell, 2008; Hannam et al., 2006). Earthquakes prompt increased movement both immediately, as individuals seek safety, and more long-term, including migration away from damaged cities and complex rebuilding processes. Disasters set physical and emotional barriers around mobility. Damaged roads and collapsed buildings in Christchurch created substantial mobility challenges for residents temporarily. The emotional costs of physical immobility had long-term impacts on emotional and psychological recovery.

Mobility and the Remaking of Gender and Sexual Diverse Identities

Geographical queer theory interrogates how space impacts the construction, performance and practice of sexual identities (Bell & Valentine, 1995; Knopp, 1995). This includes a particular interest in the spatial construction of heterosexuality as normative; the spatial practices through which LGBT individuals create a sense of belonging; and the uneven power relations that determine access to and use of space (see Browne et al., 2007; Gorman-Murray & McKinnon, 2015). Mobility is intimately involved with the making and remaking of gender and sexual diverse identities through their connections to place and space. Gender and sexual diverse identity determine access to both mobility systems and mobility practice. This includes the vulnerability of LGBT people moving through disaster spaces, as well as the forms of mobility practiced in Christchurch.

For many gender and sexual diverse people, home represents a 'safe place' where they can be their authentic selves. Disasters disrupt this and force them to move through spaces in which discrimination is unavoidable (Gorman-Murray et al., 2014). Disasters force a 'coming out' for individuals who otherwise choose not to publicly disclose their identities (Yamashita et al., 2017). Consequently, some gender and sexual diverse individuals choose to stay, or feel they have no choice but to stay, in their homes during disaster. This increases their vulnerability by inhibiting their mobility and use of evacuation shelters (This collection, Chap. 3). For many queer people who have lost their homes, rebuilding safe, ontologically secure domestic spaces is necessarily a mobile process (Gorman-Murray et al., 2014).

Urban 'gayborhoods' are declining globally, prompted by fewer queer businesses, choices by LGBTIQA+ people to make their homes outside the gayborhood (Reynolds, 2009), generational changes in understandings of sexual identity and increasing visibility beyond gay villages enabled by social and political gains (Gorman-Murray & Waitt, 2009). This fragmentation and dissolving of unique gayborhoods presents issues concerning mobility post-disaster (see 'Results and Discussion').

Arguing against binary definitions which label these spaces as either stable or declining, Nash and Gorman-Murray (2014) adopt a mobilities approach arguing.

> that gay villages and new or alternative neighbourhoods are interlinked, interrelated and mutually constituted through the varied mobilities of people, narratives and knowledges within fluid networks embedded in shifting motilities. (p. 757)

Their argument explores sexual and gender identity as determining factors in the decision to move and the scalar experience of movement. Displays of affection while walking along streets go unmarked between heterosexual couples but place same-sex, transgender and gender non-conforming individuals/couples at high levels of public harassment and violence, including when using public transport.

Intersecting factors such as race and indigeneity also determine access to LGBT space and mobility or motility. Māori LGBT people, for example, many of whom have adopted the identity *Takatāpui* (Kerekere, 2017), face racism from within gender and sexual diverse communities and anti-*Takatāpui* discrimination from without. As Johnston (2010) argues, in relation to post-colonial settler societies, 'sexual subjectivities are always entangled with race, ethnicity, and postcoloniality' (citing Aiello et al., 2013, p. 100).

Research Context and Methods

We interviewed LGBT residents to understand how mobility and the desire for mobility figured in their post-disaster efforts to re-establish, maintain or re-invent links to LGBT community. These mobility recovery practices were taking place in the context of social changes within LGBT communities that were in a state of flux before the earthquake. The earthquake simply accelerated these changes, including the disappearance of queer neighbourhoods, the choice to move into non-queer residential areas and loss of queer venues and safe hang-out spaces (Nash & Gorman-Murray, 2014, 2015).

University Ethics clearance was obtained (H9754). The principle guiding this research was not to cause or trigger any further trauma for participants. Consequently, an ethical, trauma-informed methodology was employed (Voith et al., 2020; Drozdzewski & Dominey-Howes, 2021). Nineteen semi-structured interviews were conducted with LGBT residents in 2015. Interviewees self-identified their gender and sexual identities (six female, twelve male and one genderqueer). Eleven identified as gay, six as lesbian, one as bisexual and one 'not straight.'

Interviewees were recruited via LGBT press advertisements, contact with LGBT agencies and snowballing, lasted up to ninety minutes and were conducted at locations of interviewees' choosing. Interviews opened with questions about life as a Christchurch LGBT person before the earthquake and followed a chronological path from the earthquake (onwards) including recovery processes. Interviews were transcribed, and thematically coded for issues related to mobility. We use pseudonyms to protect identities.

Results

The analysis of recovery mobility was divided into three *temporal* phases: the *impact phase*, the day and week of the earthquake; the *short-term recovery phase* of six months; and the *longer-term recovery phase* covering four years following the earthquake, from mid-2011 until interviews were conducted in 2015. These temporal categories are not fixed or discrete. Their boundaries are porous and varied according to personal capacities and experiences. These temporal phases allow us to explore changing forms of mobility undertaken by LGBT individuals as they sought safety and shelter; began to make plans for re-establishing their lives in a vastly changed city; and attempted to re-construct a longer-term sense of recovery, community and identity.

Impact Phase

Fixed places become *instantly mobile* during earthquakes (Adams-Hutcheson, 2017). Buildings rock, shake and collapse. Furniture is displaced and structures representing safety threaten life. Patrick (20s, bisexual male) was at work and explained: 'My most vivid memory was the very large photocopier getting thrown across the room and the walls warping.' The means of mobility which individual's control are lost. Bodies shake with fear, stumble and fall. Transport vehicles rock, sway and fall into gaps in upturned ground.

When cities are so transformed, they become spaces in which mobility is disrupted *and* needed (Cook & Butz, 2016). As argued by Hannam et al. (2006, p. 7), 'Urban disasters bring to the fore the astounding fragility of complex mobility systems.' The immediate reactions of the interviewees were to check that the people near them were safe and to locate, or make contact, with loved ones. Damaged or destroyed roads and overloaded telecommunications facilities challenged this. Journeys were made difficult because of liquefaction (a geological process whereby the soil turns to a liquid that objects—like houses—sink into).

For those without a car, disrupted public transportation systems inhibited movement across the city. Karl (50s, gay male) recalled: 'I decided at about four o'clock to go over to my son's house. ... That usually was a five-minute trip. It took me over an hour to get there.' Claire (50s, lesbian) stated:

> I remember driving down ... the main street of town over there, feeling like, it was all brown and the road was covered in all liquefaction, and it was drizzling and I just kept thinking, 'Is this the end of the world?'

For many Christchurch residents, the normalities of urban life instantly became dangerous or inaccessible, forcing movement across damaged spaces for safety. Locating and being with family were both the impetus for moving and the means of support and healing. The definition of 'family' for our interviewees was variable. Some defined family as biological relatives and their relationships within these family groups often revealed happy acceptance of various gender and sexual diverse identities. For some, same-sex partners were actively included as 'family' and, for others, the difference between same-sex and opposite-sex relationships in terms of familial support and inclusion was negligible.

Sean (20s, gay male) lived with his boyfriend in an inner-city apartment severely damaged. After the earthquake, Sean's entire family, including his boyfriend, parents, siblings and siblings' partners, gathered at his grandmother's house, an example of family inclusion not available to all gender and sexual diverse people. Subsequently, the family gathered in Sean's parents' house: 'We actually just spent a few days just, I don't know, just chilling out and listening to the radio and were very much quite a tightknit family group there.' The interrelationship between post-disaster mobility and gender and sexual diversity is apparent in this act of temporary, collective place-making. The family travelled from different places to meet at the family home and participate in the co-creation of physically and emotionally supportive spaces, including Sean and his boyfriend.

For one interviewee, the challenge of moving through the damaged and unstable city was exacerbated by homophobia. Annabelle (50s, lesbian) is the mother of three teenagers. She was working during the earthquake and immediately attempted to cross the city to find her partner and children. For part of the journey, Annabelle 'hitched a ride' from a stranger who proceeded to blame LGBT people for the earthquake: 'She explained to me that this was Jesus … it was all about the gay stuff and Jesus is not happy.'

Post-earthquake, residents negotiated the immobility instituted by destroyed or damaged roads and continuing aftershocks via alternate mobility systems. LGBT support networks negotiated unsafe urban landscapes by relying on re-established telecommunications systems. Julie lives alone but received support from lesbian friends. Julie explained:

> [W]e'd just text each other or talk on the phone and sometimes all night we'd just stay on the phone and keep in touch with someone. You can't drive to each other's houses because of course with a big earthquake you keep off the roads … so we would stay at home and use text and phone a lot.

While aftershocks continued, Julie was able to comfort and be comforted. Caring voices reassured her that she was not alone and that she did not need to travel through dangerous spaces to reconnect (Rushton & Scarlett, 2023; Larkin, 2019).

Short-Term Recovery

Short-term recovery saw LGBT people renegotiating relationships with people and places. The damaged city re-centred focus on home and a re-making of meaning (Gorman-Murray et al., 2014). Regions of Christchurch were off-limits due to damage and red zoning. Although movement was challenging, everyday mobilities were necessary, for work and essential activities like shopping. These acts of urban mobility became difficult, requiring careful thought and psycho-social stress. Ongoing tension in moving about the city persisted for months, especially as demolition of parts of the city progressed, changing the very nature and recognition of previously known places.

Many former LGBT social places (e.g., LGBT or LGBT-friendly clubs and bars) were destroyed. The Christchurch CBD contained a small collection of LGBT-friendly venues that comprised an inner-city LGBT precinct. This was not a 'gay village' or 'gayborhood' of the kind found in larger cities but was an LGBT focal point. The earthquake destroyed these urban queer sites and some forms of LGBT mobility. Moving towards, within or home from an LGBT bar ceased to happen because these LGBT spaces no longer existed or were unable to operate. Interviewees noted that shifts in LGBT social practices and connections to urban space in Christchurch were underway prior to 2011 and the earthquake accelerated them. Discussing what he saw as a decrease in the number of LGBT venues and an increase in the use of mobile dating apps by gay men, Peter (40s, gay male) stated:

> That was happening anyway, but the earthquake forced a lot of it to happen in Christchurch faster than it possibly would have happened. You could say, 'Well over time the bars probably would have started to die out as that's happening anyway in other places.' [But] it would have been a slow, drawn-out process. Whereas what we had was it just happened.

LGBT venues or neighbourhoods are far from the only urban spaces significant to the construction or performance of LGBT identities (Bell & Valentine, 1995). For some LGBT people, gay bars are not perceived to be safe, welcoming and inclusionary for reasons including age, race or indigeneity, gender and bodily ability (Doan, 2007; Caluya, 2008). In describing the 'losses' of these spaces for the LGBT community, it is important to acknowledge ways in which many in the community had never felt any ownership of, or attachment to them, or for whom these spaces represented exclusion more than community. For many, creating a domestic space in which their LGBT identities could be safely performed and from which the intrusion of discriminatory forces can be excluded carries greater importance than a bar or nightclub (Gorman-Murray et al., 2014).

As the earthquake transformed city streets and public spaces, it also transformed the physical structures and emotional meanings of home. The disaster changed movement within, as well as in and out of gender and sexual diverse people's homes. This change prompted lives of constant mobility, without a stable location to re-establish routines, or in which to make plans. The earthquake disrupted routine mobilities, altering domestic atmospheres in ways ranging from the inconvenient to the traumatic. Craig (20s, gay male) had lived with his partner in an inner-city apartment. He said: 'To go from just living by yourself with your partner a while to then go into staying in people's houses and feeling like a burden was just really difficult.'

Others had homes that remained habitable but were without utilities for weeks. The search for new accommodation or visits to support networks to use showers and other facilities necessitated complex mobility. Catherine's home was without water for six weeks and sewerage for six months. She and her partner developed a 'roster' with friends to use their homes for showering. Temporary toilet facilities were in their back garden. The home was no longer a place in which simple acts like showering were possible. Such acts required mobile planning and ever-changing pathways across Christchurch. For younger LGBT people, socialising with friends remained deeply important with new mobile practices established to maintain friendship networks—focussed around and within domestic spaces. There was, therefore, both a reduction in the sites of sociality and a re-making of the ways in which socialising took place within and *between* places. These may be thought of as 'new forms of mobile practice' representing resilience, coping and adaptation—necessary in the face of slow government and private sector-led recovery. Just prior to the earthquake, Sophie (20 s, lesbian) had moved to Christchurch to be with her girlfriend. Sophie commented:

> Most of the friends that we had, or that we have actually, are gay couples and so I remember doing more, like not just going over there, but like actually going over and spending the night with people. I don't know if that was just

> because we were a bit younger and had the energy for, you know, being in a sleeping bag on someone's floor or what, but, yeah, it was maybe we hung on a little tighter.

Two interviewees' homes (Zach and Chris—both gay) served as sanctuaries for LGBT youths struggling with complex family relationships regarding their sexual identities and unsafe familial homes (Rodó-de-Zárate, 2015). These pre-disaster support mechanisms were re-enacted post-disaster (Gorman-Murray et al., 2017; Ong, 2017). Chris's home was badly damaged, without power and eventually demolished. But for weeks after the disaster, it was the destination of young people who understood it as a space offering them safety and care linked to their marginalised identities and as a safe 'family of choice.' For these youths, 'home' was not a fixed site in which they lived but was instead a place to travel towards during crisis.

For those with the time and financial means, temporary travel, or mobility away from Christchurch, became a recovery strategy. Interviewees travelled to locations where everyday facilities still functioned and in which generally mundane mobilities—shopping or taking a shower—required little thought. Sean stayed with his partner's mother in a town several hours north of Christchurch and stated: 'I mean even just like the running water and usable toilets and that sort of thing was quite attractive.' He returned home 'feeling a bit more recharged.' Annabelle stayed in her family's 'quite basic' holiday cabin outside Christchurch and said: '[T]he water was working out there and so that was really, really healing.' By temporarily removing themselves (being mobile) from Christchurch's challenges, interviewees were able to return with increased coping capacity and resilience.

While mobility is part of recovery practices for some, immobility or lack of access to these practices for those with financial or carer ties to damaged Christchurch limited or delayed their recovery. As Peter Adey (2006, p. 77) has argued: '[T]here is never any absolute immobility, but only mobilities which we mistake for immobility, what could be called relative immobilities.' Being tied to Christchurch did not mean being immobile, but rather being unable to avoid the disrupted and often emotionally distressing mobility systems post-earthquake.

Longer-Term Recovery

Interviewees re-established connections to broader friendship networks and re-made their communities. However, disaster recovery—at the individual and community level—did not follow an uninterrupted chronological path of gradual improvement. Interviewees' mental and emotional states shifted frequently from stabilisation and happiness to depression and anxiety. Robyn described: '[A]n initial fall over was about a year-and-a-half after the quake, and then I was fine, and then I had another series of processes where things got tougher and tougher and tougher.' Robyn linked her recovery to the challenges of mobility in what she described as a 'broken' city, stating:

> The culture around Christchurch was still very stressful. Some days it took 15 minutes to drive to work. Some days it took an hour because you were stuck on detours and traffic and shit like that. … Yeah, the detours have detours.

Desires to psychologically 'move on' post-disaster were hindered by aftershocks, destruction of unstable buildings, distress of living in damaged homes awaiting repairs and the challenges of moving from place to place. Craig told us that he had experienced a period of depression from late 2012 to mid-2013. Craig explained:

> It was the constant drain of just dealing with the post-disaster environment. … Whether it was just that a road's closed and it takes you slightly longer to get home, these things just grind on you and grind on you.

Another challenge was walking through the city at night. The re-opening of the CBD in mid-2013 resulted in the opening of bars and restaurants. As with roadways, footpaths were damaged and under repair, meaning former routes were no longer accessible or required walking (practicing mobility) across cracked and unstable ground). Large city blocks were made up of empty buildings awaiting demolition or of gravel carparks where buildings had once stood. Large areas were deserted and poorly lit. This issue exacerbated the risk, fear or memory of prior public abuse for LGBT individuals. Risks to physical and emotional safety produced by the earthquake and the possibility of discriminatory abuse while moving through urban spaces became entwined, each escalating the other. Amid attempts to adapt to changed urban mobilities lay continued regret for what was lost.

Sophie described an event, prior to the earthquake, when she and her girlfriend went for a walk in their neighbourhood:

> There were a bunch of guys leaving a party and I was holding hands with my partner, and they came out and were like, 'Ooh, lesbians, dykes.' One of them was making indecent gestures and stuff like that.

Peter remembered, 'I used to go into town. You could go to a bar, and you could wander around the place. You could go to the sauna. It was just a normal town and that was all just gone.' Karl stated:

> The thing I missed the most was I used to love going out dancing. To this day, I still don't know where to go anymore. ... All those regular places—'Let's meet at such and such' or 'Let's go to dinner here.' [Now] it's like, 'I wonder if that place is still open?'

Describing changes in LGBT neighbourhoods, Nash and Gorman-Murray (2014, p. 768) argue: 'Social life is now mobile through new temporal and spatial logics, such as mobile clubs, events and temporary locations, with implications for the transformations of urban spaces.'

Mobility practices, instituted in response to the sudden lack of 'fixed' sites and yet hindered by the challenges of moving through post-earthquake Christchurch, meant LGBT individuals began to establish new, more obviously mobile communities. Groups such as LYC (or 'Love Your Condom') and UpRising began to host LGBT nights at normally 'straight' venues. Rather than a specific and permanent LGBT venue, or even a regular night at one specific 'straight' club, events were held at varying locations. Adoption of straight locations for queer events/nights came out of necessity due to the destruction of queer-only spaces. There was some disparity between those who happily adopted—and, indeed, celebrated—this new understanding of a mobile community and those who longed for what had been lost. This variability of queer-friendly spaces and venues represents another form of emergent mobility that the LGBT community needed to adapt too that either supported, or, hindered personal recovery depending on the perspective of the interviewee. For Jim, 'I actually think for me personally there's been more going on since the quake. … I mean, I think LYC kept the community going more in Christchurch.'

Jim's comment acknowledges that continuing connections to community through processes of deliberate mobility contributed to LGBT recovery. Following withdrawal and focus on domestic spaces, many found a renewed and active interest in connecting with a broader LGBT community in the longer-term recovery phase by moving to and between those places. As Christchurch residents re-made their city, the practice of 'taking over' a straight space for a night was specifically designed to foster a sense of belonging and participation for LGBT locals. While acknowledging the positive benefits of these practices, other interviewees expressed sadness at the lack of LGBT sites that were understood as more reliably fixed in place. Peter reflected: 'I'd much rather have the couple of bars that we had and the environment that we had back. … You could just wander around and it was really nice.'

As a white, cis-gender man, Peter had easier access to these spaces than other LGBT people, and not everyone shared his nostalgic fondness for lost spaces. His sense of loss reveals the tension between the enactment of mobile practices as a form of resilience and the vulnerability and emotional distress produced through the loss of once important urban sites. Locating the new temporary and mobile LGBT places required planning around opening times and navigating sometimes unfamiliar and unsafe urban pathways which, for some, caused further distress.

Reaching the safety and comfort of LGBT events required navigating spaces within neighbourhoods that were confronting. The mobility challenges created by the earthquake and LGBT discrimination were entwined and co-produced. Interviewees stated the earthquake had forced an abrupt reduction in community size. The form of post-earthquake mobility chosen by many Christchurch residents was permanent relocation elsewhere. Although there is no data available on the size of the LGBT community before the earthquake, many people, including LGBT individuals, left afterwards (Bellamy, 2014). This had varying emotional impacts on those who had left (Adams-Hutcheson, 2015; Hogg et al., 2016) and those remaining.

Catherine's closest friend moved away from Christchurch and the increased distance between them was one of the more significant longer-term impacts of the disaster on Catherine's emotional well-being. Many interviewees had seriously considered leaving Christchurch. Reasons for staying included attachments to family and friends and work commitments. Interviewees expressed the tension between attachments to place and the desire to move away by making regular recuperative trips to other New Zealand places. Peter explained: 'I'm going up to Hamilton this weekend and it'll just be normal up there. There's a party at a friend's place and we might go to a bar, or we might go to the movies, and everything is just normal.'

Peter describes 'normal' as a travel destination, rather than as descriptive of life in Christchurch. Post-earthquake, Peter had developed a friendship network in Hamilton and Auckland. In response to the changes in LGBT community in Christchurch, he had developed his own set of mobile practices through which he could, at least temporarily, remove himself from the disaster-impacted city and become a participant in what was, to him, a 'normal' LGBT community.

Mobilising Disaster Justice

Lukasiewicz (2020, p. 8) details ten distinct but overlapping types of disaster justice. While she does not flesh out what is specific to each type, at their shared core is a commitment to equitable and just procedures and governance. Our empirical results demonstrate how five of the ten types of disaster justice that Lukasiewicz proposes provide a way of both framing and responding to the injustices experienced by our participants: intergenerational justice, temporal justice, spatial justice, corrective justice and relational justice.

Lukasiewicz's notion of temporal justice is evident in our participants' accounts and our three temporal recovery phases. Different people bore the burdens of the disaster across different time periods. For example, some individuals did not have access to running water or toilets for weeks to months after the earthquake. For others, they were not affected in this way, representing variable temporal injustices. Other participants talked about the long-term psycho-social impacts related to returning to a normal recovered position post-earthquake which was easier for some LGBT individuals who had the means of mobility to travel to 'safe' spaces for recuperation and rest.

Perhaps the strongest of the disaster subtypes in our data was relational justice, recognising and addressing the importance of relationships and social networks in the distribution of benefits and burdens. Our participants utilised, created and maintained new systems of communication to share information and stay connected during post-disaster recovery. These are examples of building community networks, and mobilising social capital, resulting in increased resilience. As noted, some residents chose to leave the city temporarily as a way of coping with ongoing, long-term disaster consequences. However, this recovery practice was not uniformly available to all LGBT residents.

Intergenerational justice refers to the fair and equitable intergenerational distribution of benefits and burdens and our data demonstrate examples experienced by younger and older LGBT individuals. For example, younger LGBT people were more likely to rent housing. Older LGBT people were more likely to have a mortgage or own their own home. Governance and policy responses to those that rented versus homeowners, or those with a mortgage, varied enormously representing significant intergenerational injustice. Younger LGBT people tend to be on lower incomes and share properties and have fewer financial resources with which to compete in a competitive rental market emerging after disaster representing an example of intergenerational injustice. Conversely, older LGBT people with parental responsibilities and single LGBT people living alone regardless of age and disconnected from their biological families were disadvantaged in terms of the distribution of available government benefits.

There were clear examples of spatial injustice. Spatial injustice may be divided into three categories: private, public and commercial. Home represented the most obvious examples of spatial injustice—the loss of homes and their damage and the need for people to share basic amenities such as showers and bathrooms. For others, being in public spaces like shelters represented spatial injustice where their private LGBT identities became publicly known. Lastly, the loss of LGBT community spaces such as clubs, bars and other safe spaces, where LGBT could meet and socialise, involved significant spatial injustices.

An example of corrective justice (injustices caused by wrongful actions or events) involved one of our lesbian participants who told us of her and her partner's discomfort in moving about public places. Their sexual identity made them potential victims of verbal abuse leaving them feeling unsafe. Such discomfort in public spaces was noted by others, where not all members of the LGBT community felt safe or welcome in gay bars or clubs.

Conclusion

Understanding mobility in LGBT disaster recovery requires examination of transformed environments and embodied experiences of mobility and of the ways in which individual and community identities are enacted and produced. Equally important is an understanding of processes of change already underway when disaster strikes, and the ways in which disaster may accelerate change. In a small city of the Global North, the LGBT community of Christchurch re-made itself in ways that reflect broader shifts in LGBT identities, connections to place and mobile practices. While some LGBT community members mourn what was 'lost,' for others, post-disaster recovery meant creating new, mobile ways of being within a permanently changed city.

Intersecting with the positive benefits of mobile practices on individual and community recovery were challenges of moving through post-earthquake spaces that left many people emotionally and physically vulnerable. Damage to buildings, roads and infrastructure ensured that travelling across or within the city was distressing. The disaster produced new tensions of mobility. Individual mobility practices varied across space and time within and away from the city, according to variations in scale. Processes of mobility were not equally accessible, limiting the recoveries of individuals tied to the city and to the continuing pace of recovery. Ongoing marginalisation of LGBT people exacerbated inequitable access to safe mobility within and through Christchurch. Issues of safety from abuse or discrimination while moving through Christchurch were an element of post-earthquake mobilities.

The earthquake disrupted domestic boundaries. For many, home had provided safety, routine and order, but the earthquake forced new pathways which, at times, left already traumatised individuals vulnerable to anti-LGBT discrimination. Damaged and lost spaces, experiences of discrimination and inequitable opportunities for recuperative travel suggest the heightened challenges for LGBT mobility created by disaster. Yet, mobile practices were equally a means by which the community actively re-made itself within Christchurch and enacted recovery. The earthquake led to transformations in both the meaning of home and LGBT sociality that were more explicitly mobile. For many LGBT residents these transformations were vital to rebuilding community and new ways of connecting to their urban environment.

Our research traces the role of mobility/immobility as elements of recovery from a highly disruptive and traumatic disaster while remaining cognisant of the ways in which mobile practices were already in use by a community renegotiating attachment to place. Due to the disaster, the need to move through the earthquake-impacted city produced specific forms of vulnerability for LGBT people. However, queer people enacted mobile practices as a form of resilience. As time passed, tensions in mobility could be seen in the uncertain desire to return to a community that was reliably fixed in place and reveals that the pre-disaster community may already have been on the move.

The empirical data shed light on aspects of disaster [in]justice differentially experienced by LGBT people in Christchurch post-earthquake. These injustices reflected a variety of underlying socio-economic, demographic and other structural processes. The results suggest more work is needed in a variety of post-disaster settings to understand how disaster justice plays out during recovery post-disaster.

Conflict of Interest This research was funded by an Australian Research Council Discovery Grant, DP130102658, and none of the authors have any conflict of interests.

References

Adams-Hutcheson, G. (2015). Voices from the margins of recovery: Relocated Cantabrians in Waikato. *Kōtuitui: New Zealand Journal of Social Sciences, 10*(2), 135–143.

Adams-Hutcheson, G. (2017). Embodied vibrations: Disastrous mobilities in relocation from the Christchurch earthquakes, Aotearoa New Zealand. *Transfers, 7*(3), 3–37.

Adey, P. (2006). If mobility is everything then it is nothing: Towards a relational politics of (Im)mobilities. *Mobilities, 1*(1), 75–94.

Aiello, G., et al. (2013). Here, and not yet here: A dialogue at the intersection of queer, trans, and culture. *Journal of International and Intercultural Communication, 6*(2), 96–117.

Bell, D., & Valentine, G. (1995). *Mapping desire: Geographies of sexualities.* Routledge.

Bellamy, P. (2014). *Social effects of Canterbury earthquakes.* New Zealand Parliamentary Library. Available at: https://www.parliament.nz/en/pb/research-papers/document/00PlibC51211/social-effects-of-the-canterbury-earthquakes

Browne, K., Lim, J., & Brown, G. (2007). *Geographies of sexualities: Theory, practices and politics*. Ashgate.

Caluya, G. (2008). "The Rice steamer": Race, desire and affect in Sydney's gay scene. *Australian Geographer, 39*(3), 283–292.

Canterbury Earthquake Recovery Authority. (2016). *Canterbury Land Information Map*. Available at: http://ceraarchive.dpmc.govt.nz/sites/default/files/Documents/land-information-map-20110623.pdf

Cook, N., & Butz, D. (2015). The dialectical constitution of mobility and immobility: Recovering from the Attabad Landslide disaster, Gojal, Gilgit-Baltistan, Pakistan. *Contemporary South Asia, 23*(4), 388–408.

Cook, N., & Butz, D. (2016). Mobility justice in the context of disaster. *Mobilities, 11*(3), 400–419.

Cresswell, T. (2008). Understanding mobility holistically: The case of hurricane Katrina. In S. Bergmann & T. Sager (Eds.), *The ethics of mobilities: Rethinking place, exclusion, freedom and environment* (pp. 129–140). Ashgate.

Cretney, R. M. (2017). Towards a critical geography of disaster recovery politics: Perspectives on crisis and hope. *Geography Compass, 11*, e12302. https://doi.org/10.1111/gec3.12302

Cretney, R. M., & Bond, S. (2017). Shifting relationships to place: A relational place-based perspective on SES resilience. *Urban Geography, 38*(1), 8–24. https://doi.org/10.1080/02723638.2016.1139865

Doan, P. (2016). You've come a long way baby: Unpacking the metaphor of transgender mobilities. In G. Brown & K. Browne (Eds.), *The Routledge companion to geographies of sex and sexualities* (pp. 237–246). Routledge.

Doan, P. (2007). Queers in the American city: Transgendered perceptions of urban space. *Gender, Place and Culture, 14*(1), 57–74.

Dominey-Howes, D., Gorman-Murray, A., & McKinnon, S. (2014). Queering disasters: On the need to account for LGBTI experiences in natural disaster contexts. *Gender Place and Culture, 21*(7), 905–918.

Drozdzewski, D., & Dominey-Howes, D. (2021). Researcher trauma: Considering the ethics, impacts and outcomes of research on researchers. In S. Henn, J. Migglebrink, & K. Horschelmann (Eds.), *Research ethics in human geography* (pp. 168–181). Routledge.

Easthope, H. (2009). Fixed identities in a mobile world? The relationship between mobility, place and identity. *Identities, 16*(1), 61–82.

Fernandez, G., & Ahmed, I. (2019). Build back better: Approach to disaster recovery: Research trends since 2006. *Progress in Disaster Science, 1*, 100003. https://doi.org/10.1016/j.pdisas.2019.100003

Gorman-Murray, A., & Waitt, G. (2009). Queer-friendly neighbourhoods: Interrogating social cohesion across sexual difference in two Australian neighbourhoods. *Environment and Planning A: Economy and Space, 41*(12), 2855–2828.

Gorman-Murray, A., McKinnon, S., & Dominey-Howes, D. (2014). Queer Domicide? LGBT displacement and home loss in natural disaster impact, response and recovery. *Home Cultures, 11*(2), 237–262.

Gorman-Murray, A., et al. (2017). Listening and learning: Giving voice to trans experiences of disasters. *Gender, Place and Culture, 25*(2), 166–187.

Gorman-Murray, A., et al. (2018). Problems and possibilities at the margins: LGBT experiences in the 2011 Queensland floods. *Gender Place and Culture, 24*(1), 37–51.

Gorman-Murray, A & McKinnon, S. (2015). Queer geography: rethinking sexuality and space. *International Encyclopedia of the Social and Behavioral Sciences, 19*, 759–764.

Hannam, K., Sheller, M., & Urry, J. (2006). Editorial: Mobilities, immobilities and moorings. *Mobilities, 1*(1), 1–22.

Hogg, D., et al. (2016). The effects of relocation and level of affectedness on mood and anxiety symptom treatments after the 2011 Christchurch earthquake. *Social Science and Medicine, 152*, 18–26.

Johnston, L. and Longhurst, R. (2010). *Space, place and sex: geographies of sexualities*. Plymouth, UK: Rowman and Littlefield Publishers.

Kerekere, E. (2017). *Part of the Whānau: The emergence of Takatāpui identity he Whāriki Takatāpui*. PhD thesis, Victoria University of Wellington. Available at: http://researcharchive.vuw.ac.nz/xmlui/bitstream/handle/10063/6369/thesis_access.pdf?sequence=1

Knopp, L. (1995). Sexuality and urban space: A framework for analysis. In D. Bell & G. Valentine (Eds.), *Mapping desire: Geographies of sexualities* (pp. 149–161). Routledge.

Larkin, B. (2019). Pride and prejudice: LGBTIQ community responses to disaster events worldwide. *Australian Institute for Disaster Resilience, 34*(4), 60–66.

Lukasiewicz, A. (2020). The emerging imperative of disaster justice. In A. Lukasiewicz & C. Baldwin (Eds.), *Natural hazards and disaster justice: Challenges for Australia and its neighbours* (pp. 3–24). Palgrave Macmillan.

Myrdahl, T. M. (2016). Visibility on their own terms? LGBTQ lives in small Canadian cities. In G. Brown & K. Browne (Eds.), *The Routledge companion to geographies of sex and sexualities* (pp. 37–44). Routledge.

Nash, C., & Gorman-Murray, A. (2014). LGBT neighbourhoods and "new mobilities": Towards understanding transformations in sexual and gendered urban landscapes. *International Journal of Urban and Regional Research, 38*(3), 756–772.

Nash, C., & Gorman-Murray, A. (2015). Lesbian spaces in transition: Insights from Toronto and Sydney. In P. Doan (Ed.), *Planning and LGBTQ communities: The need for inclusive queer spaces* (pp. 181–198). Routledge.

Ong, J. (2017). Queer cosmopolitanism in the disaster zone: "My Grindr became the United Nations". *The International Communication Gazette, 79*(6–7), 656–673.

Reynolds, R. (2009). Endangered territory, endangered identity: Oxford street and the dissipation of gay life. *Journal of Australian Studies, 33*(1), 79–92.

Rodó-de-Zárate, M. (2015). Young lesbians negotiating public space: An intersectional approach through places. *Children's Geographies, 13*(4), 413–434.

Rushton, A., & Scarlett, J. (2023). The blame game: Disaster, queerness and prejudice. *Disaster Prevention and Management, 32*(2), 352–363. https://doi.org/10.1108/DPM-11-2022-0222

Truebridge, N. (2017). Govt spent $1.5b acquiring Christchurch red zone land that's now worth just $21m. *Stuff,* 28 August. Available at: https://www.stuff.co.nz/national/96144297/govt-spent-15-billion-acquiring-christchurch-red-zone-land-thats-now-worth-just-21m

Voith, L., et al. (2020). Using a trauma-informed, socially just research framework with marginalised populations: Practices and barriers to implementation. *Social Work Research, 44*(3), 169–181.

Yamashita, A., Gomez, C., & Dombroski, K. (2017). Segregation, exclusion and LGBT people in disaster impacted areas: Experiences from the Higashinihon Dai-Shinshai (Great East-Japan disaster). *Gender, Place and Culture, 24*(1), 64–71.

CHAPTER 7

'...You Used to Be One of the Boys'

Kate Roy

'You used to be one of the boys' was a comment made to me at work by my manager after I raised concerns about sexist and homophobic behaviours in the workplace. The manager identified as a heterosexual white male, and the workplace was part of Australia's emergency management sector. This statement is symbolic of real-world discriminatory and hostile behaviours I experienced whilst working in several emergency management organisations. Most of the behaviours were casual and generalised, occurring in numerous settings, meetings, and by email and text messages. It seemed an acceptable part of the culture and was pervasive throughout daily interactions. It felt like discriminatory behaviours were socially reinforced and normalised by others and were part of the organisation's culture, power structures and inequalities. In this chapter, I outline my personal experiences working in the emergency management sector in Australia.

I am a queer cisgender woman living in Australia. I had taken up a new position in a state government agency when I experienced a series of sexist and homophobic incidents. I was angry and fought back. It felt like the organisation didn't appreciate me admonishing them about their obligations to eliminate such behaviours nor, critically, believe me when I told

K. Roy (✉)
BHRM, MALP, Emergency Management Professional, Sydney, NSW, Australia

D. Dominey-Howes et al. (eds.), *Queering Disasters, Climate Change and Humanitarian Crises*, Sustainable Development Goals Series, https://doi.org/10.1007/978-981-96-3857-4_7

them that these behaviours had occurred. I had earned the epithet of 'troublemaker' with my strong opinions on the matter, 'causing disruption' within the team. Soon after, management threatened terminating my employment.

Confrontations like these remind me of similar moments in contemporary Australian society when problematic identity-based discrimination occurs when people do not comply with dominant societal expectations. I am reminded of the moment when the then Australian of the Year (2021) Grace Tame, a prominent female activist and advocate for survivors of sexual abuse, was criticised for not smiling at our (then) Prime Minister (Karp, 2022). Victims and other marginalised communities are often publicly admired and commended for speaking out about their experiences. Yet there is an expectation that they will 'politely' engage in public discourse and only speak out civilly (Cameron, 2022). This demand to be 'civil' usually means that women must control and disguise their emotions while silently putting-up with men's behaviour (Remeikis, 2022). Fighting back is not considered civil. Neither is anger. Indeed, other research shows that when marginalised communities speak out they are rarely afforded the brief period of public admiration but are vilified from the outset (Gardiner, 2018, Southern, 2019, Trott, 2021). Challenging the status quo isn't acceptable at all, no matter how politely or civilly.

Like other women who speak out I, too, was sharply reminded that I 'lack an internal grace or beauty of character, that once adopted, would allow me to move graciously through the world without anger and confrontation' (Blair et al., 1994, p. 383). I was being told, not so subtly, to 'go along to get along'. Team harmony, it was implied, depended on me suppressing the anger I felt as a result of my colleagues' homophobic slurs. Cautions 'to be nice' are frequently used by men, often cisgender and heterosexual, to 'keep us in our place'[1] (Blair et al., 1994, p. 383). It seemed casual racism, sexism and homophobia were part of the culture and I, as a threat to this casual collective, was swiftly and easily dealt with. I was put in my place, which resulted in me feeling like I had little choice but to leave the organisation.

Ironically, my workplace experience was all the more galling because it occurred at a time when sexism and the mistreatment of women (and by default 'other') were centre stage in Australia and central to the #MeToo

[1] 'Us' in this context referring to people who are not cisgender, heterosexual males.

movement.[2] Victims who have long been silenced had found their individual and collective voice. #MeToo and similar movements had ignited a global discussion about sexual harassment and gender inequality (Hill, 2021). It prompted a social reckoning in Australia in 2020 (McKinnon, 2022), especially focusing on the (mis)treatment of women by prominent male figures. And yet, instead of the men in my workplace stepping up and doing something to promote better workplace practices, they left the heavy lifting to women. This experience is not unique. As Maricourt and Burrell (2022) contend, feminist-led changes in gender relations that challenge the status quo have led to an anti-feminist backlash from those who see these changes as a challenge to their power and privilege. It was my experience that men and in particular those who identified as cisgender and heterosexual, persisted in defending their behaviour and policing mine because admitting and accepting they needed to change made them uncomfortable (Maricourt & Burrell, 2022).

My career in emergency management began in 2011 working on a strategy created in response to a Parliamentary Inquiry investigating claims that bullying and harassment were common staff behaviours in a statewide emergency service (Parker, 2008). Major issues that had plagued employees in the organisation included depression, anxiety, self-harm and even suicide (Parker, 2008). A common complaint about management raised during the Inquiry was the existence of a nepotistic 'old boys club' (Parker, 2008, p. xi) culture that pervaded the organisation—not unlike my own experiences where being 'one of the boys' was advantageous or indeed an unspoken requirement. One submission from the Inquiry stated, '[T]he top heavy, boys-club-mentality of Management needs to be expunged. We need to start from scratch' (Parker, 2008, p. 275).

Of significant concern to staff was the negative workplace culture and the way in which management failed to handle these matters. Numerous

[2] The expression 'me too' was used in 2006 by African-American activist Tarana Burke as a private act of solidarity with young black girls who were survivors of sexual violence. Burke developed a non-profit called Just Be and the expression 'me too' was used to raise awareness of the pervasiveness of sexual abuse and assault in society (Trott, 2021). In 2017, the expression became a viral hashtag #MeToo when American actress Alyssa Milano publicly shared her experience of sexual abuse on social media. This prompted millions of people to publicly share their experiences of sexual abuse and harassment in an expression of global vulnerability using the hashtag #MeToo (Trott, 2021). In 2021, in Australia, the era of #MeToo became seismic when multiple allegations of sexual misconduct occurred in Australia's Parliament House (Hill, 2021).

submissions suggested that drawn-out grievance and complaints-handling processes within the organisation had exacerbated many of these situations.

This seeming exposure to serious workplace culture problems within at least parts of the emergency management sector didn't seem to be isolated to one organisation. My experiences in other emergency management roles since have been equally bleak.

On another occasion in a different emergency management organisation, a senior ranking male implied that sexual favours were required for situational information on a disaster event. A female colleague and I travelled to the disaster site with this senior officer and a call was answered in the vehicle on speaker phone. One of his junior staff boasted that he had just got off the phone from his 'Missus' and had told her to 'strap a mattress to her back' as he would be returning from deployment that evening. The senior officer had replied, 'Attaboy'.

When I joined a different emergency management agency as a volunteer and was getting to know my peers, I had asked the question, 'why do you volunteer?' The unanimous response was, 'because it gets us time away from the Missus'. Women, it seems, are not considered whole, composite beings in their own right, but rather tools to be used or annoyances to avoid. As feminist scholars have shown this way of positioning women is common to communities and workplaces around the world (Summers, 1975; Keller, 2021; Cameron, 2022).

Developing and progressing my career in a hypermasculine male-dominated sector, I was often automatically regarded at meetings as the tea lady or the secretary, responsible for nothing more than men's comfort. 'Can I have a cup of tea, love?' and 'Are you taking the minutes?' were the norm. As a consequence of these gendered expectations, it felt like reactions to my professional emergency management contributions were often heard and dismissed or simply not heard at all. Behaviours and comments like these signified to me at best a lack of awareness of my skills and expertise and at worst blatant misogyny. Gender dynamics can play out in many ways during meetings, which provide arenas where key social processes embedded within organisations unfold (Hemshorn de Sánchez, 2020).

A more recent example took place when I attended a meeting with a senior-ranking white male member of a third emergency management agency, who identified as cisgender and heterosexual, where I was directed to put down my pen (because I had been relegated to taking the meeting minutes). Upon putting down my pen he continued with instructions pertaining to an upcoming meeting in which he directed that a male consultant be engaged because the cohort of predominately older men would not respect a female consultant. He did not want that specific statement about gender recorded in the meeting minutes, and yet it can be assumed he felt comfortable with this discrimination and did not fear repercussions. I felt that now, I was resentfully complicit in the organisation's structural, institutional and cultural dimensions of gender inequality.

Sexism, misogyny and gender norms are the basis of male authority and power within the emergency sector. Flood (2005) argues that wilful ignorance of these forms of systemic discrimination and inequality and resistance to change can be summed up in one word: privilege. Male privilege allows (cisgender) men to reinforce their dominance, acknowledging that sexism and misogyny exist while at the same time wilfully allowing it to continue unchallenged (Flood, 2005).

I have been seen through men's construct as a woman, a simpering secretary, tea lady and troublemaker.

I have been seen through men's construct as lesbian, objectified, fetishised and subjected to sexual inquiry.

The intersectionality[3] of queer+ woman has resulted in much of my experience within emergency management workplaces being on a continuum of discrimination, sexual inquiry and objectification. I have been asked many times '[W]hen did you "come out"?', and the answer to that is not a single instance. It is multiple (tiring) instances, over and over again, with almost every new colleague. In some contexts, I am perceived as heterosexual which circumvents these inquisitions. Whilst this alleviates the banality, I simultaneously feel guilty for perpetuating or allowing the assumption that I am heterosexual to go unquestioned. Some gender non-conforming and non-binary people in these situations may not have the option of passing as heterosexual, or being their authentic selves may carry additional risks. Heteronormativity infuses everyday interactions. Sometimes it is just easier to be what they think you are and what they want you to be, one of the boys.

[3]The term emerged from legal black feminist scholarship, specifically Crenshaw (1989, p. 149) who provided the following definition of intersectionality: 'a metaphor for understanding the ways that multiple forms of inequality or disadvantage sometimes compound themselves and create obstacles that often are not understood among conventional ways of thinking'.

Pease (2010), McLean (2008) and Adams et al. (2017) have written, in detail, about these and other similar experiences. In my workplaces I felt and continue to feel compelled to disclose my sexuality to my close colleagues to be my authentic self, though self-exposure is not without its dangers (Pease, 2010). In doing so, I was subjected to demeaning homophobic epithets—'lesbo' and 'homo'. Unwanted sexual innuendos came thick and fast. Repeated and intrusive questions about my sex life and sexually explicit images being sent to my email and mobile phone were unwelcome occurrences.

It can feel like heterosexual people generally do not think of themselves as having a sexual identity in the same way as LGBTQIA+[4] people. They simply regard themselves as 'normal' as their sexuality is rarely, if ever, questioned (Pease, 2010). These discourses have significant implications for queer people who feel they already do not fit into heterosexual spaces. When heterosexuality is accepted as normative (which can be seen in the hypermasculinised emergency management sector), it becomes the standard for all legitimate sexual behaviour (Pease, 2010). For heterosexuals to see themselves as normal, queer people must then remain abnormal. Being 'othered' is already hard. Being 'othered' in a hypermasculinised workforce is even harder. For example, introductions of '(t)his is my lesbian friend' reduces me to one element of who I am, ignoring and devaluing my multiple, intersecting identities. Inquiries into sexual activity and sexual acts followed by references to the phallus were consistent jokes made by the boys in 'good humour'. Unwanted jokes, innuendos and other harmful behaviours were the norm. What offended me most was the thought that because I was attracted to women, they assumed that I would want to participate in their sexism and misogyny, like them, one of the boys.

I describe these experiences not as an exercise in self-indulgence but rather to draw attention to some big issues in the emergency management sector.

First, moments of casual misogyny, sexism and homophobia are still common and illustrate how cultural assumptions of misogyny and heterosexuality can characterise everyday interactions (Adams et al., 2017). They

[4] There are a number of different terms for these populations. In this chapter I use the terms LGBTQIA+ (Lesbian, Gay, Bisexual, Transgender, Queer, Intersex, Asexual) and 'queer' to encompass any individual that may be marginalised by cisgendered heteronormativity.

(un)expectedly happen in a variety of times and places, from business-as-usual meetings in corporate headquarters to deployments in the field. Parkinson et al. (2022) and Leonard et al. (2022) document the impacts of these hypermasculine, emergency sector work cultures. They identify the many barriers LGBTQIA+ individuals face in participating in emergency service work and their lived experiences of poor inclusion, marginalisation and discrimination. Both papers report research which considers how emergency service organisations' work cultures and practices negatively affect LGBTQIA+ people.

Poor inclusion has been found to have 'harmful cognitive, emotional, behavioural, and health outcomes' (Shore et al., 2011, p. 1265). The consequences of these hetero-gendered and -sexualised patterns in workplaces can be severe (Hemshorn de Sánchez, 2020) and include reduced safety, well-being and organisational effectiveness (Eriksen, 2019).

My experiences show how emergency management workplaces are distressingly 'heterosexed' (England, 1994), revealing the dominance and impacts of homophobic and sexist views, behaviours and strategies. Dominey-Howes et al. (2016, 2022) identify religious exemptions under Australian Equal Opportunity legislation as another barrier to inclusive emergency service provision and employment. In Australia, faith-based organisations that provide response and recovery services on behalf of state governments to disaster-affected individuals and communities are afforded exemptions to Commonwealth[5] and State[6] anti-discrimination laws, meaning they can lawfully refuse service to LGBTQIA+ people. Not only are these examples discriminatory, they also demonstrate the failure of public policy by enabling the exclusion of marginalised people.

Second, what does not get measured and reported does not get managed. An analysis of NSW legislation, reports, policies and plans by Dominey-Howes et al. (2016) in relation to their inclusivity of LGBTQIA+ people is discouraging. For this chapter I conducted a similar analysis using the same combined keyword[7] search within the two government documents that govern emergency management in NSW: the NSW State Emergency and Rescue Management Act 1989 and NSW State Emergency Management Plan. In 2023, eight years after Dominey-Howes et al.'s (2016) publication, there were again zero hits.

[5] Sex Dis-crimination Act 1984 (Commonwealth).

[6] Anti-Discrimination Act 1977 (NSW).

[7] Keywords: Sexuality, gender, identity, minority, lesbian, transsexual, transgender, bisexual, gay, homosexual, queer, LGBTI, diversity, inclusion.

In the context of workplace policy, the Australian Workplace Equality Index is considered the definitive national benchmark on LGBTQIA+ workplace inclusion and encompasses the country's largest and only national employee survey designed to gauge the overall impact of inclusion initiatives within organisations. The index drives Australia's best practices in LGBTQIA+ initiatives and sets a comparative benchmark for Australian employers across all sectors. To date, the NSW Police Force is the only NSW emergency management organisation to participate.

LGBTQIA+ personnel within the emergency management sector are not supported and their experiences are not being heard. Some emergency management organisations have diversity and inclusion policies or plans. Conducting a combined keyword[8] search of these documents specifically relating to sexual and gender diversity yielded dismal results. The very documents that were designed solely for diversity and inclusion still exclude LGBTQIA+ people. The scarcity of any meaningful policy aimed at ensuring sexual and gender diverse workers within the sector are supported is grim.

To make a bad situation worse, at the time of writing, Sydney is host to World Pride 2023, an event that promotes visibility and awareness of LGBTQIA+ issues on an international level. Numerous Australian State and territory emergency management organisations are proudly displaying their inclusivity at this event, with pride logos, participation in various high-profile events such as Fair Day[9] and the Mardi Gras parade.[10] Some organisations unveiled assets with pride decals[11] as mobile displays of inclusivity and even defended these when criticised by the public on social media. And yet, it appears that the same organisations are unable to enact real change at a day-to-day level or adequately support staff who report experiencing discrimination and marginalisation.

My experiences, combined with the curiosity of inquiry, have led me to conduct my own research guided by intersectional feminism into the

[8] Keywords: Sexuality, identity, minority, lesbian, transsexual, transgender, bisexual, gay, homosexual, queer, LGBTI, orientation.

[9] Fair Day is part of the Sydney Gay and Lesbian Mardi Gras. The annual all-day celebration marks the start of Mardi Gras. It is a free community event comprising of community stalls, live music, entertainment, food and beverage outlets hosted in a public park in Sydney, Australia.

[10] The Mardi Gras parade is part of the Sydney Gay and Lesbian Mardi Gras. Marching groups and floats parade down Oxford Street to showcase the beauty and strength in pride and self-expression as the LGBTI community continue the fight for equality. It is an annual parade and attended by hundreds of thousands of people from around the world.

[11] NSW Rural Fire Service fire truck, NSW Fire and Rescue fire truck, NSW State Emergency Service high clearance vehicle.

experiences of queer personnel in the Australian emergency management sector—the good, the bad and the ugly. Feminist research has been extremely influential in disasters research, drawing attention to how 'intersecting social characteristics, including gender, sexuality, race, class, and bodily ability each contributes to the social construction of disaster' (McKinnon, 2022, p. 1).

Australia is entering a new phase of disaster risk reduction policy and practice as climate change increases the frequency and intensity of disasters, urgently requiring new and alternative approaches (McKinnon, 2022). To highlight and critique the practices and norms in the emergency management sector that organisational heads would prefer to believe are relics of the past (Leonard et al., 2022) is also now an urgent task. Investigating the structure and practice of emergency management organisations is not novel. However, a gap does exist as it relates to LGBTQIA+ personnel experiences. Australians expect workplaces to be safe, respectful and inclusive. Unfortunately, that is not always the case. Before these incidences I might not have had the occasion or courage to explore the topic. To be clear, revisiting the distress of these sexist and homophobic incidences is difficult. But if the outcomes of my sharing and analysis of my experiences can be used to make the sector safer for others, then all the better.

Conflict of Interest The author has no conflict of interests.

References

Adams, T. E., ellis, C. & Jones, S. H. (2017). Autoethnography. Hoboken, NJ, USA: John Wiley & Sons, Inc.

Blair, C., Brown, J. R., & Baxter, L. A. (1994). Disciplining the feminine. *The Quarterly Journal of Speech, 80*, 383–409.

Cameron, D. (2022). Women, civility and the language of politics: Realities and representations. *The Political Quarterly (London, 1930), 93*, 25–31.

Crenshaw, K. (1989). Demarginalizing the Intersection of Race and Sex: A Black Feminist Critique of Antidiscrimination Doctrine, Feminist Theory and Antiracist Politics. University of Chicago Legal Forum, 139–67.

Dominey-Howes, D., Gorman-Murray, A., & McKinnon, S. (2016). *Emergency management response and recovery plans in relation to sexual and gender minorities in New South Wales.International Journal of Disaster Risk Reduction, 16*, 1–11.

Dominey-Howes, D., Mckinnon, S., Gorman-murray, A. & Eriksen, C. (2022). Sexual and gender minorities in disasters. In: (EDS.), I. T. K. M. E. C. P.-R. (ed.). *Routledge Handbook of Environmental Hazards and Society.*

England, K. V. L. (1994). Getting personal: Reflexivity, positionality, and feminist research. *The Professional Geographer, 46*, 80–89.

Eriksen, C. (2019). Negotiating adversity with humour: A case study of wildland firefighter women. *Political Geography, 68*, 139–145.

Flood, M., & Pease, B. (2005). Undoing men's privilege and advancing gender equality in public sector institutions. *Policy and Society, 24*, 119–138.

Gardiner, B. (2018). 'It's a terrible way to go to work:' What 70 million readers' comments on the Guardian revealed about hostility to women and minorities online. *Feminist Media Studies, 18*, 592–608.

Hemshorn De Sanchez, C., & Meinecke, A. (2020). *Managing meetings in organizations.* Ringgold Inc.

Hill, J. (2021). How #MeToo is changing Australia. *Quarterly essay*, 1–131.

Karp, P. (2022). 'I can just think it': Peter van Onselen admits controversial Grace Tame column was 'probably unnecessary. *The Guardian.*

Keller, P. (2021). Objectified women and fetishized objects. *Journal of Ethics and Social Philosophy, 19.*

Leonard, W., Parkinson, D., Duncan, A., Archer, F., & Weiss, C. O. (2022). Under pressure: Developing lesbian, gay, bisexual, transgender and intersex (LGBTI) inclusive emergency services. *Australian Journal of Emergency Management, 37*, 52–58.

Maricourt, C. D., & Burrell, S. R. (2022). MeToo or #MenToo? Expressions of backlash and masculinity politics in the #MeToo era. *The Journal of Men's Studies, 30*, 49–69.

McKinnon, S. (2022). Masculinities and disaster. In *Oxford research encyclopedias of natural Hazard science.* Oxford University Press.

Parker, R. (2008). The management and operations of the NSW Ambulance Service. In: 2, N. S. W. P. L. C. G. P. S. C. N. (ed.). New South Wales.

Parkinson, D., Duncan, A., Leonard, W., & Archer, F. (2022). Lesbian and bisexual women's experience of emergency management. *Gender Studies, 39*(1), 1–24.

Pease, B. (2010). *Undoing privilege: unearned advantage in a divided world.* London, NBN International.

Remeikis, A. (2022, January 26). On the project my rage bubbled to the surface. That's what happens when women are no longer willing to make nice. *The Guardian.*

Shore, L. M., Randel, A. E., Chung, B. G., Dean, M. A., Holcombe Ehrhart, K., & Singh, G. (2011). Inclusion and diversity in work groups: A review and model for future research. *Journal of Management, 37*, 1262–1289.

Southern, R. (2019). *Othering political women: Online misogyny, racism and ableism towards women in public life.* Springer International Publishing.

Summers, A. (1975). *Damned whores and God's police: The colonization of women in Australia.* Melbourne, Allen Lane.

Trott, V. (2021). Networked feminism: Counterpublics and the intersectional issues of #MeToo. *Feminist Media Studies, 21*, 1125–1142.

CHAPTER 8

The Changemakers and the Losers

Jon Martin Larsen

Introduction

As part of my master's thesis in international journalism in 2006, I collected stories from minorities seeking to understand how minorities, migration and terror influenced Turkish society. I found that LGBTQ+ migrants were changemakers who contributed constructively to develop more openness towards LGBTQ+ minorities in Turkish society. Eighteen years later, following my PhD project on municipal minority communication (Larsen, n.d.), I returned to Türkiye looking for the refugees who are supposed to come to Norway, seeking to give recommendations to the municipalities on how to create a dialogue with their minority residents. Promises to prioritise LGBTQ+ asylum seekers were renewed by the Norwegian government in 2020 (Vadla, 2020), but this vignette highlights that Iranian LGBTQ+ refugees are caught in a situation where they cannot stay in Türkiye, cannot return to their home countries, or flee to Europe or Norway. My vignette shares the stories of Rahim and Kian, two Iranians who belong in the gender and sexuality diversity, to shed light on

J. M. Larsen (✉)
Kristiania University of Applied Sciences, Oslo, Norway
e-mail: jonmartin.larsen@kristiania.no

D. Dominey-Howes et al. (eds.), *Queering Disasters, Climate Change and Humanitarian Crises*, Sustainable Development Goals Series, https://doi.org/10.1007/978-981-96-3857-4_8

Iranian LGBTQ+ refugees in Türkiye and invite readers to consider the complexity of oppression and hostilities towards LGBTQ+ in their home countries or host-countries but also the opportunities that LGBTQ+ can bring a receiving country.

Rahim's Story (2006)

Tears are running down her beard stubble. The cheap black eyeliner is being washed away from her eyes in small rivers down her face. She tries to clean herself with a small tissue hidden in her sleeve, just like her very religious mother used to do at home in a small city in Iran. She is in her early thirties, and her name is Rahim. Originally, a masculine Iranian name given to boys, Rahim was never a boy, neither a man today. Only her physical appearance bears a man's print and casts a strong shadow as the sun sets on the bank of Lake Van in Eastern Türkiye. Rahim is an asylum seeker. In 2004, a boyfriend in Iran gave her $430.00 to be smuggled into Türkiye.

I met Rahim, her boyfriend and her best friend in the city of Van in Turkey, in 2006. Van was, and still is, home to the eastern base of the United Nations High Commissioner for Refugees (UNHCR, n.d.) and is located close to both the Iranian and Iraqi borders. In 2006, Van was said by local authorities to be home to 70,000 people in the city centre, 3000 of them refugees. The city's total population was around 300,000 without any confirmed number of refugees. Today, Türkiye is home to the world's largest refugee population. Presently, Türkiye accommodates approximately 3.6 million Syrian refugees and around 320,000 individuals from other nationalities who are considered persons of concern (UNHCR, 2024a; Kaya, 2023). Van has an estimated population that has passed 550,000 people (PopulationStat, 2024), and the total amount of refugees hosted in the southeastern part of Türkiye is considered to be more than 1.7 million refugees (UNHCR, 2024b).

Iranian LGBTQ+ Fleeing

Iranians are the fourth top country of origin with 5340 refugees and 11,976 asylum seekers in Türkiye as of February 2024 (UNHCR, 2024a), but Iranians in Türkiye are also entering on work visas or without any legal permit, both including LGBTQ+ from Iran. After finding a temporary refuge and contributing to a rise in LGBTQ+ visibility and activism during the first decade of 2000 (Selen, 2020), my fieldwork reveals that Iranian LGBTQ+ have been severely ignored the last decade, as persecution and hate speech from Turkish authorities and population in general towards any LGBTQ+ are continuously on the rise (Selen, 2020, 2024).

Iran is a state that for decades has viewed LGBTQ+ people as deviant and worthy of death. These homophobic and transphobic views have enabled violence from both citizens and the state towards LGBTQ+ individuals. Furthermore, scholars have also recently documented stories on how same-sex-attracted peoples have been strongly encouraged to undergo sex-reassignment surgeries (Alexander et al., 2023). Reports on state-enabled killings of gay, lesbian, and transgender people have been frequently published by human rights organisations such as Amnesty International and Human Rights Watch since the Iranian revolution in 1979 (ibid). In 2005, Human Rights Watch commented on the execution of two men for homosexual conduct and said that the execution highlighted a pattern of persecution of gay men in Iran. At the same time, Human Rights Watch listed other cases of persecution and execution of gay men in Iran in the years before (Human Rights Watch, 2005). Between 2004 and 2020, Iran executed at least seventy-eight men for *livat*, penetrative anal intercourse with another man (Sato and Alexander 2021, pp. 84–86). Scholars, both outside and inside Iran, have also documented discrimination, violence and sexual violence towards Iranian transgender people (Nematollahi et al., 2022).

'I escaped from people, from Muslims who couldn't tolerate me,' Rahim told me, after being smuggled out of Iran. After finding shelter in Van, Rahim immediately told the UNHCR about her life in Iran. She described her family as deeply religious and under strict control by her father, a high-ranked man in the army, and deeply committed to having things the '*normal*' way, she explained, meaning the heterosexual way.

> As a child I was dragged to doctors and psychologists many times. My father wanted them to cure me, but eventually they found out that nothing could be done. Later, when my father found out that I was trying to have a boyfriend, he tried to kill me.

After two years of waiting, Rahim secured her asylum in a new country. All the documentation from doctors and psychologists helped her. This was confirmed by a coordinator at the local lawyer association in Van who knew Rahim well: 'Gays and transgender people often carry a psychologist report or a letter from a doctor or psychologist from their home country. A piece of paper like that is easily accepted by UNHCR.'

Opening the Closet

When I interviewed Rahim, she was not crying out of happiness because she got her asylum. She cried because she would soon be half a world apart from her boyfriend and best friend, and from the first place that took her in as a woman. For Rahim, her boyfriend and her best friend, Van had become a paradise.

The three told me that some local residents thought of their presence as troublesome. Some locals were hateful, and Rahim also told me about her own experiences of violence and rape in the city. But many more in the local population, it seemed, saw the benefits from their stay. Strong testimonials from several local LGBTQ+ and straight people in Van described how Rahim, her boyfriend and best friend, had opened the local population's eyes and 'closet doors' for local LGBTQ+ people. As the leader of a local women's association put it:

> Almost out of nowhere, sometime last year, gay people began to appear in the streets. We recognized them because they were better dressed than every man and they were very sociable. Local people started to open up to them and that's remarkable. It is good for our society that they are here. We want to try and help them. We have tried to find small jobs for them in cafes and such, but they disappear fast.

A receptionist at one of the hotels said, 'Yes, I have seen them around. They are funny. They don't bother me. If something bad happened to them I would help.'

A local gay man in his thirties, who organised local gay parties, agreed:

> It must be more than two hundred gays in this city, and at least I know 99 per cent of them are prostitutes. But do you know, refugees as visible as Rahim and her best friend are doing something good for us. I believe that they are trying to change the situation. Now, some of us are accepted in daylight. But it is also my opinion that Rahim and her friend will be killed after some time. New tolerance has limitations. Afterwards it is up to us to carry on stating our presence in this society.

A Decade of LGBTQ+ Visibility

The visibility for LGBTQ+ people in Türkiye gained a momentum after current Turkish President Recep Tayyip Erdoğan spoke about LGBTQ+ rights in 2002 when he campaigned to become Prime Minister. A week before national elections were going to be held, a university student asked Erdoğan on the TV show *Genç Bakış* (Outlook of the Youth) about his plans concerning LGBTQ+ rights if he was elected as Prime Minister. He sympathetically answered, 'Homosexuals must have legal security that frames their rights and freedom' (Leydense, 2013, quoted in Selen, 2020).

Selen (2020) suggests that this moment was 'the first positive acknowledgment of homosexuality by a politician' (p. 5524). Selen found that since 2003, LGBTQ+ activism has increased the visibility of LGBTQ+ individuals in Türkiye, including activism such as pride parades, protests, and demonstrations in Istanbul and other cities. Poole (2022, p. 52) also describes 'a peak in queer visibility' and freedom since 2010.

The Shift in Erdoğan's Position

In 2013, eleven years after Erdoğan acknowledged the issue of gay rights, he shifted his position, stating 'homosexual coupling is immoral' (Eşcinsel Çift Ahlaka Ters, 2013, p. 1, quoted in Selen, 2020). Erdoğan's shift coincided with the period when he started turning his back on the West following a collapse in negotiations for membership in the European Union (EU) that started in 2005 (Çapan & Zarakol, 2019). Since 2013, the conditions for LGBTQ+ activism and visibility have worsened and pride initiatives, such as organising pride parades and events, even debates and movie screenings, have been suppressed through governmental bans, police violence, and the threats of homophobic groups (Selen, 2020).

In October 2023, Erdoğan again made it particularly clear, as part of his last election campaign, that he does not 'recognize' LGBTQ+. According to Reuters (2023), Erdoğan vowed to combat 'perverse' trends which he said aimed to destroy the institution of family in the country. Korkut et al. (2023) found that Erdoğan, throughout the election campaign, used anti-LGBTQ+ arguments to attract the conservative voters from the opposition bloc because he 'systematically directed hate speech towards homosexuals and transgender people' (p. 19).

Following Erdoğan's move towards anti-LGBTQ+ policies, many studies have been conducted over the last decade about the growing number of migrants and refugees waiting in Türkiye, and the limited, if any, offers of protections for gender and sexual minorities (Kalan, 2018). This research shed light on the precarious situation of LGBTQ+ migrants and refugees in the country, who are also often subjected to widespread violence (Kalan, 2018). At the end of 2023, the EU developed new policies regarding migration for non-Europeans, with special mention of migrants in Türkiye. They suggested that migrants who were seeking asylum in Europe could be held in closed camps in Türkiye close to the European borders (Al-Jazeera, 2023). This move or suggestion from the EU also created new barriers for Iranian LGBTQ+ refugees.

For an Iranian LGBTQ+ person, applying for asylum in Türkiye or further to a third country was, and is, a high risk. Several organisations working with Iranian LGBTQ+ people in Türkiye confirmed this in interviews for my PhD project in 2022, 2023, and 2024. Similarly to the pseudonyms I have used for safety, I will not name the organisations here as it will put their staff, volunteers, and beneficiaries at risk. Many of these organisations confirm that the current Turkish migration system, where all migrants and asylum seekers register with Turkish authorities before they are processed by the UNHCR, is not trusted by Iranian LGBTQ+ people. One reason is that Iranian LGBTQ+ people who registered years ago have not seen their cases at the UNHCR processed.

Also, documents from doctors and psychologists that were previously considered an advantage to those applying at the UNHCR, as in Rahim's case, would now be considered a risk. Asylum applications from an Iranian LGBTQ+ person describing the reason for their application as being an LGBTQ+ person could end up being a death sentence if rejected by both the Turkish authorities and by the UNHCR. Also, organisation staff have told me that a possible consequence of a rejection from Turkish authorities is a forced return to Iran where organisations confirm that a returned LGBTQ+ person may face imprisonment or execution. A manager in one of the organisations said:

> That is the reason why Iranians would never reveal their LGBTQ+ identity to Turkish authorities'.

And if the Iranian LGBTQ+ do not reveal their identity then they are presumable not picked up by UNHCR, or Norway for that matter. A staff member at an organisation for the most vulnerable migrants and refugees said, 'Some describe themselves as being stuck in Türkiye for years. We know many of them very well, and they are becoming more and more desperate. We do not know what to do for them.'

One of those who would never apply for refugee status as a gay person is Kian, an Iranian gay man who entered Türkiye first on a work visa.

Kian's Story

'We are not safe anywhere here,' Kian told me in autumn of 2023 in Türkiye. I cannot write where or when, because he is in a city with increasingly visible homophobia in what used to be one of the more liberal cities. Since the last election, he told me, people have been tagging 'LGBT is a crime' in the streets around his house and he does not know if he is the target or not.

Kian crossed the border to Türkiye in his early twenties, almost a decade ago, looking for work after completing a university degree. He succeeded right away.

Following our conversations in Türkiye, several videocalls revealed Kian's desperate situation. 'I will take the boat in a couple of days,' Kian said with his usual calm voice in a videocall in late October 2023. Even through the speaker on his mobile phone, his voice was calm, deep, and strong, a voice that one could listen to for hours—for relaxation, meditation, or deep thinking. The calmness of his voice was so far from the reality of his life.

Kian was determined to be smuggled out of Türkiye to Europe. His determination was highly influenced by the increasing homophobia in his city and the now many anti-LGBTQ+ speeches of President Erdoğan. Also, Kian was aware of another fact that organisations working with LGBTQ+ migrants in Türkiye started documenting in late 2023: the forced displacement of Iranian LGBTQ+ from more liberal cities to more conservative ones, as the Turkish authorities decide where they shall live and where they can work.

'We are not safe anywhere here,' Kian repeated.

The same month that Kian decided to leave, in October 2023, almost 6,500 people arrived by boats to the Greek Islands from Turkey according to numbers collected from refugees and volunteers by the Norwegian NGO, Aegan Boat Report. Almost 12,000 people were stopped by Turkish coast guards or others before they left Turkish waters or pushed back from the Greek Islands into Turkish territories again. This was the highest number of boats leaving Türkiye since 2019 (Aegean Boat Report,

2023). The following months in 2023 was also record high in attempts to leave Türkiye, indicating that one out of three was successful in their crossing from Türkiye to the islands or mainland elsewhere in Europe with no certain numbers of how many had drowned on their way. For Kian these odds still amounted to a bearable risk to take. Also the risk of being pushed back to Türkiye for that matter, risking jail or possible deportation to Iran where the risk was prison or even a possible death sentence.

'But couldn't you apply for a study visa in Norway or Europe?' I suggested, even though I knew a tourist visa would never be an option because he didn't have the money to apply. An Iranian passport makes it also very difficult, if not impossible, to get any kind of student visas stopping Kian's dream of doing a PhD.

Norwegian and European intelligence have reported annually over the last years that Iranians represent some of the highest risks in threat assessments. Particularly mentioned in the 2023 report is the following: 'Norwegian research and educational institutions will be exploited for illegal knowledge transfer. Actors linked to Iran as well as Russia, China and Pakistan will represent a special challenge' (Norwegian Police's threat assessment, 2023, p. 6). It means that there is almost impossible to enter Europe and Norway due to the formal limitations on migration and asylum, and almost impossible to initiate a dialogue with Turkish authorities or UNHCR about asylum for Iranian LGBTQ+.

In early November of 2023 I got a text message from Kian. 'I just heard from the smugglers. I'm leaving in a few days. Hopefully to one of the nearest Greek Islands.'

Conclusion

In this vignette, by weaving in the stories of Rahim and Kian, I have tried to highlight LGBTQ+ migrants living in-between formalities of a refugee or asylum-process without many possibilities, or with no possibilities at all, to disclose their gender and sexuality diversity. My vignette adds to literature on oppression and the multiple axes of stigma faced by LGBTQ+ refugees and asylum-seekers (García Rodríguez, 2023) that also goes beyond their queerness and their minority backgrounds (Gebrelul, 2023) and into states' political attitudes towards gender and sexual diversity. Unpacking the oppression and hostility towards LGBTQ+ people with any migrant background in every society and host-country, temporarily or permanent, is needed such as Kian's story, as well as unpacking the possibilities of changemaking for the better in a hosting community represented by Rahim's story. Positive narratives of LGBTQ+ migrants and refugees, like Rahim's story, could also lead states to navigate and work better with gender and sexuality minorities, and to develop new laws, regulations and services needed to both save lives and meet ambitions of putting priority on the most vulnerable migrants and refugees, if that is in states' policies.

Conflict of Interest The author has no conflicts of interest to declare that are relevant to the content of this chapter.

References

Aegan Boat Report. (2023). *Agean Boat Report Data Studio*. Accessed December 20, 2023, from https://lookerstudio.google.com/reporting/1CiKRl_R7-1UbMHKhzZe_Ji_cvqF7xlfH/page/oOyx

Alexander, C., Sato, M., & Zanghellini, A. (2023). State-enabled killing of same-sex-attracted people: A legal pluralist account. *Law and Social Inquiry, 48*(3), 719–747.

Al-Jazeera. (2023, December 10). *EU reaches agreement to overhaul migration system, tighten asylum rules*. https://www.aljazeera.com/news/2023/12/20/european-union-reaches-agreement-on-reforming-migration-rules. Viewed on May 14, 2024.

Çapan, Z. G., & Zarakol, A. (2019). Turkey's ambivalent self: Ontological insecurity in "Kemalism" versus "Erdoğanism". *Cambridge Review of International Affairs, 32*(3), 263–282. https://doi.org/10.1080/09557571.2019.1589419

García Rodríguez, D. (2023). Critiquing trends and identifying gaps in the literature on LGBTQ refugees and asylum-seekers. *Refugee Survey Quarterly, 42*(4), 518–541.

Gebrelul, M. (2023). *Minoriserte subjekter av begjær Identitetsforståelse blant skeive personer med minoritetsbakgrunn i Norge* (Master's thesis).

Human Right Watch. (2005, November 21). *Iran: Two More Executions for Homosexual Conduct*. https://www.hrw.org/news/2005/11/21/iran-two-more-executions-homosexual-conduct

Kalan, J. (2018). The history of human rights abuses against the LGBTI Community in Turkey. In F. Muedini (Ed.), *LGBTI rights in Turkey: Sexuality and the state in the Middle East* (pp. 29–65). Cambridge University Press.

Kaya, A. (2023). *The world's leading refugee host, Turkey has a complex migration history*. Migration Policy Institute. https://www.migrationpolicy.org/article/turkey-migration-history

Korkut, U., Basbugoglu, T., & Sahin, O. (2023). Turkey's 2023 elections: Another victory for Erdogan. *Political Insight, 14*(3), 16–19.

Larsen, J. M. (n.d.). *Rhetorical risk reduction: Exploring rhetorical communication between municipalities and their minorities seeking to improve quality of life.* [Unpublished doctoral dissertation, publication in process]. Kristiania University of Applied Sciences.

Leydense. (2013, August 4). *Recep Tayyip Erdoğan - Eşcinsellik Hakkındaki Düşüncesi [Recep Tayyip Erdoğan - His thought concerning homosexuality].*

Nematollahi, A., Farnam, F., Gharibzadeh, S., & Khoda-Khah, P. (2022). Discrimination, violence, and suicide in transgender women in Iran. *Health Care for Women International, 43*(7–8), 861–872.

Norwegian Police. (2023). *National threat assessment.* Accessed December 18, 2023, from https://www.pst.no/alle-artikler/trusselvurderinger/ntv-2023/

Poole, R. J. (2022). *Queer Turkey: Transnational poetics of desire*, 25, 259. Transcript Verlag.

PopulationStat. (2024). *Van, Turkey Population.* https://populationstat.com/turkey/van

Reuters. (2023, July 10). *Turkey's Erdogan tells supporters he does not 'recognise' LGBT.* Accessed December 18, 2023, from https://www.reuters.com/world/turkeys-erdogan-tells-supporters-he-does-not-recognise-lgbt-2023-10-07/

Sato, M., & Alexander, C. (2021). *State-sanctioned killing of sexual minorities: Looking beyond the death penalty.* Melbourne: Monash University. https://doi.org/10.26180/14069318.v3.

Selen, E. (2020). The public immoralist': Discourses of queer subjectification in contemporary Turkey. *International Journal of Communication, 14*, 5518–5536. https://ijoc.org/index.php/ijoc/article/view/12033

Selen, E. (2024). *Intersectionality and hate speech: The plight of LGBTI+ citizens and migrants in Turkey* [In publication].

UNHCR. (2024a). *Refugees and Asylum Seekers in Turkey.* Accessed May 7, 2024, from https://www.unhcr.org/tr/en/refugees-and-asylum-seekers-in-turkey

UNHCR. (2024b). Türkiye. *Fact sheet,* February 2024. Accessed April, 2024, from https://www.unhcr.org/tr/wp-content/uploads/sites/14/2024/04/Bi-annual-fact-sheet-2024-02_Turkiye_FINAL-25March24.pdf

UNHCR. (n.d.). Türkiye. *Overview.* Accessed April 1, 2024, from https://www.unhcr.org/tr/en/overview-2

Vadla, A. (2020, July 6). Regjeringen endrer retningslinjene: Vil prioritere skeive flyktninger. *Vårt Land.* Accessed April 1, 2024, from https://www.vl.no/nyheter/2020/07/06/regjeringen-endrer-retningslinjene-vil-prioritere-skeive-flyktninger/

CHAPTER 9

Applying a Disaster Justice Lens to LGBTQ+ Refugees in Complex Humanitarian Emergencies

Edward J. Alessi, Yong Gun Lee, and Sarilee Kahn

Introduction

Disaster justice is a practical framework for addressing the impacts of complex humanitarian emergencies on lesbian, gay, bisexual, transgender, and queer refugees, as well as those with other diverse sexual orientations or

E. J. Alessi (✉)
School of Social Work, Rutgers, The State University of New Jersey, New Brunswick, NJ, USA
e-mail: ealessi@ssw.rutgers.edu

Y. G. Lee
Department of Social Work and Social Administration, The University of Hong Kong, Pokfulam, Hong Kong

S. Kahn
School of Social Work, McGill University, Montreal, QC, Canada

D. Dominey-Howes et al. (eds.), *Queering Disasters, Climate Change and Humanitarian Crises*, Sustainable Development Goals Series, https://doi.org/10.1007/978-981-96-3857-4_9

gender identities (LGBTQ+).[1] This chapter critically re-evaluates our (Edward and Sara's) 2017 study exploring the experiences of LGBTQ+ refugees who fled from the Middle East, North Africa, and South Asia (MENASA) and sought asylum in Austria or the Netherlands (Alessi et al., 2018, 2020, 2021a). We begin by providing the study's rationale and by defining a complex humanitarian emergency (hereafter, complex emergency) and disaster-related justice. Then, we present the details of the study and revisit the experiences of these LGBTQ+ refugees through the lens of the *Disaster Risk Management Cycle*. Finally, we provide recommendations for ensuring disaster-related justice for LGBTQ+ refugees during complex emergencies using the *Prevent/Mitigate, Prepare, Respond, Recover* framework. Using first-hand accounts of LGBTQ+ refugees during a complex emergency provides opportunities to strengthen policy and practice recommendations for this group.

Background and Rationale

Our 2017 study was conducted on the complex emergency that began in 2015, when 1.3 million people, mainly from Syria, Afghanistan, and Iraq, fled to Europe to escape war and ongoing political conflict (Pew Research Centre, 2016). It was considered one of the first complex emergencies of the twenty-first century, and Austria and the Netherlands, similar to other European (EU) countries (e.g., Hungary, Germany, Sweden, Norway), had record numbers of asylum applications relative to their overall populations (Pew Research Centre, 2016). Official estimates on asylum claims that have been disaggregated by sexual orientation or gender identity do not exist. However, significant numbers of LGBTQ+ individuals from Syria, Afghanistan, and Iraq were believed to have been part of this complex emergency (European Union Agency for Fundamental Human Rights [FRA], 2017), including those persecuted because of their LGBTQ+ identities.

While refugee groups share similarities (e.g., the need to secure housing, find work, and adjust to a new culture), they should not be

[1] While both *refugees* and *asylum seekers* flee their country of origin due to persecution, refugees are vetted in a country of transit *before* resettlement in a host country, and asylum seekers are vetted *after* arrival. However, when the study was conducted, we used the term 'refugees,' although study participants were technically asylum seekers, because that was the term being used to refer to the influx of individuals who fled to Europe.

considered a homogenous group, since doing so obfuscates their vulnerabilities and prevents the implementation of policies that meet their diverse needs. LGBTQ+ refugees frequently have difficulty relying on diaspora communities for support due to homophobia, biphobia, and transphobia, and they also experience intersecting stigmas (e.g., xenophobia, racism/ colourism, homophobia, transphobia, classism) from host community members (Camminga, 2018; Held, 2023; Wimark, 2021).

LGBTQ+ people around the world, including in countries with more progressive attitudes and policies, face rising stigmatisation, marginalisation, and victimisation. For example, in the United States, anti-transgender bills have been adopted while transgender women of colour continue to experience violence at extraordinary rates (Arayasirikul et al., 2022; Trans Legislation Tracker, 2023). In 2020, 67 countries criminalised consensual same-sex sexual activity (Mendos et al., 2020), and research demonstrates that LGBTQ+ individuals living in such countries experience severe and prolonged trauma, including physical and sexual assault, extortion, public shaming, social ostracism, forced marriage, and conversion therapy (Alessi et al., 2017; Gowin et al., 2017; Hopkinson et al., 2017; Piwowarczyk et al., 2017; Yarwood et al., 2022; Nematy et al., 2023). Living with a stigmatised identity also negatively impacts people's ability to seek and receive support from family, community members, and police (Alessi et al., 2021b). These individuals may also lack legal protections, and usually cannot move domestically because they are still required to conceal their identities for protection (UNHCR, 2012). As a result, LGBTQ+ individuals may flee countries if same-sex sexual behaviour has been criminalised or where their rights are limited.

During a complex emergency, socio-structural conditions may adversely affect the health and well-being of LGBTQ+ individuals, not only due to the crisis but also because of stigma related to their LGBTQ+ identities (Rumbach & Knight, 2014). Considerations must be made for LGBTQ+ refugees who are fleeing alongside individuals who are homophobic/ transphobic, and who are unable to 'pass' as cisgender or heterosexual. LGBTQ+ refugees fleeing persecution find themselves in chaotic environments where they cannot be guaranteed access to diverse and inclusive services. Yet, providing such services during and following a complex emergency is vital to building their resilience and well-being.

Defining Complex Emergency and Disaster-Related Justice

According to UNHCR (2001), a complex emergency refers to a humanitarian crisis:

> in a country, region or society where there is a total or considerable breakdown of authority resulting from internal or external conflict, and which requires an international response that goes beyond the mandate or capacity of any single agency and/or the ongoing UN country programme.

The term 'complex emergency' tends to refer to a specific time point and does not always capture how the exigencies of extreme civil or political crises interact with individual life histories and pre-existing social-structural systems. For populations who are minoritised, their histories of abuse and exposure to pre-existing systemic discrimination in their countries of origin may lead to psychological trauma—effects that are magnified by humanitarian crises (Norris et al., 2011). This risk may be mitigated with the right support, highlighting the importance of using a disaster-justice lens to assist groups who remain marginalised during complex emergencies.

Disasters are similar to and different from complex emergencies. A disaster relates to a natural occurrence or human-made event, such as a flood or industrial spill, respectively (Myers & Wee, 2004; Anderson & Gerber, 2018). Underpinning both disasters and humanitarian emergencies are the complicated social processes that often exacerbate long-established social and economic disparities (Myers & Wee, 2004). A disaster becomes a humanitarian emergency when precipitated by multiple factors and when international assistance is required (Anderson & Gerber, 2018).

The connections between humanitarian emergencies and disasters indicate that a disaster justice lens is an ideal frame to examine the experiences of LGBTQ+ refugees in a complex emergency. While disaster justice continues to be an emerging concept, its aim is the fair treatment of people who experience catastrophic events (Verchick, 2012; Lukasiewicz, 2020), including complex emergencies. Disaster justice, deriving from social justice, also considers how societies allocate resources during a disaster; how disasters lead to excess vulnerability among stigmatised groups; and how social-structural conditions existing before the disaster impact responses (Lukasiewicz, 2020). Because of its emphasis on social justice, disaster justice illuminates how systems of oppression inhibit fair allocation of resources that provide protection and facilitate recovery for LGBTQ+ refugees, who have typically endured trauma and intersecting oppressions.

Study Details

The study was conducted in the early summer of 2017 and comprised 38 LGBTQ+ individuals between the ages of 18 and 53 who fled from the MENASA region to Europe. Participants had been living in Austria (n = 18) between 9 weeks and 36 months or the Netherlands (n = 18) between 7 and 48 months. Seventeen participants migrated from either Syria or Iraq, while the remainder from other countries in the MENASA region. Most participants (n = 24) fled their countries of origin due to persecution based on sexual orientation or gender identity, while those from Syria and Iraq fled due to war and political strife and/or persecution related to their sexual orientation or gender identity. Twenty participants had received asylum while 15 had claims pending, 2 were denied status, and 1 was granted subsidiary protection.

The majority of participants identified as gay (n = 24), while the remainder identified as transgender woman (n = 5), lesbian (n = 3), bisexual (n = 3), queer (n = 2), or transgender man (n = 1). We collected data by conducting qualitative interviews to understand participants' experiences prior to migration, in-transit, and while in the host country. Data from these interviews comprise the first-hand accounts in this chapter. Twenty participants spoke English, 13 Arabic, 2 Farsi, and 1 Urdu. Interviews not conducted in English included interpreters. (For a full description of the methodologies used and the results, see Alessi et al., 2018, 2020, 2021a). Participants' migration pathways to Europe varied; in sum, travel was not always directly from the country of origin to Europe. Twenty-five participants fled their countries of origin to other Middle Eastern countries first, and 20 paid smugglers to take them from Turkey to Greece. Upon arrival in Austria and the Netherlands, 28 (11 in Austria and 17 in the Netherlands) were placed in refugee camps.

Participants reported experiencing traumatic events during pre-migration, in-transit, and in the host country. Many of these events were precipitated because of their sexual orientation or gender identity. The majority also spent their lives hiding their sexual orientation or gender identity because they feared being victimised or further stigmatised. However, they also experienced events that were not directly related to their sexual orientation or gender identity (e.g., living in a war zone or a country facing political upheaval; having a family member who was murdered; being imprisoned; and experiencing chronic family conflict and abuse).

Thus, participants did not feel safe, secure, or protected before the humanitarian emergency began; this feeling may be intensified during such a crisis (Rumbach & Knight, 2014). For many, the trauma precipitated by the humanitarian emergency became part of their cumulative experiences of trauma. When we refer to trauma, we mean the 'destruction of basic organising principles by which we come to know self, others, and the environment; traumas wound deeply in a way that challenges the meaning of life' (Root, 1992, p. 229). This includes specific symptoms that emerge from severe and prolonged trauma, such as reexperiencing the trauma or having intrusive thoughts; dissociation; avoidance; difficulty connecting with others' change in cognitions (e.g., blaming oneself); experiencing negative beliefs about the self and the those around them; and hypervigilance. It can also include emotional dysregulation, impaired relationships, and disruption of one's identity and place in the world (Maercker et al., 2022).

To respond to complex emergencies among LGBTQ+ refugees in a way that ensures disaster justice for this population while also accounting for their long histories of trauma, we reflect on and resituate our study results using the *Disaster Risk Management Cycle* and the *Prevent/ Mitigate, Prepare, Respond, Recover* framework.

Disaster Risk Management Cycle

The *Disaster Risk Management Cycle* comprises four phases of humanitarian emergencies: pre-emergency, acute emergency, post-emergency, and recovery phase (Anderson & Gerber, 2018). The needs of affected individuals are said to change within and across the phases of the emergency. As Gómez and Kawaguchi (2018) argue, using a phase-based approach reflects 'the dynamic change of needs throughout a crisis and its aftermath, requiring both short-term and long-term commitments to achieve the final goal of securing humans' (p. 16).

Pre-emergency

Much has been documented about the factors driving the complex emergency affecting LGBTQ+ individuals in the MENASA region. First,

U.S. government-led wars and a history of colonialism have produced instability in Syria, Iraq, and Afghanistan (Mayblin, 2014; Fonkem, 2020). Additionally, religious and cultural prohibitions against LGBTQ+ people have been influenced by laws imported by colonial powers, in addition to more contemporary Western homophobic and transphobic discourses (Han & O'Mahoney, 2018; L, 2018; Gloppen & Rakner, 2020). Currently, individuals in the MENASA region are still subjected to laws that punish homosexuality and same-sex sexual behaviour (Mendos et al., 2020).

Many participants spoke about growing up in environments where traditional values, shaped by conservative religious beliefs, impacted their day-to-day experiences. Several participants recalled being exposed to vilification, humiliation, and denunciation, often led by religious authorities, of LGBTQ+ identities and same-sex behaviours. This pressured individuals to conceal their sexual orientation or gender identity to survive and led some to thoughts of suicide (Alessi et al., 2018, 2021a). Participants reported experiencing physical and emotional abuse by family members and some were rejected by them once they knew they were LGBTQ+ (Alessi et al., 2018, 2021a). They also contended with persecution by community members and state actors, and while these events were considered traumatic, it was the loss of family support that permeated many participants' narratives about their lives in their countries of origin.

Acute Emergency

The onset of the next phase, *acute emergency*, starts just after disaster erupts. It involves active exposure to imminent threat. For some participants, the acute emergency was present before fleeing their countries of origin. For example, they described what it was like to live in a war-torn country. Adnan, a transgender man, talked about the daily decisions he had to make for himself and his family. For instance, they could remain in their home to stay protected, but staying inside could be just as dangerous as being outside, since many people's homes were being bombed (Alessi et al., 2018). Another participant, Zayd, a gay man, discussed seeing several people murdered by gunfire, assisting those who were shot, and holding the hands of those actively dying after being shot (Alessi et al., 2018).

Various factors impact the decision to flee one's country of origin during a complex emergency, like those from Syria and Iraq in our study. As mentioned, these participants were not only fleeing war but also victimisation or stigmatisation related to their sexual orientation or gender identity to avoid harm or even death. Furthermore, participants from other countries in the MENASA region joined the population flow of refugees to seek protection in Europe based on persecution of their sexual orientation or gender identity.

The migration process for most participants was not linear. Migrants frequently travel to other countries before arriving in the host or destination country (Zimmerman et al., 2011). Twenty-five participants fled to other Middle Eastern countries first, with 22 spending time in Turkey (Alessi et al., 2018). Because of its geographical location and proximity, Turkey was considered a more accessible destination. Some thought they would live there permanently because they felt the cultural aspects of Turkey were more similar to their country of origin than countries in Europe. Others considered Turkey as a country of transit before travelling to Europe. While participants' lives were somewhat better in Turkey (i.e., they were not exposed to war-torn conditions and could find a job off-the-books), they still had to hide their sexual orientation or gender identity, which led participants, even those who thought they were staying permanently, to seek asylum in Europe.

Many participants paid smugglers to help them make the dangerous journey by boat from Turkey to Greece. Once arriving in Greece, it was easier for participants to travel to other parts of Europe. One participant, Alex, a gay man, recalled that the boat he was on sank and that he and his boyfriend had to swim for part of their journey. The traumatic experience of surviving this near-death experience was compounded by earlier traumatic experiences related to his sexual orientation. Another participant, Ram, who was afraid of water decided to walk from Turkey to Greece but got lost on his way there. He was so terrified by the experience that he suffered from PTSD symptoms in Amsterdam, exemplifying why it is also important to account for trauma among LGBTQ+ refugees that does not necessarily relate to their sexual orientation or gender identity.

The complexities of forced migration were further illustrated by participants who travelled from Greece to other parts of Europe on foot. LGBTQ+ refugees endure not only the stress encountered by the broader population of refugees but also the hypervigilance that comes with hiding their sexual orientation or gender identify from other refugees who may be homophobic or transphobic (Yarwood et al., 2022; Gottvall et al., 2023). Some participants employed various concealment strategies to protect themselves from potential harm en route. Lina, a transgender woman, wore a hijab to conceal her identity. She believed she was protected from transphobia because her wearing a hijab helped her 'pass' as a cisgender woman. However, this was not the case for another participant, Pani, who was gender nonconforming. They not only encountered other refugees who attempted to sexually assault them but also European immigration officials who did little to protect them. It was only after joining a heterosexual family that Pani felt safe on the migration trail. Walking with a family provided some protection; walking alone, however, conferred risk of harm.

Not being able to rely on government officials and state actors to protect them compounded participants' stress and trauma. For instance, Ali, a gay man, was paired with a homophobic interpreter as he tried to cross from one European country to another. The interpreter was accusing Ali of being a 'terrorist,' so European immigration officials would make the decision not to allow him to cross the border. Had Ali not realised what was occurring and not advocated for himself, he was unsure what would have happened to him (e.g., he might have been arrested or sent back to his country of origin). What Ali experienced was not a rare occurrence; evidence demonstrates that immigration officials, ancillary service providers, and interpreters can be discriminatory towards LGBTQ+ refugees (Ferreira, 2022).

Many European countries experienced significant challenges managing the complex emergency that began in 2015. For many participants, their specific needs were overlooked or not considered in the planning process. For instance, participants encountered significant challenges accessing safe accommodation. They reported being placed in migration camps with other refugees who held homophobic and transphobic attitudes, and thus once again had to conceal their sexual orientation and gender identity. Some, like Zayd, experienced threats of harm and physical violence by other refugees after he started a relationship with another man in the camp, and his attempts to report it to refugee assistance workers did not lead to repercussions for the perpetrators. Meanwhile, others reported that they were blamed by aid workers for not doing enough to hide their LGBTQ+ identities (Alessi et al., 2018), exacerbating their feelings of anxiety, fear, and isolation. Also contributing to participants' isolation was the geographic location of some of the migration camps, which were far from urban centres with visible LGBTQ+ communities. For Ram, who reported already being traumatised after fleeing a war-affected country and getting lost in the woods on his way from Turkey to Greece, the nagging sense of loneliness in the refugee camp contributed to his suicide attempt. Ram and other participants also mentioned that suicide attempts by LGBTQ+ refugees in migration camps were not uncommon (Alessi et al., 2018).

Post-emergency Phase

In the *post-emergency* phase, individuals and families who have fled will make decisions about whether it is safe to return to their country or origin or they will attempt to gain protected status in the host country (Al Husein & Wagner, 2023; Alrababah et al., 2023). All participants in our study applied for asylum in Austria and the Netherlands. Seeking refuge from political persecution, persecution based on sexual orientation or gender identity, or both served as common bases for an asylum application. Nevertheless, as demonstrated in previous studies (Lee & Brotman, 2011; Kahn & Alessi, 2018; Borges, 2020; Gottvall et al., 2023), participants described multiple challenges, including long wait times, confusing processes and procedures, and difficulties of having to prove, based on white/American/Eurocentric conceptualisations of sexuality and gender, that they were LGBTQ+.

Moreover, 10 participants were seriously impacted by the Dublin Regulation, also referred to as Dublin. There have been multiple iterations of Dublin, and the most recent one was in 2013 (Council Regulation [EC] No 604/2013). Dublin purportedly makes it easier for European Union (EU) member states to determine which country is responsible for handling an individual's asylum claim. In sum, it mandates that if a person seeks asylum in one EU country, they cannot have been granted a visa from or passed through and fingerprinted in another EU country first. If they have, they must seek asylum in that particular country (Alessi et al., 2020). We have previously critiqued Dublin's response, especially for the problems it caused LGBTQ+ refugees (see Wieland & Alessi, 2021), but overall, this legislation negatively affected participants' ability to recover following the complex emergency. On their migration journeys to Austria and the Netherlands, which oftentimes involved walking or being smuggled, some participants had been fingerprinted in other EU countries (e.g., Hungary, Croatia). Others had previously obtained a visa to another EU country before arriving in Austria or the Netherlands. Dublin mandated that these participants return to these countries to seek asylum. However, these participants did not feel comfortable seeking asylum in these countries because LGBTQ+ rights were not affirmed there or particular services were not available (e.g., gender-affirming healthcare).

Those who were mandated to return to an EU country that was affirming of LGBTQ+ people did not necessarily want to return there either. For example, one participant had family in Austria. She felt she would lose her main source of support if she had to seek asylum in another European country for which she had initially obtained a visa before fleeing her country of origin. Another participant was scared for her life if she was to stay in the EU country she had already obtained a visa to enter, because she had family there who had threatened to harm her for being in a relationship with a woman. Dublin left these participants in an even more precarious situation, as the foundation for what any refugee needs to integrate into a new society was put in jeopardy: protected immigration status (Ager & Strang, 2008). For example, Iwan, a gay man, reported that he was affected by Dublin and, as a result, was denied asylum in the Netherlands. Rather than move to the EU country where he had initially obtained a visa for, he stayed in the Netherlands because his social support system was already there. However, he was denied asylum and thus found it difficult to obtain a job to support himself while living as an undocumented person. He engaged in sex work to make a living. While sex work comes with risks (e.g., exploitation, of HIV and STIs, interpersonal violence) for LGBTQ+ refugees (Yarwood et al., 2022; Gibbons & Best, 2023), it can also provide opportunities for displaced individuals to support themselves financially and to increase agency in their life (Global Network of Sex Work Projects, 2024; Rosati et al., 2024).

Unlike other refugee groups, LGBTQ+ refugees often receive little support, as they may experience or anticipate stigma and discrimination from both diaspora and host communities (Lee & Brotman, 2011; Fox et al., 2020). Thus, community organisations, especially those focused on supporting this population are critical for LGBTQ+ refugees' survival, assisting them with meeting basic needs, finding housing, and applying for asylum. For most participants, interacting with these organisations provided a corrective experience; that is, it was the first time they had felt safe, seen, and heard since arriving in Europe, providing a built-in community for many who had felt isolated, alienated, and alone. For instance, several participants in Vienna and Amsterdam expressed feelings of relief and gratitude to be part of LGBTQ+ refugee organisations in their respective cities (Alessi et al., 2018, 2020).

Nonetheless, there were participants who felt that, while the organisations assisted them in accessing essential resources, they still felt disconnected from other individuals. As Mohammed, a gay man explained, there was a misconception that he would automatically connect with other refugees who were gay since they shared the same sexual orientation. But for him this was not the case and left him wondering whether he would ever be able to connect with anyone, since he felt this way in his country of origin, too. Mohammed's situation is common among people who have experienced severe and prolonged interpersonal trauma. This form of trauma frequently leaves LGBTQ+ refugees feeling unable to trust others; overestimating interpersonal threat even in situations where the risk is low; seeing themselves as not contributing much to friendships or romantic relationships; and dealing with their anxiety by avoiding interpersonal relationships altogether (Alessi et al., 2016). This also suggests that the assistance provided to LGBTQ+ refugees during a humanitarian should be provided in a trauma-informed manner, which accounts for how experiences of interpersonal trauma can impact help seeking (Kahn et al., 2018).

In addition to LGBTQ+ focused community organisations, participants also relied on the help of community members, some who were strangers, to assist them with integration. When looking for accommodation, Marwa, a queer woman, reported that potential roommates offered her a discount because of her situation. Although community members can help ease the integration of LGBTQ+ refugees, Lukasiewicz et al. (2017) argue that assistance to refugees should be primarily the responsibility of federal, state, and local governments. This could include vetting community organisations and providing them with funding to help vulnerable populations in accordance with the organisation's mission. Without such oversight, LGBTQ+ refugees might be blamed for not being strong or resilient enough to help themselves when structural barriers, including stigma and discrimination, frequently drive the social and health inequities that prevent them from integrating into host societies.

Participants faced discrimination from diaspora community members when accessing vital services for refugees. For example, Betty, a transgender woman, reported being threatened in language classes and not receiving the protection from resettlement workers when members of the diaspora community harassed her, leaving her anxious and scared. It also prevented her from obtaining the language skills needed to integrate and undermined her ability to support herself in the formal labour market. Participants also encountered racism and xenophobia from other LGBTQ+ people and cis-hetero individuals who they felt either exoticized or looked down on them for being refugees, Muslim, or from the MENASA region. Some gay male participants reported experiencing rejection based on a bias against physical features presumed to be associated with being Muslim (e.g., being circumcised) or anticipating rejection based on their perceived lack of capacity to be involved in a serious, long-term relationship as a refugee (Alessi et al., 2020). Additionally, several gay men who identified as a racial minority and Muslim recounted experiences of overt and covert forms of discrimination, including being denied entry to gay venues in Vienna and being rejected by other gay men because they were refugees. This made integrating more challenging and thwarted feelings of belongingness that might otherwise come from affiliating with grassroots LGBTQ+ organisations that provide assistance to refugees.

Recovery Phase

The fourth phase, *recovery*, signals a change in priorities. In the context of LGBTQ+ refugees from the MENASA region seeking asylum in Austria and the Netherlands, recovery entails being able to successfully integrate into host countries while also being able to cope and manage the trauma encountered during pre-migration, in-transit, and in settlement. Although the results of our research were limited by a cross-sectional study design, they nevertheless suggest that sources of support and development could provide the structure that promotes resilience and recovery in this group. These sources of support and development included, but were not limited to, accessing mental health treatment and counselling, linking with affirmative community organisations to assist with making social connections and meeting basic needs (including support with asylum claims), and participating language lessons in safe environments. Because the Prevent/Mitigate, Prepare, Respond, Recover (PPRR) framework (Resilient Community Organisations, 2015) also includes a focus on recovery, it is included in the discussion and Recommendations below.

Recommendations

In this chapter, we adapt the PPRR framework and incorporate recommendations outlined by Rumbach and Knight (2014), Alessi and Kahn (2017), and Alessi et al. (2021b) to delineate specific actions that policymakers and practitioners can take towards ensuring disaster-related justice for LGBTQ+ refugees during complex emergencies (see Table 9.1 for a summary).

Table 9.1 Prevention, preparation, response, and recovery to ensure the protection of LGBTQ+ refugees in humanitarian emergencies

Prevention	*Preparation*	*Response*	*Recovery*
• Work to ensure international human rights protection for LGBTQ+ people. • Fund LGBTQ+ organisations in countries where rights are limited or non-existent. • Hold religious organisations accountable for exporting homophobia and transphobia to other countries. • Create policies that reduce structural violence (e.g., ensure trans people can access documents that match their gender identity).	• Conduct ongoing LGBTQ+ training for humanitarian aid and refugee assistance workers, interpreters. • Develop policies and plans that provide LGBTQ+ refugees with safe accommodation. • Create linkages with community organisations; train them to provide legal and psychological support to LGBTQ+ refugees. • Elicit the feedback of LGBTQ+ refugees in all phases of the preparation process.	• Provide for basic needs (housing, food, healthcare). • Provide access to safe accommodation and guarantee access to single stall restrooms. • Engage LGBTQ+ refugees in ways that do not out them in humanitarian settings. • Facilitate referrals to affirmative health, mental health, and legal organisations. • Explore ways to strengthen social and community ties for LGBTQ+ refugees.	• Facilitate governmental and community collaboration with LGBTQ+ refugees to ensure their social integration. • Address lingering symptoms of trauma and to help ease the integration of LGBTQ+ refugees. • Create opportunities for LGBTQ+ refugees to work in groups to build community and navigate the challenges of integration, including intersecting stigmas (homophobia, transphobia, racism, xenophobia).

Conclusion

A disaster justice lens is especially useful in guiding and planning safe and equitable paths to recovery for LGBTQ+ refugees affected by complex emergencies. The first phase, *prevention*, emphasises macro-level approaches for strengthening basic protections for LGBTQ+ people to minimise their risks during complex emergencies. The second phase, *preparation*, requires policymakers, in tandem with LGBTQ+ refugees, to build organisational capacity to ensure disaster-related justice for LGBTQ+ people. In the third phase, *response*, where post-crisis needs arise, recommendations include creating safe zones for LGBTQ+ refugees where they can safely receive crucial resources; providing access to safe accommodation and facilities; facilitating referrals to organisations that provide mental health, psychosocial, and legal supports that are affirmative and trauma- and culturally-informed; and working closely with LGBTQ+ refugees so they can feel safe when accessing community services. The fourth phase, *recovery*, requires ongoing governmental and community collaboration with LGBTQ+ refugees to ensure their survival, protection, and social integration.

When complex emergencies occur, Disaster Risk Management and PPRR-based responses can minimise risks for LGBTQ+ refugees while also protecting and empowering them. LGBTQ+ refugees themselves should be involved in every phase of disaster prevention, planning, and management, as this will bring policymakers and aid workers closer to ensuring disaster-related justice for this population during a complex emergency.

Conflict of Interest The authors have no conflict of interests.

References

Ager, A., & Strang, A. (2008). Understanding integration: A conceptual framework. *Journal of Refugee Studies, 21*(2), 166–191.

Al Husein, N., & Wagner, N. (2023). Determinants of intended return migration among refugees: A comparison of Syrian refugees in Germany and Turkey. *International Migration Review, 57*(4), 1771–1805.

Alessi, E. J., & Kahn, S. (2017). A framework for clinical practice with sexual and gender minority asylum seekers. *Psychology of Sexual Orientation and Gender Diversity, 4*(4), 383–391.

Alessi, E. J., Kahn, S., & Chatterji, S. (2016). 'The darkest times of my life': Recollections of child abuse among forced migrants persecuted because of their sexual orientation and gender identity. *Child Abuse & Neglect, 51*, 93–105.

Alessi, E. J., Kahn, S., & Van Der Horn, R. (2017). A qualitative exploration of the premigration victimization experiences of sexual and gender minority refugees and asylees in the United States and Canada. *The Journal of Sex Research, 54*(7), 936–948.

Alessi, E. J., Kahn, S., Woolner, L., & Van Der Horn, R. (2018). Traumatic stress among sexual and gender minority refugees from the Middle East, North

Africa, and Asia who fled to the European Union. *Journal of Traumatic Stress, 31*(6), 805–815.
Alessi, E. J., Kahn, S., Grennfield, B., Woolner, L., & Manninig, D. (2020). A qualitative exploration of the integration experiences of LGBTQ refugees who fled from the Middle East, North Africa, and Central and South Asia to Austria and The Netherlands. *Sexuality Research and Social Policy, 17*(1–2), 13–26.
Alessi, E. J., Greenfield, B., Kahn, S., & Woolner, L. (2021a). (Ir)reconcilable identities: Stories of religion and faith for sexual and gender minority refugees who fled from the Middle East, North Africa, and Asia to the European Union. *Psychology of Religion and Spirituality, 13*(2), 175–183.
Alessi, E. J., Cheung, S., Kahn, S., & YU, M. (2021b). A scoping review of the experiences of violence and abuse among sexual and gender minority migrants across the migration trajectory. *Trauma Violence Abuse, 22*(5), 1339–1355.
Alrababah, A., Masterson, D., Casalis, M., Hangartner, D., & Weistein, J. (2023). The dynamics of refugee return: Syrian refugees and their migration intentions. *British Journal of Political Science, 53*(4), 1108–1131.
Anderson, M., & Gerber, M. (2018). Introduction to humanitarian emergencies. In D. Townes (Ed.), *Health in humanitarian emergencies* (pp. 1–8). Cambridge University Press.
Arayasirikul, S., Turner, C., Trujillo, D., Sicro, S. L., Scheer, S., Mcfarland, W., & Wilson, E. C. (2022). A global cautionary tale: Discrimination and violence against trans women worsen despite investments in public resources and improvements in health insurance access and utilization of health care. *International Journal for Equity in Health, 21*, 32. https://doi.org/10.1186/s12939-022-01632-5
Borges, S. (2020). "We have to do a lot of healing": LGBTQ migrant Latinas resisting and healing from systemic violence. *Journal of Lesbian Studies, 24*(2), 94–109.
Camminga, B. (2018). "Gender refugees" in South Africa: The "common-sense" paradox. *Africa Spectrum, 53*(1), 89–112.
European Union Agency for Fundamental Human Rights (FRA). (2017). *Current migration situation in the EU: Lesbian, gay, bisexual, transgender and intersex asylum seekers.* European Union Agency for Fundamental Human Rights.
Ferreira, N. (2022). Utterly unbelievable: The discourse of "fake" SOGI asylum claims as a form of epistemic injustice. *International Journal of Refugee Law, 34*(3–4), 303–326.
Fonkem, A. (2020). The refugee and migrant crisis: Human tragedies as an extension of colonialism. *The Round Table, 109*(1), 52–70.
Fox, S. D., Griffin, R. H., & Pachankis, J. E. (2020). Minority stress, social integration, and the mental health needs of LGBTQ asylum seekers in North America. *Social Science and Medicine, 246*, 112727. https://doi.org/10.1016/j.socscimed.2019.112727

Gibbons, J. L., & Best, D. L. (2023). Gender and migration: Women and nonbinary individuals at risk. In G. J. Rich, J. Kuriansky, U. P. Gielen, & D. Kaplin (Eds.), *Psychosocial experiences and adjustment of migrants*. Academic Press.

Global Network of Sex Work Projects. (2024). *Policy brief: sex workers in conflict zones and humanitarian crises*. Accessed March 15, 2024, from https://www.nswp.org/sites/default/files/pb_sws_in_conflict_zones_english_prf02.pdf

Gloppen, S., & Rakner, L. (2020). LGBT rights in Africa. In *Research handbook on gender, sexuality and the law*. Edward Elgar Publishing.

Gómez, O. A., & Kawaguchi, C. (2018). A theory for the continuum: Multiple approaches to humanitarian crises management. In A. Hanatani, O. A. Gómez, & C. Kawaguchi (Eds.), Crisis management beyond the humanitarian-development nexus (pp. 15–36). Routledge. https://doi.org/10.4324/9781351006828

Gottvall, M., Brunell, C., Eldebo, A., Johansson Metso, F., Jirwe, M., & Carlsson, T. (2023). Post-migration psychosocial experiences and challenges amongst LGBTQ+ forced migrants: A meta-synthesis of qualitative reports. *Journal of Advanced Nursing, 79*(1), 358–371.

Gowin, M., Taylor, E. L., Dunnington, J., Alshuwaiyer, G., & Cheney, M. K. (2017). Needs of a silent minority: Mexican transgender asylum seekers. *Health Promotion Practice, 18*(3), 332–340.

Han, E., & O'Mahoney, J. (2018). *British colonialism and the criminalization of homosexuality: Queens, crime and empire*. Routledge.

Held, N. (2023). "As queer refugees, we are out of category, we do not belong to one, or the other": LGBTIQ+ refugees' experiences in "ambivalent" queer spaces. *Ethnic and Racial Studies, 46*(9), 1898–1918.

Hopkinson, R. A., Keatley, E., Glaeser, E., Erickson-Schroth, L., Fattal, O., & Nicholson Sullivan, M. (2017). Persecution experiences and mental health of LGBT asylum seekers. *Journal of Homosexuality, 64*(12), 1650–1666.

Kahn, S., Alessi, E. J., Kim, H., Woolner, L., & Olivieri, C. J. (2018). Facilitating mental health support for LGBT forced migrants: A qualitative inquiry. *Journal of Counseling and Development, 96*(3), 316–326.

Kahn, S., & Alessi, E. J. (2018). Coming out under the gun: Exploring the psychological dimensions of seeking refugee status for LGBT claimants in Canada. *Journal of Refugee Studies, 31*, 22–41.

L, A. (2018). How homosexuality became a crime in the Middle East: Colonialism, culture wars and fundamentalist politicians have restricted sexual freedom. *The Economist*. https://www.economist.com/open-future/2018/06/06/how-homosexuality-became-a-crime-in-the-middle-east

Lee, E. O. J., & Brotman, S. (2011). Identity, refugeeness, belonging: Experiences of sexual minority refugees in Canada. *Canadian Review of Sociology/Revue canadienne de sociologie, 48*(3), 241–274.

Lukasiewicz, A. (2020). The emerging imperative of disaster justice. In A. Lokasiewicz & C. L. Baldwin (Eds.), *Natural hazards and disaster justice: Challenges for Australia and its neighbours* (pp. 3–24). Palgrave Macmillan.

Lukasiewicz, A., Dovers, S., & Eburn, M. (2017). Shared responsibility: The who, what and how. *Environmental Hazards, 16*(4), 291–313.

Maercker, A., Cloitre, M., Bachem, R., Schlumpf, Y. R., Khoury, B., Hitchcock, C., & Bohus, M. (2022). Complex post-traumatic stress disorder. *Lancet, 400*(10345), 60–72.

Mayblin, L. (2014). Colonialism, decolonisation, and the right to be human: Britain and the 1951 Geneva convention on the status of refugees. *Journal of Historical Sociology, 27*(3), 423–441.

Mendos, L. R., Botha, K., Lelis, R. C., De La Pena, E. L., Savelev, I., & Tan, D. (2020). *State-sponsored homophobia: global legislation overview update.* Accessed March 15, 2024, from https://ilga.org/downloads/ILGA_World_State_Sponsored_Homophobia_report_global_legislation_overview_update_December_2020.pdf

Myers, D., & Wee, D. (2004). *Disaster mental health services: A primer for practitioners.* Routledge.

Nematy, A., Namer, Y., & Razum, O. (2023). LGBTQI + refugees' and asylum seekers' mental health: A qualitative systematic review. *Sexuality Research and Social Policy, 20*(3), 636–663.

Norris, A. E., Aroian, K. J., & Nickerson, D. M. (2011). Premigration persecution, postmigration stressors and resources, and postmigration mental health: A study of severely traumatized U.S. Arab immigrant women. *Journal of American Psychiatric Nurses Association, 17*(4), 283–296.

Pew Research Centre. (2016). *Number of refugees to Europe surges to record 1.3 million in 2015* [Online]. Pew Research Centre. Accessed June 1, 2023, from https://www.pewresearch.org/global/2016/08/02/number-of-refugees-to-europe-surges-to-record-1-3-million-in-2015/

Piwowarczyk, L., Fernandez, P., & Sharma, A. (2017). Seeking asylum: Challenges faced by the LGB community. *Journal of Immigrant and Minority Health, 19*(3), 723–732.

Resilient Community Organisations. (2015). *Emergency management: Prevention, Preparedness, Response and Recovery* [Online]. Accessed June 1, 2023, from https://resilience.acoss.org.au/the-six-steps/leading-resilience/emergency-management-prevention-preparedness-response-recovery

Root, M. P. (1992). Reconstructing the impact of trauma on personality development: A feminist perspective. In L. S. Brown & M. Ballou (Eds.), *Personality and psychopathology: Feminist reappraisals* (pp. 229–266). Guilford Press.

Rosati, F., Pistella, J., Coletta, V., & Baiocco, R. (2024). Racialized migrant transgender women engaged in sex work: Double binds and identifications with the community. *Archives of Sexual Behavior, 53*(3), 1153–1168.

Rumbach, J., & Knight, K. (2014). Gender and sexual minorities in humanitarian emergencies. In L. W. Roeder Jr. (Ed.), *Issues of gender and sexual orientation in humanitarian emergencies: Risks and risk reduction*. Springer International Publishing.

Trans Legislation Tracker. (2023). *2023 anti-trans bills tracker* [Online]. Accessed June 1, 2023, from https://translegislation.com/

UNHCR. (2001). *Coordination in complex emergencies* [Online]. Accessed June 1, 2023, from https://www.unhcr.org/publications/coordination-complex-emergencies

UNHCR. (2012). *UNHCR guidelines on international protection no. 9: Claims to refugee status based on sexual orientation and/or gender identity*. Accessed February 27, 2024, from https://www.unhcr.org/media/unhcr-guidelines-international-protection-no-9-claims-refugee-status-based-sexual-orientation

Verchick, R. (2012). Disaster justice: The geography of human capability. *Duke Environmental Law and Policy Forum, 23*, 23–71.

Wieland, R., & Alessi, E. J. (2021). Do the challenges of LGBTQ asylum applicants under Dublin register with the European Court of Human Rights? *Social and Legal Studies, 30*(3), 405–425.

Wimark, T. (2021). Homemaking and perpetual liminality among queer refugees. *Social and Cultural Geography, 22*(5), 647–665.

Yarwood, V., Checchi, F., Lau, K., & Zimmerman, C. (2022). LGBTQI+ migrants: A systematic review and conceptual framework of health, safety and wellbeing during migration. *International Journal of Environmental Research and Public Health, 19*(2), 1–22.

Zimmerman, C., Kisss, L., & Hossain, M. (2011). Migration and health: A framework for 21st century policy-making. *PLoS Medicine, 8*(5), e1001034.

CHAPTER 10

Gender and Sexual Diversity and the Urban Climate Crisis: Intersectional Injustices of Housing and Livelihoods in Kampala, Uganda

Katie McQuaid and Neil J. W. Crawford

Introduction

Climate change and disaster justice are intersectional gender justice issues. To better understand how, we have conducted intersectional feminist research with gender and sexual diverse communities in Uganda. Foregrounding queer lived experiences of climate change forces us to confront the complex and situated gender politics and systemic power inequalities that disproportionately burden some communities over others. We employ postcolonial queer scholarship (Hawley, 2001; Spurlin, 2006;

K. McQuaid (✉)
School of Geography, University of Leeds, Leeds, UK
e-mail: K.R.V.McQuaid@leeds.ac.uk

N. J. W. Crawford
Division of History, Heritage and Politics, University of Stirling, Stirling, UK

D. Dominey-Howes et al. (eds.), *Queering Disasters, Climate Change and Humanitarian Crises*, Sustainable Development Goals Series, https://doi.org/10.1007/978-981-96-3857-4_10

Boone, 2015; see also Gaillard, 2022, p. 151) to better understand gender beyond the ('Western') binary; address power inequities; embrace diversity; and centre local epistemologies and identities. We locate our knowledge of climate change impacts within local experiences that subvert hegemonic narratives of gender and climate change. This pushes us to confront the heteronormative, technocratic, top-down, and colonial models that dominate knowledge production and policymaking in climate and disaster studies.

Bringing marginalised voices and priorities away from the periphery of disaster studies—indeed gender studies—to the centre of our research, without perpetuating narratives of victimhood, is critical if we are to translate deeper complex approaches to gender into inclusive and sustainable climate and disaster justice strategies (Arora-Jonsson, 2017; Buckingham & Le Masson, 2017; Michael et al., 2023). Only then can we have strategies that are responsive to the needs and capacities of diverse populations and inclusive of gender and sexual diverse people. As Rydstrom and Kinnvall (2020, pp. 5–6) argue: 'A damaging event interlocks with gender specific inequalities which underpin ordinary times and might, in doing so, not only fuel but even exacerbate gendered susceptibility to short-term and long-term disaster-related hazards and ramifications.' This is arguably even more pronounced in the case of gender and sexual diverse people.

In Uganda, climate change impacts are most apparent not in disasters and shocks (though these occur), but through the slow onset and accumulation of crises. These include increasing heat, erratic weather, unpredictable seasons; rising food and water insecurity; increasing costs of living; and declining physical and mental health. Slow onset events evolve from small changes over extended periods of time, increasing the frequency and intensity of regularly occurring events, most having dire community consequences (Head, 2016). Our focus on slow-onset, intersecting crises offers opportunities to understand *the everyday* gender ramifications that (will) shape responses to future disasters. This directly acknowledges that 'gendered patterns help to produce disasters and part of what is disastrous is patriarchy' (Luft, 2016, p. 8). Understanding how social, political, and environmental burdens intersect in different ways in non-disaster (but accumulative crises) conditions can highlight the complex and messy workings of gendered power in perpetuating the gendered inequalities and injustices that will inevitably shape disaster justice.

Our applied research demonstrates how a person's sexual orientation, gender identity and expression, and sex characteristics (SOGIESC) actively shape the kinds of livelihoods they can pursue and their access to social, state, and humanitarian support networks, especially family support networks critical to responding to, and recovering from, shocks and crises. They also influence their access to healthcare and identity documentation; education and vocational training; the ways they can move in and around the city (this collection, Chap. 4); the neighbourhoods people can afford to live and find peace within; and the standard (and safety) of housing they can access. All these intersect to directly shape the ways in which they experience and respond to climate change impacts in a country where state-sanctioned anti-queer and trans attitudes, policies, and actions are pervasive. All these factors are deeply interlinked. We cannot begin to understand vulnerability or resilience to disasters, climate change, and humanitarian crises without addressing the messy workings of everyday gendered power.

Understanding the fast-developing political-legal context in Uganda for gender and sexual diverse people is critical. The Anti-Homosexuality Act, 2023, was signed into law by President Yoweri Museveni on 26 May 2023 and passed by parliament with only one opposing vote. At the time of writing, legal challenges to overturn the law have been unsuccessful. The law follows on from other legal attacks on gender and sexual diverse people, including the overturned Anti-Homosexuality Act of 2014 and the vetoed Sexual Offences Bill of 2021. The current Act has outlasted previous efforts. Rights groups have noted that the Act has already had negative consequences including a spike in violence (HRAPF, 2023). The Act targets those who 'aid and abet homosexuality' with direct implications for housing and livelihoods. For example, Section 11(d) permits the imprisonment for someone who rents property to gender and sexual diverse people (discussed later). Gender and sexual diverse people have been criminalised ever since Uganda was a British Protectorate—with discriminatory features of the penal code retained when Uganda gained Independence in 1967. The latest Act has drawn international attention and condemnation which has made headlines previously for its staunch efforts to further criminalise and punish gender and sexual diverse people.

Applied Methodology

The project, on which this chapter draws, examines the disproportionate impacts of climate change and environmental injustice on marginalised groups, including gender and sexual diverse communities in Uganda.

These impacts are myriad and overlapping. For the purposes of this chapter, we focus on two vital (and intersecting) issues: housing and livelihoods. We draw on ethnographic research and foreground the lived experiences of respondents as they were shared with us between December 2021 and April 2023 mainly in Kampala, Uganda's capital and largest city. Our fieldwork comprised a mixed-methods approach, including extensive participant observation, 30 ethnographic life story interviews, two multi-day applied arts workshops, and three group discussions with shelter residents. Our methods also included a series of 15 creative workshops involving over 200 participants that combined drawing and discussion and involved 15 civil society organisations and artists led by (and serving) gender and sexual diverse communities as co-facilitators, hosts, and research partners. This approach was adopted to ensure a multiplicity of insights from across a wide and diverse community, ensuring that data did not come exclusively from dominant organisations or key community figureheads.

To better understand how systemic and structural oppressions affect gender and sexual diverse people, we deliberately sought to capture and highlight the diversity of the broader 'community' (as it is informally referred to) of gender and sexual diverse people in Kampala while offering safe spaces for different communities to share their unique experiences individually and collectively. Workshops focused on different, though often overlapping, parts of the community, such as transgender men, refugees, and asylum-seekers, or those with experiences of incarceration. Some individuals fell within or across the purview of different workshops, with some people participating in more than one. Our research was co-designed and conducted in collaboration with the community. This allowed us to reflect on local priorities, build ownership, and resist and subvert historical and colonial power dynamics of research by adopting a collaborative methodology that enabled the community itself to determine what and how this research was produced and applied. This included shaping the questions and how they were asked, and in the analysis and application of the knowledge produced. Here, we draw from a dataset that includes field notes, transcriptions from one-to-one interviews and the creative workshops, and visual and creative data, including collage and poetry.

Two initial workshops involved invited participants who were in leadership positions within gender and sexual diverse community-based organisations. These provided a broad overview of issues and allowed for greater access and collaboration. The following workshops centred on the experiences of specific sections of the broader gender and sexual diverse community. They included individual workshops with gay, bisexual, and queer men; gay, bisexual, and queer women; transgender and gender-nonconforming women; and transgender and gender-nonconforming men. They also included workshops with gender and sexual diverse older people; refugees and asylum-seekers; people living with disability; and other gender and sexual diverse minorities. These workshops were all co-organised with one to three local organisations and generally held at one of the organisations' offices, in a space familiar, comfortable, and safe for participants.

The workshops also reflected the makeup of Uganda's—and in particular, Kampala's—gender and sexual diverse civil society. This civil society is segmented, made up of different organisations that, while working collaboratively, have their own specific memberships that serve different parts of the community or focus on specific issues. Most of our research was conducted in Kampala and surrounding locations as this is where the majority of relevant organisations are based. Doing so also allowed us to delve into the specifics of one key urban site—Uganda's largest city by some margin, where many urban inequalities and environmental hazards are most pronounced. The research was undertaken following full ethics approval by the University of Leeds and extensive consideration, in discussion with all local partners, of ongoing safety considerations amidst a fractious national context.

Climate Change and Urbanisation in Uganda

Climate change is and will disproportionately impact countries in the Global South who, in addition to having historically contributed the least to carbon emissions, have the least resources to be able to install costly mitigation and adaptation efforts (Crawford et al., 2023). This is the case for Uganda, which is projected to experience up to 5 °C temperature rise during the twenty-first century (Twinomuhangi et al., 2021). Such changes will likely lead to a decrease in rainfall, particularly in parts of the country located near Lake Victoria, including Kampala, resulting in increased droughts, extreme heat, erratic and extreme rainfall, and more flooding (Twinomuhangi et al., 2021). Many of these consequences are already playing out and are set to increase. They are especially felt by those living in informal settlements who often lack access to (inadequate) services, are characterised as the 'urban poor', and make up most of the city

(Lwasa, 2010). These climate impacts are occurring against a backdrop of rising urbanisation, with urban populations expanding faster than infrastructure or housing allows. Uganda's urban population has risen from 15 to 21% between 2000 and 2021 and in 2019 the Parliament of Uganda awarded city status to 15 towns and municipalities in anticipation of a more urban future, as well as attempts to ease pressure on Kampala. Such crises meet continuing political instability in the region that is forcibly displacing refugees and asylum-seekers across its borders in significant numbers. This leads, inevitably, to rising urbanisation of forced displacement as harsh environmental conditions and expanding populations increasingly push refugees and asylum-seekers out of rural settlements towards 'self-settlement' in its towns and cities (Crawford, 2021; Ronald, 2022; Tshimba, 2022).

How these climate impacts are experienced, and the environmental injustices that exacerbate them, is deeply gendered, as are the ways in which people can and do respond to these accumulating crises. How gender and sexual diversity shapes and (often) limits the capacities of certain communities to anticipate, respond to, and recover from these interlocking crises is the empirical focus of this chapter. We deliberately take an intersectional, inclusive, and feminist approach to better understand the lived experiences of these climate impacts. This approach foregrounds gender and sexual diversity and in so doing advances our understanding of how climate and disaster justice are inherently gendered issues. Here, at the queer frontline of climate change, we can begin to better comprehend how environmental, social, and climate injustices intersect and layers of disadvantage accumulate and play off each other in diverse ways.

Where the Environmental, Social, and Climatic Intersect

(Gendered) Housing and Livelihoods

Gender and sexual diverse people in Kampala are frequently forced to live in poor neighbourhoods where the housing is dense and amenities and facilities (including water, drainage, infrastructure, waste management, etc.) are inadequate. In these neighbourhoods, public services are under-resourced and over-subscribed, and pollution, flooding, and heat are concentrated. In some cases, the density of urban housing and business of urban life offers safety through anonymity, particularly for people who have moved to the city in response to homophobia and transphobia in the towns and villages where they grew up (as has been noted in other studies of queer experiences of urban space; see Doderer, 2011). In other cases, closely packed living conditions increase social surveillance and risks of public identification and 'outing'. In turn, this leads to increasing housing precarity and risks of being unhoused as people are evicted, suffer harassment and violence, or invite the attention of hostile authorities. Housing precarity was something shared across the whole community, and as one of our respondents, Sandra, a lesbian activist, reminds us, it takes many forms:

> With housing, as much as we'd say we have a lot of homeless members within the community in shelters, but then homelessness does not stop at shelters, because if we're staying in a one-roomed house and we're six people, we are still homeless, and we don't talk about that.

Participants drew a direct line between their gender identity and expression, precarious livelihoods, and their housing situation. Julie, who describes herself as a masculine-presenting lesbian, notes that 'if you have money, you can afford privacy'. Susan, a lesbian woman, agreed:

> [I]f we find work then we get better shelter, however when they see 'studs' [masculine-presenting women] and our dress code, they not give you a job. We miss out on jobs because of our expressions. The challenge of jobs make us sleep in bad places. Most of our community, women, stay in such areas as Katanga [an informal settlement in Kampala] where the houses are cheaper.

She highlights how gendered social norms around dressing, expression, and mannerisms are keenly socially 'policed', restricting socio-economic opportunities. As Michael, a transgender man, puts it:

> studs have been mislabelled for being trans… that the way we talk, the way they present us… how we compose ourselves and how we talk… We have to learn how we approach every office you go to and how we compose ourselves.

Andrew, a transgender man, reflected on how these norms drive socio-economic exclusion:

[W]e all want to do business, we want to work, we want to get money but sometimes it's not that easy. You can't do everything… I'm professionally a counsellor, but I can't be a counsellor in any other [non-LGBTQI+] organizations because of the way they look at counsellors. They believe counsellors are supposed to be: if you're a woman, you're supposed to dress like a woman to be respected. I worked in a village and dressing in [trousers] is an abomination. No-one is going to come to that clinic if you dress like a man and you're a woman. They will not come. I had to resign from that job because of that. So they think your appearance matters more than the brains that you have, the services that you offer.

For others, especially those who are perceived to be cisgender or heterosexual, it is less challenging to find employment or secure livelihoods. Put simply, Julie argues, 'it's better when femme.' But for the most part, socio-economic exclusion begins early in life. Angela, a non-binary person, notes that 'many aspects of our life are about our orientation: we are pushed out of school, no job, that financial frustration'. Janepher, a transgender woman and sex worker, points out:

We all understand that LGBTI people or probably the gender and sexual minorities people in Uganda have not had higher education levels. So we are talking about members that are working in restaurants. We're talking about members that are working to sell water in the streets. We are talking about members that are selling clothes on the roads. We are talking about members that are actually selling sweets on some vendor streets. We are talking about these members that are actually doing sex work… So these are the members that are basically affected by the climatic changes […] You realise we have seasons when it is sunny, and it is not favourable for someone sitting on the street to actually be able to sell their good or probably it is not favourable for a sex worker that is going to be on the street at night and then be able to give services when it is raining. So it is deteriorating the income and so that leads to a lot of major health issues and other issues like drug abuse like insistent on gender based violence, insistent social economic justice, whereby people are not involved.

Family rejection plays a significant role here in exacerbating disadvantages, pushing young people out of homes and education. Jo, a gender-nonconforming young person who moved to Kampala on their own as a teenager, reflects:

> I had sex when I was still young. I had slept with a bunch of, you know, old men. And I feel like no-one could ever date me… From a person who is just you know looking at the world moving, looking at people being happy… But I am not always, you know, some people are always seeing me smiling or being happy but deep down in my heart I am like, 'oh my God.' You know, it happens, you miss your Mum, but when I'm missing my Mum she doesn't love me. Then I am missing my Grandmother, she's down in the ground, she's already dead. My Uncle no longer wants me.

Stephen describes arriving in Kampala as a young gay man. He describes being 'shunned by neighbours', how 'young men in Uganda' are especially not 'economically stable', and how landlords will not rent a property to someone whom they know or suspect to be gay.

Already facing challenges in renting properties, gender and sexual diverse people in Uganda face new challenges in the wake of the passing in May 2023 of the new Anti-Homosexuality Act, which actively disincentivises landlords from renting homes to people they may perceive to be 'homosexual'. Section 9 states that a person who 'knowingly allows any premises to be used by any person for purposes of homosexuality' could face 'imprisonment for a period not exceeding seven years'. Section 14 declares that anyone with 'reasonable suspicion that a person has committed or intends to commit the offence of homosexuality' has a 'duty' to report it, with failure to do so potentially resulting in a significant fine or up to five years in prison (Government of Uganda, 2023).

Heightened Vulnerabilities Among Older and Transgender People

Housing and livelihood challenges are even more profound when gender and sexual diversity intersect with age. Speaking about the expectations, assumptions, and experiences of older bisexual, gay, and queer men, Robert shares:

> [W]hen it comes to homelessness, it really affects the older people more than it would affect the young people […] They need to live independently. They cannot come and stay in such a house where young people are, they will be segregated, they cannot tell friends that at this age, I don't have where to stay. So they're really affected to the extent that many older people, they will get 5,000 [Shillings], go to a bar and sleep in a bar drinking or trying to make a night just because they don't want to come to [an organisation] and ask: 'do you have shelter and I come spend a night?' They will feel

embarrassed and stigmatized. Elderly people are really affected, now when it rains, imagine what happens when they're outside in a bar? They are vulnerable.

Geoffrey, an older gay man, reflects on his circumstances:

when you are at a certain age, you're supposed to be having kids. At a certain age, you're supposed to be having a wife, you're supposed to be having a home and everything. And so, when you are not in that position and with the climate [making agriculture difficult] and everything like that, you find that you're like an outcast, you're not really part of the community. So it affects the economic prospects of the older LGBT because they cannot fit in.

The link between precarious livelihoods and housing can be most pronounced for transgender people. Reflecting on the post-COVID period, Janepher shares:

money worries are on the rise for LGBTI persons, but most especially transwomen sex workers that probably live alone most often because they are rejected from their families. And this completely brings a feeling of isolation and fear… is very much harder for trans people to actually be able to have a stable economic structure.

Rejection from families, stigma from neighbours, and lack of access to religious institutions cut many off from the social safety nets and support many Ugandans rely upon when crises hit. As Derrick, the leader of an organisation working with young gay, bisexual, and queer men, remarks:

members of the LGBT community, even before the disaster, have not been having the same action just as equal as the general population. This is because of the exclusion, the isolation, the restricted social networks… this here affects the resilience of how a person might experience the climate shocks or the weather changes.

Philip, a transgender man, describes the isolation of transgender people *within* the wider gender and sexual diverse population:

they couldn't even go back to their homes, back in their families. Because remember, most of them were chased away. They're not accepted. So challenges of homelessness were so much, and if you get to know about trans issues, trans people are never comfortable. Yes, I understand there are different shelters in Uganda that are helping the LGBT community. But trans people weren't comfortable in those shelters.

Such shelters, which tend to be inhabited by young, unhoused members of the community who have run out of other choices, have frequently been the target of police raids, such as those that happened under the pretext of COVID-19 safety measures (Ghoshal, 2020). The raids further demonstrate the *everyday* entrenched links between insecurity, violence, housing, age, and economic means that constitute the status quo for gender and sexual diverse communities before any climate shocks or disasters.

Water Security, Urban Heat, and Health Challenges

Better understanding the lived experiences of these everyday challenges—and how they manifest across and within the community—is critical if we are to develop grounded and inclusive approaches to disaster justice. During a multi-arts workshop Julie described her living situation, highlighting experiences common within the community. The room she rented was small and barely lit by a bare bulb hanging from the ceiling. Temperatures during the dry season were stifling, but the room had no ventilation beyond a small window that was closed and locked. She could not afford a fan and only had intermittent electricity. When the dry season extends, as is increasingly common, the dust from local unpaved roads combines with rising temperatures to make the small room where she both lives and sleeps feel suffocating. As a masculine-presenting lesbian, Julie describes why she cannot open the window: it is too much of a security risk to invite the gaze of her neighbours. As she explains: 'people get into your business in these congested spaces. Where you would have had your privacy, it make the person who hasn't noticed you, notice you. They notice you *because you're different*.' Once 'noticed', word rapidly spread throughout these dense neighbourhoods. For Julie, over the course of five mornings in a row she had found 'GAY WOMAN' written in charcoal along the wall next to her front door.

At the time of our discussion, the onset of the rains had been delayed and Julie's tap had run dry, which was common for national water during the height of the dry season. She was therefore going through a process of weighing up how to get water into the house. To go to the standing pipe and queue up during the daylight seemed too risky, especially as politicians, religious leaders, and educators alike were stoking stigma and moral hysteria around gender and sexual diversity. Even before the Anti-Homosexuality Bill 2023 was tabled, she would only leave her house after dusk to mitigate against the risk of attack due to the way she dresses. However, to go to the standing pipe late at night, when it would be the most anonymous and quiet, increased her vulnerability to theft or gender-based violence, risks facing many women on Kampala's poorly lit streets. This left a relatively short period in the day when going to get water was perceived to be the least risky. Another option many of her neighbours took was to pay someone to collect her water and deliver the jerrycans; however, with every sip of water, she described wondering if it might be poisoned, on top of the anxiety of opening her door to someone she could not trust.

Water insecurity had further profound sanitation implications for people who menstruate. As Julie notes:

> as women we are affected so much. They think we [masculine-presenting women] don't experience menstruation, but we do. During the hot season we are feeling hot, need comfort but can't go outdoors. [We] need water for cleaning those pads, but it's not good at all, we are humans but people don't take us as such.

Many of these neighbourhoods' poor-quality rented rooms do not have their own bathrooms and as Blessing, a non-binary person, recounts: 'you have to get out [of your house] every time to take a bath, to change your sanitary towel', exposing yourself to neighbours, and in many cases when it is hot and/or the social risk of going out during the daytime is deemed too high, people are forced into poor menstrual hygiene. '[Y]ou are getting sick,' said Blessing, 'which gets back to economic aspect of going to hospital: can you afford? Which gets back to your mental health. When you get to the hospital, the doctor says: "is this a man or woman?"'

While petty crime and insecurity affect nearly all inhabitants of these low-income areas, these risks intersect with stigma for gender and sexual diverse people, including gender-based violence, harassment, and vandalism, but also an inability to report crimes to the police or seek justice or redress for fear of police stigma or even arrest. Nansubi, a lesbian woman, describes how the impacts of crime and social isolation in her poor neighbourhood further intersect with the climate:

> when I get to think of rain and I'm living in Bwaise [an informal settlement in the north of Kampala], the crime rate is high when it's raining. *Omusaja aja nakuba akayumba'ko ejinja, akaji nekagwa'eli* [a thief will come with a huge stone and throw it at your door and it falls down], so [the] crime rate is really high when it's raining. Because you can't, you can't even [raise the] alarm.

When the heat and pressure of stress and fear all begin to exact a significant toll on people's mental health, there is no recourse to nature, no tree or green space in which to seek solace or to breathe freely. Blessing, like many others, could not afford the transport to less dense neighbourhoods or greener open spaces. Many compounds and thoroughfares in Kampala's densely packed poor neighbourhoods have been picked bare of tree cover and vegetation, further exacerbating levels of heat and dust during the dry seasons with implications for both physical and mental health. This returns us to the impacts of precarious livelihoods and socio-economic status. Many of Kampala's greener areas are found within the walls of hotels and restaurants, all of which require money to access, and even with money, they may still be inaccessible to gender and sexual diverse people due to discrimination. Julie describes her situation as a stud as equating to 'emotional homelessness' with nowhere to go when the house becomes unbearably hot, with no sense of belonging—emotional or physical—and a very real danger of eviction hanging over all gender and sexual diverse people. Perhaps Florence, a lesbian and artist, captures this layering of disadvantages best:

> The very marginalised groups become victims again and again... as housing becomes a challenge due to the fact that many areas are flooding, people like us become further victimised. When resources are less, there'll be other people who are considered before others. And as LGBTI people we often come at the lower.

Housing Challenges for Refugees and Asylum-Seekers

Julie's case was not unique. Ilana, a transgender woman and Congolese refugee, spoke of keeping to herself and remaining indoors. Like Julie, she recounted a deep sense of insecurity when inside her home: 'I normally have to lock myself in the house, 'cause sometimes I can even be sleeping and I even think like, let me lock my window they might even throw a bottle of fire in my house and it will burn me in.' Their neighbours could mitigate against climate and environmental impacts, for example, by leaving their windows open to let in fresh air and a breeze or sitting outside to cook and seek cooler temperatures. Julie and Ilana, however, like many other gender and sexual diverse people, were forced to remain inside small and poorly ventilated housing. As other studies have indicated, isolation indoors is a common protective measure for trans people in the region (Camminga, 2020; Marnell, 2023). This social isolation was exacerbated for refugees and asylum-seekers we spoke to by their fear of accessing humanitarian offices that are staffed by Ugandans and the perceived risks of encountering discrimination and/or reporting and arrest. This perpetuates a series of violent continuities that further marginalise gender and sexual diverse refugees and asylum-seekers (McQuaid, 2014, 2020) within the broader host and refugee communities.

When considering housing injustices, gender and sexual diverse refugees and asylum-seekers often have little recourse but to move into poor-quality homes in Kampala, ones with metal or tin roofs that leak during heavy rainfall. These urban refugees and asylum-seekers often lack access to sufficient funds to move into better quality homes, with rent money, often from small grants from aid organisations, and livelihoods eked out from the informal economy. However, in contrast to many poor urban Ugandans in similar positions, refugees and asylum-seekers also face additional concerns and barriers to renting. Landlords may choose not to rent their properties to refugees and asylum-seekers due to assumptions about their inability to pay, their lack of permanence, or more generalised prejudice against refugees, intentionally increasing rent to exploit, intimidate, or push out tenants. This can force people to remain in substandard housing due to fear or inability to access better quality places or concern about losing rental time paid in advance. Such discrimination and stigma in the housing market were further exacerbated in cases where refugees and asylum-seekers were also gender and sexual diverse:

> We used to rent a house and then it rained a lot and then sometimes there would be too much wind that would blow the roof off. Then we went and reported to the landlord... Then he was like: 'I don't want you in my house, you can leave.' We were asking him what we had done and he told us that we used to shout making noise in the room, like shouting for others. They also used to report... that we are homosexual guys living in that house. Then we tried to hide... The neighbours started just to hate us then he

> called all authorities... he told them: 'I don't want those people to stay in my house'... you know, getting the advance for paying rent for three months is not easy.

Conclusion

This chapter demonstrates how the layers of disadvantages and environmental burdens facing gender and sexual diverse communities accumulate across multiple intersecting variables. By focusing on housing and livelihoods in a contemporary African city, we glimpse the myriad everyday impacts of these multiple, intersecting variables, including physical and mental health; water in/security; immobility; isolation and social exclusion; vulnerability to extreme weather and inability to mitigate against it; housing precarity and risks of being unhoused; and vulnerability to violence and crime. Without attention to how gendered power works to perpetuate and exacerbate these disadvantages in the everyday and efforts to transform these inequities through intersectional and inclusive models of climate and disaster justice, inequitable and systemic sexual and gender dynamics, power, and relations will continue to expose gender and sexual diverse people to far greater burdens.

Such injustices cannot be understood in isolation. It is critical that we approach disaster justice not just through the lens of gender, but with broader intersectional consideration of the environment, politics, colonial history, global capital, urbanisation, homonationalism, forced displacement, and more. For example, we cannot advance gender justice without considering social justice and the intersection of gender and sexual diversity with social factors, including class, citizenship, ethnicity, race, disability, and age, among others (Mikulewicz et al., 2023). Similarly, we must consider environmental (in)justice and the increasing environmental challenges facing rapidly growing urban populations, and how these disproportionately burden people with disabilities; people who have experienced displacement; the financially less well-off; or those who are otherwise socially marginalised. Lukasiewicz (2020, p. 9) positions disaster justice 'at the confluence of environmental, climate and social justice'. Our chapter, however, has argued for a more explicit reckoning with gender justice, which highlights the disproportionate distribution of burdens before, during, and after crises, on those with the least power, voice, access, recognition, and resources. As Janepher, a transgender woman and sex worker, put it: 'climate change is, in effect, an inclusion into the socio-economic and cultural norms, or probably the cultural dissatisfaction that LGBTI persons are going through.' We must, she reminds us, look carefully at 'our economic status, our socio-economic justice, and… our political intervention'.

Conflict of Interest The authors have no conflict of interest.

References

Arora-Jonsson, S. (2017). Gender and environmental policy. In S. MacGregor (Ed.), *Routledge handbook of gender and environment* (pp. 289–303). Routledge.

Boone, J. (2015). *The Homoerotics of orientalism*. Colombia University Press.

Buckingham, S., & Masson, V. (2017). *Understanding climate change through gender relations*. Routledge.

Camminga, B. (2020). Encamped within a camp: Transgender refugees and Kakuma refugee camp (Kenya). In J. Bjarnesen & S. Turner (Eds.), *Invisibility in African displacements* (pp. 36–52). Zed Books.

Crawford, N. J. W. (2021). *The urbanization of forced displacement: UNHCR, urban refugees, and the dynamics of policy change*. McGill-Queen's University Press.

Crawford, N. J. W., Michael, K., & Mikulewicz, M. (2023). *Climate justice in the majority world: Vulnerability, resistance, and diverse knowledges*. Routledge.

Doderer, Y. (2011). LGBTQs in the City, queering urban space. *International Journal of Urban and Regional Research, 35*(2), 431–436.

Gaillard, J. C. (2022). *The invention of disaster: Power and knowledge in discourses on Hazard and vulnerability*. Routledge.

Ghoshal, N. (2020). *Uganda LGBT shelter residents arrested on COVID-19 pretext*. Human Rights Watch.

Government of Uganda. (2023). *The anti-homosexuality Act, 2023*. Available at: https://www.parliament.go.ug/sites/default/files/The%20Anti-Homosexuality%20Act%2C%202023.pdf

Hawley, J. (2001). *Post-colonial, queer: Theoretical intersections*. SUNY Press.

Head, L. (2016). *Hope and grief in the Anthropocene*. Routledge.

Human Rights Awareness and Promotion Forum (2023). One Month After: Increasing Cases of Violence and Violations Based on Real or Presumed Sexual Orientation and Gender Identity in the First Month of the Enforcement of the Anti-Homosexuality Act, 2023. 14 July. Kampala: Human Rights Awareness and Promotion Forum.

Luft, R. (2016). Racialized disaster patriarchy: An intersectional model for understanding disaster ten years after hurricane Katrina. *Feminist Formations, 28*(2), 1–26.

Lukasiewicz, A. (2020). The emerging imperative of disaster justice. In A. Lukasiewicz & C. Baldwin (Eds.), *Natural hazards and disaster justice: Challenges for Australia and its Neighbours* (pp. 3–24). Palgrave Macmillan.

Lwasa, S. (2010). Adapting urban areas in Africa to climate change: The case of Kampala. *Current Opinion in Environmental Sustainability, 2*(3), 166–171.

Marnell, J. (2023). City streets and disco beats: Recentring the urban in queer and trans migration studies. *Urban Forum*, 14 March. https://doi.org/10.1007/s12132-023-09483-1

McQuaid, K. (2014). Violent continuities: Telling stories of one sexual minority life in the African Great Lakes region. *Identities: Global Studies in Culture and Power, 21*(5), 570–585.

McQuaid, K. (2020). "There is violence across, in all arenas": Listening to stories of violence amongst sexual minority refugees in Uganda. *International Journal for Human Rights, 24*(4), 313–334.

Michael, K., Mikulewicz, M., & Crawford, N. J. W. (2023). Introduction: Climate justice beyond the minority world—Towards Decolonial knowledges. In N. J. W. Crawford, K. Michael, & M. Mikulewicz (Eds.), *Climate justice in the majority world: Vulnerability, resistance, and diverse knowledges* (pp. 1–17). Routledge.

Mikulewicz, M., Caretta, M., Farhana, S., & Crawford, N. J. W. (2023). Intersectionality and climate justice: A call for synergy in climate change scholarship. *Environmental Politics*, 15 February. https://doi.org/10.1080/09644016.2023.2172869.

Ronald, M. (2022). An assessment of economic and environmental impacts of refugees in Nakivale, Uganda. *Migration and Development, 11*(3), 433–449.

Rydstrom, H., & Kinnvall, C. (2020). Introduction: Climate hazards, disasters, and gender ramifications. In C. Kinnvall & H. Rydstrom (Eds.), *Climate hazards, disasters, and gender ramifications* (pp. 1–25). Routledge.

Spurlin, W. (2006). *Imperialism within the margins: Queer representation and the politics of culture in Southern Africa*. Palgrave Macmillan.

Tshimba, D. (2022). *Asylum in urban spaces: The case of refugees in cities in Uganda*. EU Trust Fund for Africa (Horn of Africa Window) Research and Evidence Facility.

Twinomuhangi, R., Sseviiri, H., Mulinde, C., Mukwaya, P., Nimusiima, A., & Kato, A. (2021). Perceptions and vulnerability to climate change among the urban poor in Kampala City, Uganda. *Regional Environmental Change, 21*(2), 39. https://doi.org/10.1007/s10113-021-01771-5

PART III

Sins of Omission: Procedural Vulnerabilities and the (Re) Marginalisation of Queer People in Disaster Response and Recovery

CHAPTER 11

Impact of the 2010 Chilean Earthquake and Tsunami on Gender and Sexual Minorities

Borja Velez and Supriya Akerkar

Introduction

Humanitarian gender policies have focused on reducing the vulnerability of women and girls (Sweetman, 2012). However, heterosexist assumptions in Disaster Risk Reduction (DRR) and recovery strategies marginalise and discriminate against sexual and gender minorities (SGM) including gender diverse people who do not conform to the gender binary. The discrimination experienced by SGM is underpinned by a cis-heteropatriarchal system where political, economic and social relations favour cisgender, heterosexual men. This discrimination leads to the invisibility of SGM in DRR and recovery strategies (Gaillard et al., 2017b; Rushton et al., 2019). It also places transgender women and gender

B. Velez (✉) • S. Akerkar
Centre for Development and Emergency Practice, Oxford Brookes University, Oxford, UK

D. Dominey-Howes et al. (eds.), *Queering Disasters, Climate Change and Humanitarian Crises*, Sustainable Development Goals Series, https://doi.org/10.1007/978-981-96-3857-4_11

diverse individuals in extreme situations, post disasters, which dramatically increase their vulnerability and reduce their recovery capacity (Rushton et al., 2020).

This chapter explores the impacts of the 2010 Chilean earthquake-tsunami (hereafter, '2010 disaster') on SGM, focusing on transgender women and gender diverse individuals. It is the first study to offer analyses of the impacts of a Chilean disaster on SGM, providing a unique contribution to queer disaster studies. The specific vulnerabilities and capacities of Chilean SGM are highlighted, and official recovery documents analysed to review if and how SGM are included in DRR planning and programmes.

Data obtained through an online questionnaire and interviews showed instances of harassment and discrimination based on individuals' diverse sexual orientations, gender identities, gender expressions or sexual characteristics (SOGIESC). The data show how SGM were systematically excluded from official DRR and recovery strategies in 2010 and were not included in the development of official disaster prevention plans and policies. Our analyses use an intersectional lens to show how interacting forms of discrimination based on gender, sexuality and social class increased the vulnerability of many female transgender sex workers.

Absence of Gender and Sexual Diversity in DRR and Recovery Policies

Our review shows that official DRR and recovery policies and planning documents do not include SGM and their disaster experiences and needs. The impacts of this lack of inclusion are two-fold: (1) there is no consideration of the forms of discrimination unique to SGM and how they increase SGM's disaster risk and vulnerability; and (2) there are no policies and programmes that aim to empower SGM communities, increasing their disaster resilience and recovery capacity.

The Impacts of Heteropatriarchal Discrimination

DRR and recovery policies (and research) seek to address gender and gender inequality but assume a binary focused almost exclusively on women's and girls' increased vulnerabilities and coping capacities (Akerkar, 2022a). Minorities that fall outside binary gender norms have been excluded. This is reflected in both the limited research on SGM within disaster studies and their systemic exclusion from DRR and recovery policies and practices (Gaillard et al., 2017a; Rushton et al., 2019).

None of the following official disaster-related plans, policies, and guidelines mention SGM, let alone address systemic heteropatriarchal discrimination and its impacts on SGM: *Hyogo Framework for Action 2005–2015* (UNISDR, 2005); *Sendai Framework for DRR 2015–2030* (UNISDR, 2015); *UNISDR Strategic Framework 2016–2021* (UNISDR, 2016); and *UNISDR's Policy and Practical Guidelines to make DRR Gender-Sensitive* (UNISDR, UNDP and IUCN, 2009). The brief mentions of SGM's vulnerabilities in the latest version of the Sphere handbook and in a UNISDR document on minority groups in disasters only refer to SOGIESC diverse people living with HIV/AIDS. Clearly, these 'mentions' are far from adequate (IGLHRC/SEROVie, 2011; UNISDR, 2017; Sphere Project, 2018).

SGM Capacity and Resilience

DRR and recovery policies and plans should not only protect vulnerable communities but also increase their capacity and resilience to deal with disasters—social capital being one significant mechanism. Social capital refers to a community's level of trust, social norms, and networks that help reduce the disruption caused by a hazard or disaster (Nakagawa & Shaw, 2004).

Three types of social capital increase people's recovery capacity (Nakagawa & Shaw, 2004; Aldrich, 2010). These are *bonding*, which refers to the ties between family members, close friends, or neighbours; *bridging*, which defines the relations between people from different social groups and backgrounds that share common interests and activities; and *linking*, which refers to the ties between community groups and people in positions of influence which can help the community group move forward with recovery.

The research shows that by adopting DRR policies and relief and recovery programmes that build SGM's social capital, humanitarian agencies would not only increase these minorities' resilience but also benefit the mainstream (e.g., cisgender heterosexual) society (Gaillard, 2011; McSherry et al., 2015; Dominey-Howes et al., 2016; Gaillard et al., 2017a; Dwyer & Woolf, 2018; Gorman-Murray et al., 2018; Rodriguez & Suvianita, 2020). Both the vulnerabilities and capacities identified inform the scope of this study and establish the baseline for a comparative analysis of the data. We identify how transgender women and gender diverse individuals' 'self-organising' have enabled them to mobilise their collective agency and meet the challenges and suffering they experienced during the 2010 disaster.

Research Design, Methodology and Analytic Framework

Research Design

On February 27, 2010, a magnitude of 8.8 earthquake off the Chilean coast triggered South-America's deadliest tsunami in 40 years (Fritz et al., 2011). The disaster left 562 casualties and over 2.6 million affected people, with 800,000 people rendered homeless (EM-DAT & CRED, 2010). Chile was chosen as the site for this research because of the large impact of the disaster and because Chile had been proposed as a country for further studies (Dominey-Howes et al., 2014, p. 4). The following research questions were formulated:

1. What were the experiences of Chilean SGM during and after the disaster?
2. What factors contributed to their vulnerabilities and capacities in recoveries after disasters?
3. How were SGM included in the official relief and recovery programmes?
4. What improvements in Chilean DRR plans and policies to include SGM have been made since the 2010 disaster?

Methodology

Data was collected using both qualitative methods and secondary literature from Chile. Qualitative data was also obtained through a combination of an online questionnaire and semi-structured online interviews with five people directly affected by the disaster, including representatives of SOGIESC diverse organisations (e.g., *Organización de Transexuales por la Dignidad de la Diversidad*, Organisation for the Dignity of Diversity, [TODD] and *Transgéneras por el Cambio Talca*, Transgenders for a

Change Talca, [TCT]). Social media was used to recruit participants for both qualitative components. Eighteen people answered the online questionnaire, and one online interview was undertaken with each of the five participants affected. These participants were either transgender women or members of organisations that worked closely with this community.

The study was carried out during the COVID-19 pandemic. Questions about SGM experiences during the pandemic were also asked during interviews to assess whether the State had made any improvements in DRR and Recovery plans and policies since the 2010 disaster. COVID-19 is conceptualised as a disaster by UNDRR[1] which affected people in unequal ways (see also the IFRC's World Disaster Report 2022). The report highlighted that LGBTQ and gender diverse persons were disproportionately impacted with increased violence and abuse, both physical and psychological, and discrimination in access to basic relief and health care. The data was collected between May and September 2021 and Ethics Approval was obtained from Oxford Brookes University. Representatives from leading organisations granted permission to share the names of their institutions.

Secondary literature sources were explored to understand the specific vulnerabilities and capacities of SGM in Chile during the disaster. This research engaged, primarily, with grey and policy literatures and builds on the study of Yamashita (2012). Reports from 2010 onwards by the *Movimiento de Integración y Liberación Homosexual* (MOVILH), the main Chilean LGTBQIA+-focused organisation, were reviewed to check all information related to disasters and SGM. The following keywords (translated from Spanish) were searched for in each document: disaster, hazard, earthquake, tsunami, emergency, and February 27. The search identified 11 documents (reports between 2010 and 2020) of which only 2 included information specific to the disaster (MOVILH, 2011b, 2012). Additionally, the same keyword search was carried out on MOVILH's website's search engine to find articles related to the disaster. Of the 35 entries matching any of the keywords, only 3 mentioned the 2010 disaster (MOVILH, 2010a, 2010b, 2011a).

Legislative and policy frameworks, plans and strategies for emergency management published by the ONEMI (National Office for Emergencies of the Home and Public Security Office) from Chile were also consulted using the same keywords. From the 19 documents analysed, only 4 included the word 'gender' (ONEMI, 2018a, 2018b, 2018c, 2018d). Of these, only one referred to gender equity between men and women, and one strategic plan to 'sexual diversity' (ONEMI, 2020).

Analytic Framework: Disaster Justice and Intersectionality

We have used an intersectional disaster justice lens to analyse the experiences of SGM following the 2010 disaster (Lukasiewicz & Baldwin, 2020). Disaster justice highlights the role that pre-existing, systemic discrimination plays in dramatically increasing SGM's vulnerability to disasters (Gorman-Murray et al., 2014; Rushton et al., 2019; Seglah & Blanchard, 2021). Disaster justice acknowledges that while disasters involve natural hazards their causes and effects are related to existing systemic inequalities—social, political, economic, institutional, cultural, temporal and spatial.

Intersectionality is also a key element of our use of disaster justice. Applying an intersectional disaster justice lens to our analysis reveals that the different layers of oppression faced by gender diverse people, and especially transgender women, directly increase their vulnerabilities within disaster contexts. Crenshaw (1991) explains how the experiences and discrimination experienced by minorities are compounded by the interactions among different types of identity-based structural discrimination, including ethnicity, gender, sexuality, social class, age, disability and more. In turn, these intersectional discriminations may vary depending on geographical location, power relations and personal experiences (Daigle & Woodroffe, 2017).

Few scholars have applied the concept of intersectional discrimination to DRR and recovery programmes. The majority have prioritised analysis of one social identity to the exclusion of others in their respective studies (Fordham, 1999; Schuller, 2015; Lovell et al., 2019). To address this gap, we and scholars and practitioners in crisis and disaster management advocate the adoption of an intersectional analytical framework that facilitates a nuanced understanding of the vulnerabilities and resilience within minority groups (Vickery, 2018; Kuran et al., 2020; Akerkar, 2022b).

Findings

The analysis of the primary qualitative data highlighted two main themes: (1) the increased vulnerabilities of SGM during and after disaster due to the exacerbation of entrenched and overlapping intersectional discriminations and stigmatisations; and (2) the strategic building of social capital through self-organisation by SGM to increase their resilience and collective agency. In the absence of state support, self-organising became much more important to SGM's recovery.

Specific Vulnerabilities of SGM During the 2010 Chilean Earthquake and Tsunami

SOGIESC-Based Harassment and Discrimination

In 2010, Chile was led by a conservative right-wing party and had not yet passed any bills or acts to reduce SGM discrimination (MOVILH, 2011b). Interviews with transgender participants showed both patriarchal and transphobic attitudes among the authorities and the public in the aftermath of the 2010 disaster. One of the interviewees said transgender women's identities were further marginalised compared with other SGMs:

> It's painful to confirm that within the gender and sexual dissent, trans people lose the status of human beings [during emergencies]. ... It is also shocking to realize how the patriarchal model is replicated within the trans community. I was heard and listened [to] by the authorities just for being a man, even as a trans man, while many of my colleagues who are trans women were never listened to because they were perceived as sluts, crossdressers, or fagots.

An MOVILH's annual report (2012, p. 75) highlighted how, after having lost their houses because of the disaster, 14 transgender women were victims of a transphobic attack. One victim reported:

> We managed to get No.6 temporary shelters where we could live and work. However, 6 months after we got the temporary shelters someone set them on fire in a transphobic attack. Thankfully, no one passed away, but we needed psychological support after that and required shelter once again.

Another participant noted the severe marginalisation experienced by transgender women who lived in the red-light district and received no official State aid:

> The State was providing water to the neighbouring houses, but they did not come into the red-light districts. They [the transgender women] said: 'they don't come here because they just see us as prostitutes.'

One transgender interviewee reiterated the lack of support for transgender people: '[T}he disaster impacted us severely. We as a trans community were helpless, we did not have any benefit nor support at neither local nor national level.'

Given the systemic discrimination suffered by this community from the State, the TODD gathered relief items to support them as much as possible. Reflecting on their relief operations during our interview, the Executive Director of TODD also highlighted the community's extreme vulnerability:

> There was [another] answer [between the members of the trans community] that was quite common: 'if your organisation would not have brought us food, we would not have had anything to eat.' [Another trans woman] said: 'they will let us die as dogs just for being trans.'

SGM Hiding Their SOGIESC for Safety and Protection When Accessing Aid

SOGIESC-based harassment and discrimination is not the only vulnerability that SGM faced in the aftermath of the 2010 disaster. Some interviewees chose to hide their SOGIESC identities to prevent any forms of violence. '[I]n 2010, Chile was still a very homophobic country,' said one participant:

> [A]nd the questions about being married or having children were quite common and were always underpinned by heteronormative assumptions. When accessing aid, pretending to be heterosexual was common.

This statement, together with the following comment by another interviewee, shows how individuals often hide their SOGIESC identities to feel secure when accessing aid:

> Some of them [referring to a group of transgender women] started dressing as men to avoid being harassed in the streets or discriminated against when asking for water. In men's clothes, they could walk one or two blocks and get to the water truck to obtain the daily water they needed.

Loss of SOGIESC Safe Spaces

Information regarding the impact of the 2010 disaster in local SOGIESC communities was shared at MOVILH's website (MOVILH, 2010b). The organisation announced that the Frida Kahlo Lesbian Cultural Centre in Chillán and the Pub Vunker in Paine had been destroyed, together with the office of *Acción Gay*, Gay Action, local SOGIESC NGO (GA), in Talca. The Headquarters of GA in Valparaíso were also severely affected. The Director of Public Management at GA shared how the loss of their offices affected the local SOGIESC community:

> However, moving to this new house [in Valparaíso] took around 6 months, which obviously affected the [public policy, sexual education, human rights, and HIV prevention programmes] that we were providing. ... The office in Maule was destroyed and we did not rebuild it ... so we stopped our work in this region, which was the most affected area.

SOGIESC safe spaces in Chile provided both emotional and material support that had to be discontinued after the disaster. Furthermore, the loss of the Lesbian Cultural Centre and the Pub Vunker, where SGM could freely express themselves, meant that individuals were suddenly exposed to discrimination and harassment.

Restricted Access to Healthcare Services

Two study participants indicated that they had suffered discrimination when accessing healthcare services. One outlined that '[T]he LGBT community does not feel secure, the healthcare system is not kind to LGBT individuals.' A member of TCT stated that 'we as trans women are generally discriminated against ... and even our access to certain healthcare services is restricted.'

When asked about the impact of the COVID-19 pandemic on the Chilean SOGIESC community, two interviewees confirmed that access to healthcare was compromised:

> As an example, hormone replacement treatment for trans people have been suspended due to the pandemic as the Government stated that there was a lack of resources. For instance, the healthcare system in Talca run out of HIV treatment during the pandemic and some of our women living with HIV are relying on the treatments that 'AIDS Maule' [HIV-focused local NGO] has provided.

The interruption of certain healthcare services during disasters, especially for people in continuous treatment such as HIV/AIDs and transgender hormone therapy, has been identified as a key vulnerability of SGM in disaster contexts and has a direct impact on their health (IRIN News, 2014; The National LGBT Health Education Centre, 2016; Knight & A, 2020).

Limited Employment Opportunities
Interviewee information and secondary source data highlighted a specific vulnerability for transgender women after disaster—namely, lack of employment opportunities, directly raising concerns about the intersectional vulnerability and overlapping discriminations that they face. The intersectional discrimination and vulnerability of female transgender sex workers is rooted in heteropatriarchal norms present across society, which stigmatise and marginalise them based on their gender and sexuality (woman and transgender), which do not align with the dominant societal categories.

Many transgender women engage in sex work, often as a result of structural discrimination (Nadal et al., 2014). The socio-economic class that they belong to is another layer of oppression that comes into play when evaluating their vulnerability. In other words, female transgender sex workers' experiences of discrimination were mediated through intersections of structural, identity-based discriminations including discrimination based on gender, sexuality and social class. These findings are supported by those of a previous study, which found:

> [T]he destruction of factories and businesses in the area [referring to Rancagua, Chile after the 2010 earthquake] left thousands of people unemployed, which in turn left many transgender sex workers unemployed. (IGLHRC/SEROVie, 2011, p. 9)

Two participants shared the difficulties they faced when their employment opportunities were reduced by the 2010 disaster:

> Before the disaster, it was already harder for me to get a job as a trans woman. With the reduced employment opportunities in the aftermath of the disaster, I started struggling economically, reaching a point of not having what to eat. The combination of all these facts led me to carry out a third suicide attempt.

The situation after the earthquake was very insecure for us to continue with our routines and with our jobs [as sex workers] as the neighbourhood was disrupted and we could not walk alone due to the lack of protection we would face.

Another participant highlighted the impact of COVID-19 on employment opportunities available:

> During the pandemic … trans women that carry out sex work have suffered especially. Mostly because they have not been able to work due to the curfews and other restrictions.

These findings show how the concepts of social vulnerability, intersectional discrimination and disaster injustice can help understand and address SGM's increased vulnerabilities in the aftermath of disaster. In contrast, the following section demonstrates how vulnerabilities can be reduced by empowering these minorities with the support of local LGBTIQA+-focused organisations.

Building Social Capital Among SGM

Developing Resilience of SGM During and After the 2010 Disaster

When the 2010 disaster occurred, the TODD offered support to the smaller organisation of transgender women in Talca (TCT) who had been severely affected. The staff at the former organisation noticed that many of the affected transgender women had stopped wearing their 'more feminine' clothes either due to the loss of such items or because of the increased protection and security that wearing men's clothes granted them. Aiming to help improve their mental health, the organisation provided both psychological support and supported a beauty session including make-up, manicure and haircut services. Referring to this programme, one interviewee stated:

> They were depressed and even dressing with what would be categorized as male clothes, and suddenly they flourished and started wearing their female clothes again. … They even outlined that we had brought back their will to live through that beauty session.

Additionally, TODD applied for international funding so that they could support TCT after the 2010 disaster. One interviewee who worked for TODD outlined how 'this funding also allowed them [TCT] to organize and set [up] their own organization and even acquire some financial skills through capacity-building workshops that we organized.' The 'new' organisation was registered in January 2012, although it had been running, informally, since 1996–1997. This also demonstrates how, when designed effectively, relief and recovery operations can build social capital to increase disaster resilience and fight for social change.

Bonding

Respondents also suggested that emotional support between members of the Chilean SOGIESC community was helpful in coping with the disaster's aftermath. When asked if they received any type of support from their SOGIESC networks, many participants answered that they received emotional support from other members of the community. The staff at TODD collected all relief items from their own families and from SOGIESC networks. One interviewee reported: 'The first aid we provided to survivors was obtained from [mostly SOGIESC] friends or family members. We prioritized water ... oil, sugar and toiletries.' A member of TCT explained that before the disaster their organisation used to provide guidance and shelter to transgender women who planned to relocate to Talca. They said that even when the women from the organisation lost their houses, they kept helping other transgender women who had come from different cities. The interviewee from TCT stated:

> When TODD came with all the relief items they had collected, we shared it with all the other trans women who were not part of our organisation but were staying in the city. And when we applied for the temporary shelter, we helped everyone else to apply as we knew what was required and how the procedures in our region were.

Bridging

Our participants did not offer many examples of bridging between people from different social groups after the 2010 disaster. However, one interviewee shared a valuable insight they gained from working with a group of transgender women:

> The trans women we worked with also looked after other women by sharing the aid we distributed, especially food. They shared not only with other trans women but also with older cisgender women who used to be sex workers.

The older cisgender women were facing a lack of income before the disaster due to their age, and despite their already vulnerable situation, the transgender women decided to share their relief items with the former group. Even though both groups belonged to the same social class and background of sex workers, differences in age and gender divided these groups into two similar but different communities. *Bridging* between the groups provided opportunities for building solidarity and mutual support, links from which they could benefit in the future.

Linking

After the 2010 disaster, the support provided by TODD to the transgender women of TCT not only empowered the latter but also enabled them to *link* with government bodies. The TCT managed to officially set up a Housing Committee which helped them to eventually receive support from the Government for a temporary shelter. The success of this project encouraged these minorities to *link* with State actors and eventually empowered two transgender women to seek careers within Government bodies to keep fighting for SGM equality. According to one of these transgender women:

> I became the first trans woman at a national level to run for city councillor and there was much excitement around it. ... I did not get to it in the end, but I would have loved to be part of the municipality to have more influence and be able to change things at a structural level.

One participant also highlighted the great work developed by this group of transgender women:

> In Talca, they [referring to the group of transgender women of TCT] organized themselves so good that they created a Housing Committee that eventually allowed them to apply for post-disaster housing. 2 or 3 of them managed to move out from the temporary shelters into these new houses. Two of these women even got a job within the Government.

The examples above evidence how disaster resilience can be built by strengthening social capital to reduce the impact of disasters in vulnerable communities. Additionally, the involvement of local SGM organisations is critical to ensure that adequate support and guidance is given to these minorities to build and use their social capital to overcome adverse situations. This is particularly important in the absence of dedicated State support to SGM, as in Chile.

SGM in Official DRR and Disaster Recovery Policies and Plans

SGM in DRR and Disaster Recovery Policies and Plans Before 2010

From the 19 legislative and policy documents analysed, only 1 strategic plan by the ONEMI (2020)—published 10 years after the disaster—referred to 'sexual diversity.' Specific support to transgender individuals was practically absent in the state policies and frameworks in 2010. One participant highlighted the exclusion of transgender sex workers from state-supported recovery:

> Some gay men and lesbians could opt to receive it as they lived integrated within the rest of the population, but the trans community was concentrated in one physical place, the red-light district, and so it was easy to marginalize. ... State support was given to the population living in the surrounding streets, but it did not get to the three streets conforming the red-light district.

Further, the Executive Director of TODD met the Mayor of Concepción at that time to gain material support for the transgender community as they were not receiving any aid. The Mayor responded that 'she would first look after the people from her own community ... and then she would take care of other things.' Neither the Mayor nor Concepción's local government provided any support. Furthermore, two participants when asked if they felt included within the State's DRR and recovery strategies stated: '[T]hey did not seem designed for me or people like me' and that 'there was not aid focused towards trans people.' This demonstrates the lack of inclusion of SGM within Chilean Relief and Recovery Programmes during the 2010 disaster.

SGM in DRR and Disaster Recovery Policies and Plans After 2010

Only one (ONEMI, 2020) document mentioned 'sexual diversity,' and stated that ONEMI will introduce SOGI-related non-discriminatory practices for children and teenagers within their DRR plans and strategies. No details on how such measures would be implemented were provided. In contrast, the National Standards for Emergency Response published by both the ONEMI and UNDP (2017) raises no SOGIESC-related issues. This document is based on the 2011 version of the Sphere Handbook (The Sphere Project, 2011), which did not have any SOGIESC-related information. None of the remaining ONEMI documents reviewed in this study made any reference to SGM.

Interviewees were asked if they had perceived any SOGIESC-related improvements in the State's DRR and Recovery strategies since 2010. Their responses implied that things had not progressed much in the last decade:

> No, I have not perceived any change at a national level. There are few well-intended people in the Government as a result of many years of LGBT sensibilization. But this has not been yet reflected into public policy.

Last year, the Government set an advisory team to work on reducing violence against the LGBT population; however, there were not any campaigns to work towards [SOGIESC] inclusiveness or protection [of the SOGIESC diverse community]. There is no real interest in changing things, or at least is not shown effectively.

This suggests that although the Chilean government has taken the first step towards SOGIESC diverse inclusion and protection, it still excludes the transgender community in its DRR response and recovery plans and strategies. Lastly, when interviewees spoke about COVID-19 and its impacts on the SOGIESC community, it was evident that no effective measures had yet been implemented to protect SGM:

> [The trans community could not] access proper healthcare services. … Additionally … the Judicial system is swamped and there have been many hate crimes towards trans women who did not receive legal attention.

The government is still marginalising the trans community in all their policies. Most of the women from the organisation are sex workers, and now due to the pandemic and the curfew, it has been quite difficult for them to keep going.

Discussion

The stories from our participants speak to the need for the development of an intersectional disaster justice framework. The stigmatised and marginalised intersecting social identities of SGM resulted in high levels of inequality and social exclusion and placed SGM in extremely vulnerable situations after the 2010 disaster. Many of the specific vulnerabilities highlighted in earlier queer disaster studies were also found in Chile (D'Ooge, 2008; Pincha & Krishna, 2008; IGLHRC/SEROVie, 2011; Yamashita,

2012; Dominey-Howes et al., 2018; Rushton et al., 2019). Additionally, a new vulnerability in Chile linked to the lack of employment opportunities for transgender women in disasters was identified, resulting from the intersectional discriminations experienced.

This study contributes new knowledge to the existing literature by raising the importance of applying an intersectional lens to explore the various layers of oppression in disaster contexts. This study found that the intersections of gender, sexuality and social class faced by female transgender sex workers, together with the prevailing norms of the heteropatriarchal system, furthered their social exclusion and structural discrimination following the 2010 disaster. During COVID-19, female transgender sex workers faced increased economic hardships because of government restrictions on work and movement. Our study confirmed these findings, highlighting how female transgender sex workers are often socio-economically disadvantaged during disasters due to a lack of savings or alternative forms of income (Jebin, 2018; Carpenter et al., 2020; Koch et al., 2020). We contend that female transgender sex workers and gender diverse people were not included in DRR and recovery programmes after the 2010 disaster due to their socio-economic class and the government's lack of understanding of their stigmatisation and its impacts on their increased vulnerabilities to disasters.

The role of SOGIESC organisations in building and mobilising different forms of social capital is critical for the resilience of SGM in post-disaster contexts. After the 2010 disaster, TODD and TCT worked closely together supporting the needs of local SGM and strengthening *bonding* between the SOGIESC community members by promoting solidarity and support. These organisations also empowered the affected transgender community to *link* with the authorities, which eventually led some SGM to seek careers within Government bodies to further fight for SOGIESC equality. These findings share similarities with the capacities and agency shown by SGM to respond to previous adversities such as the AIDS/HIV pandemic in the 1980s and 1990s and the self-organising of SOGIESC minorities in Indonesia following the 2010 Mount Merapi eruption and, later, during *COVID-19* (Balgos et al., 2012; Halkitis, 2019; Rodriguez & Suvianita, 2020).

The beauty sessions and provision of adequate clothing by one of the local organisations are not to be disregarded as a powerful strategy to improve the mental health of transgender women in disasters. As Glynn et al. (2016) mention in their study, 'gender affirmation' among transgender women, who are a group at high risk of psychological distress, could reduce adverse mental health outcomes for these individuals.' 'Gender affirmation' can be achieved by someone wearing clothes they associate with their gender identity; through medical or surgical treatments, such as hormone use or breast augmentation; or through legal means to change 'sex at birth' on a birth certificate or a person's name. Therefore, both the beauty-related strategies put in place by TODD and ensuring SGM's access to healthcare services are critical to reducing the vulnerabilities of these minorities in disasters and to affirming their identities as transgender women.

Overall, our study shows the importance of laying and furthering the disaster justice framework through both procedural and distributive justice as argued by Lukasiewicz and Baldwin (2020). The lack of recognition of SGM in disaster policies and guidelines has ensured that there is no procedural guidance on how to establish SOGIESC inclusive disaster responses or management systems. Ignoring or silencing the voices of SOGIESC organisations in disaster response, as was done by the Mayor of Concepción, also highlights the need for procedural justice.

Disaster justice also focuses on the importance of policy and governance. Like Dovers (2020), Lukasiewicz and Baldwin (2020) argue that to address the roots of social injustice and discrimination against SOGIESC people in disasters involves reviewing not only emergency management plans and policies but also wider political and legal systems and processes. As such, Chile's cis-heteropatriarchal system directly influences the country's disaster management policies, resulting in a lack of SOGIESC-sensitive DRR and recovery plans (from 2010 to Chile's more recent response to COVID-19). This procedural injustice has direct impacts on distributive justice. The outcomes of such skewed policies and politics are SGM's disenfranchisement and exclusion from the available resources for disaster recovery that ought to be their right as Chilean citizens. Consequently, lack of recognition and invisibility of SGM within both mainstream politics and disaster management systems increases their inequalities and places them in extremely vulnerable post-disaster situations perpetuating the cycle of inequalities before, during and after disasters.

Conclusion

This study has analysed the impacts of the 2010 Chilean earthquake and tsunami and has developed new insights that contribute to the very sparse literature on SGM and intersectional discriminations in disasters. The analysis of our data shows that in 2010, both the vulnerabilities and capacities of Chilean SGM were systematically excluded from official DRR and recovery strategies. The response to COVID-19 a decade later demonstrates DRR and recovery plans are still underpinned by heteronormative assumptions that focus on the female-male binary. They continue to ignore the needs, vulnerabilities and capacities of SGM. The inclusion of these minorities within the States' plans and policies is urgently required to reduce their suffering and to affirm the equal human rights of one of the most marginalised social groups in disasters. Furthermore, consultations with the State and its support to local organisations working with SGM are encouraged to develop programmes with a stronger focus on building social capital. Such programmes will empower transgender women and gender diverse people and help them self-organise prior to disasters, increasing their resilience and affirming their agency in disasters.

Conflict of Interest The authors have no conflict of interests.

References

Akerkar, S. (2022a) 'Disaster risk reduction and furthering women's rights', in Akerkar, S., *Oxford research encyclopedia of natural hazard science.* Oxford University Press. https://doi.org/10.1093/acrefore/9780199389407.013.406.

Akerkar, S. (2022b). *Gender and older people.* United Nation Department of Economic and Social Affairs (UN DESA). Available at: chrome-extension://efaidnbmnnnibpcajpcglclefindmkaj/https://www.un.org/development/desa/ageing/wp-content/uploads/sites/24/2022/03/Gender-and-Older-People-Supriya-AKERKAR.pdf

Aldrich, D. P. (2010). Fixing recovery: Social capital in post-crisis resilience. *Journal of Homeland Security*, Forthcoming [preprint]. Available at: https://ssrn.com/abstract=1599632

Balgos, B., Gaillard, J. C., & Sanz, K. (2012). The Warias of Indonesia in disaster risk reduction: The case of the 2010 Mt Merapi eruption in Indonesia. *Gender and Development, 20*(2), 337–348. https://doi.org/10.1080/13552074.2012.687218

Carpenter, C. S., Eppink, S. T., & Gonzales, G. (2020). Transgender status, gender identity, and socioeconomic outcomes in the United States. *ILR Review, 73*(3), 573–599. https://doi.org/10.1177/0019793920902776

Crenshaw, K. (1991). Mapping the margins: Intersectionality, identity politics, and violence against women of colour. *Stanford Law Review, 43*(6), 1241. https://doi.org/10.2307/1229039

D'Ooge, C. (2008). Queer Katrina: Gender and sexual orientation matters in the aftermath of the disaster. In B. Willinger (Ed.), *Katrina and the women of New Orleans* (pp. 4–22). Tulane University.

Dominey-Howes, D., Gorman-Murray, A., & McKinnon, S. (2014). Queering disasters: On the need to account for LGBTI experiences in natural disaster contexts. *Gender, Place and Culture, 21*(7), 905–918. https://doi.org/10.1080/0966369X.2013.802673

Dominey-Howes, D., Gorman-Murray, A., & McKinnon, S. (2016). Emergency management response and recovery plans in relation to gender and sexual minorities in New South Wales, Australia. *International Journal of Disaster Risk Reduction, 16*, 1–11. https://doi.org/10.1016/j.ijdrr.2016.02.004

Dominey-Howes, D., Gorman-Murray, A., & McKinnon, S. (2018). On the disaster experiences of sexual and gender (LGBTI) minorities: Insights to support inclusive disaster risk reduction policy and practice. *Australian Journal of Emergency Management, Monograph, #3*, 59–67.

Dovers, S. (2020). Public policy and disaster justice. In C. Baldwin & A. Lukasiewics (Eds.), *Natural hazards and disaster justice: Challenges for Australia and its neighbours* (pp. 51–72). Springer Singapore. https://doi.org/10.1007/978-981-15-0466-2

Dwyer, E., & Woolf, L. (2018). *Down by the river: Addressing the rights, needs and strengths of Fijian gender and sexual minorities in disaster risk reduction and humanitarian response*. Oxfam Australia.

EM-DAT and CRED. (2010). *2010–0091-CHL*. UC Louvain. Accessed July 1, 2021, from www.emdat.be

Fordham, M. (1999). The intersection of gender and social class in disaster: Balancing resilience and vulnerability. *International Journal of Mass Emergencies and Disasters, 17*(1), 15–36. https://doi.org/10.1177/028072709901700102

Fritz, H. M., et al. (2011). Field survey of the 27 February 2010 Chile tsunami. *Pure and Applied Geophysics, 168*(11), 1989–2010. https://doi.org/10.1007/s00024-011-0283-5

Gaillard, J. C. (2011). *People's response to disasters: Vulnerability, capacities and resilience in Philippine context*. Angeles City: Centre for Kapampangan Studies, Holy Angel Univ.

Gaillard, J. C., et al. (2017a). Beyond men and women: A critical perspective on gender and disaster. *Disasters, 41*(3), 429–447. https://doi.org/10.1111/disa.12209

Gaillard, J. C., Gorman-Murray, A., & Fordham, M. (2017b). Gender and sexual minorities in disaster. *Gender, Place and Culture, 24*(1), 18–26. https://doi.org/10.1080/0966369X.2016.1263438

Glynn, T. R., et al. (2016). The role of gender affirmation in psychological well-being among transgender women. *Psychology of Sexual Orientation and Gender Diversity, 3*(3), 336–344. https://doi.org/10.1037/SDG0000171

Gorman-Murray, A., et al. (2014). The LGBTI community in the 2011 Queensland floods: Marginality, vulnerability and resilience. *LES Online, 6*(1), 4–20.

Gorman-Murray, A., et al. (2018). Listening and learning: Giving voice to trans experiences of disasters. *Gender, Place and Culture, 25*(2), 166–187. https://doi.org/10.1080/0966369X.2017.1334632

Halkitis, P. N. (2019). The stonewall riots, the AIDS epidemic, and the public's health. *American Journal of Public Health, 109*(6), 851–852. https://doi.org/10.2105/AJPH.2019.305079

IGLHRC/SEROVie. (2011). *The impact of the earthquake, and relief and recovery programs on Haitian LGBT People*. Available at: http://www.iglhrc.org

IRIN News. (2014). Lost in the chaos—LGBTI people in emergencies. *The New Humanitarian* [Preprint]. Accessed February 21, 2021, from https://www.thenewhumanitarian.org/fr/node/254807

Jebin, L. (2018). Status of transgender people in Bangladesh: A socio-economic-analysis. *South Asian Journal of Policy and Governance Gender Mainstreaming in Development, 42*(1), 50–63.

Knight, K., & A, T. (2020). The most vulnerable people in the world right now. *The Advocate* [Preprint]. Accessed February 21, 2021, from https://www.advocate.com/commentary/2020/5/07/most-vulnerable-people-world-right-now

Koch, J. M., et al. (2020). The cost of being transgender: Where socio-economic status, global health care systems, and gender identity intersect. *Psychology and Sexuality, 11*(1–2), 103–119. https://doi.org/10.1080/19419899.2019.1660705

Kuran, C. H. A., et al. (2020). Vulnerability and vulnerable groups from an intersectionality perspective. *International Journal of Disaster Risk Reduction, 50*, 101826. https://doi.org/10.1016/j.ijdrr.2020.101826

Lovell, E., Twigg, J., & Lung'hai, G. (2019). *Building resilience for all: Intersectional approaches for reducing vulnerability to natural hazards in Nepal and Kenya*. ODI.

Lukasiewicz, A., & Baldwin, C. (Eds.). (2020). *Natural hazards and disaster justice: Challenges for Australia and its neighbours*. Springer Singapore. https://doi.org/10.1007/978-981-15-0466-2

McSherry, A., et al. (2015). From deviant to Bakla, strong to stronger: Mainstreaming gender and sexual minorities into disaster risk reduction in The

Philippines. *Forum for Development Studies, 42*(1), 27–40. https://doi.org/10.1080/08039410.2014.952330

MOVILH. (2010a). *Colectivos LGBT de España y Chile viajan al sur en ayuda de víctimas del terremoto y del tsunami* [Spanish and Chilean LGBT colectives travel to the South to assist earthquake and tsunami survivors], 28 March. Accessed March 27, 2021, from https://www.movilh.cl/colectivos-lgbt-de-espana-y-chile-viajan-al-sur-en-ayuda-de-victimas-del-terremoto-y-del-tsunami/

MOVILH. (2010b). *Datos de utilidad y ayuda tras terremoto, tsunami y réplicas en Chile* [Useful and helpful information for a post-earthquake and tsunami context]. Accessed March 27, 2021, from https://www.movilh.cl/datos-de-utilidad-y-ayuda-tras-terremoto-tsunami-y-replicas-en-chile/

MOVILH. (2011a). *Investigan incendio intencional que destruyó viviendas de transexuales en Talca* [Begins investigation of intentional fire that destroyed transgenders' homes in Talca]. Accessed July 30, 2021, from https://www.movilh.cl/investigan-incendio-intencional-que-destruyo-viviendas-de-transexuales-en-talca/

MOVILH. (2011b). *'IX Informe Anual: Derechos Humanos de la Diversidad Sexual Chilena' (Hechos 2010)*. Movimiento de Integración y Liberación Homosexual.

MOVILH. (2012). *'X Informe Anual: Derechos Humanos de la Diversidad Sexual Chilena' (Hechos 2011)*. Movimiento de Integración y Liberación Homosexual.

Nadal, K. L., Davidoff, K. C., & Fujii-Doe, W. (2014). Transgender women and the sex work industry: Roots in systemic, institutional, and interpersonal discrimination. *Journal of Trauma and Dissociation, 15*(2), 169–183. https://doi.org/10.1080/15299732.2014.867572

Nakagawa, Y., & Shaw, R. (2004). Social capital: A missing link to disaster recovery. *International Journal of Mass Emergencies and Disasters, 22*(1), 5–34.

ONEMI. (2018a). *Plan para la Reducción del Riesgo de Desastres—Región de La Araucanía* [Disaster Risk Reduction Plan—Región of La Araucania]. ONEMI. Accessed July 25, 2021, from https://www.onemi.gov.cl/region/lib-gral-bernardo-ohiggins/

ONEMI. (2018b). *Plan para la Reducción del Riesgo de Desastres—Región de Valparaíso* [Disaster Risk Reduction Plan—Region of Valparaiso]. ONEMI. Accessed July 25, 2021, from https://www.onemi.gov.cl/region/valparaiso/

ONEMI. (2018c). *Plan para la Reducción del Riesgo de Desastres—Región del Biobío* [Disaster Risk Reduction Plan—Region of Biobio]. ONEMI. Accessed July 25, 2021, from https://www.onemi.gov.cl/region/biobio/

ONEMI. (2018d). *Plan para la Reducción del Riesgo de Desastres—Región del Maule* [Disaster Risk Reduction Plan—Region of Maule]. ONEMI. Accessed July 25, 2021, from https://www.onemi.gov.cl/region/maule/

ONEMI. (2020). *Política Nacional Para la Reducción del Riesgo de Desastres. Plan Estratégico Nacional 2020–2030* [National Policy for Disaster Risk

Reduction. 2020–2030 National Strategic Framework]. ONEMI. https://www.onemi.gov.cl/

ONEMI and UNDP. (2017). *Estándares nacionales para la respuesta a emergencias en Chile [National standards for emergency responses in Chile]*. UNDP.

Pincha, C., & Krishna, H. (2008). Aravanis: Voiceless victims of the tsunami. *Humanitarian Practice Network, 41*(1), 41–43.

Rodriguez, D. G., & Suvianita, K. (2020). How Indonesia's LGBT community is making a difference amid *COVID--19. UCL News*, 26 June. Accessed February 21, 2021, from https://www.ucl.ac.uk/news/2020/jun/analysis-how-indonesias-lgbt-community-making-difference-amid-*Covid--19*

Rushton, A., et al. (2019). Beyond binary: (re)defining "gender" for 21st century disaster risk reduction research, policy, and practice. *International Journal of Environmental Research and Public Health, 16*(20), 3984. https://doi.org/10.3390/ijerph16203984

Rushton, A., et al. (2020). The gendered body politic in disaster policy and practice. *International Journal of Disaster Risk Reduction, 47*, 101648. Available at:. https://doi.org/10.1016/j.ijdrr.2020.101648

Schuller. (2015). "Pa Manyen Fanm Nan Konsa": Intersectionality, structural violence, and vulnerability before and after Haiti's earthquake. *Feminist Studies, 41*(1), 184. https://doi.org/10.15767/feministstudies.41.1.184

Seglah, H. A., & Blanchard, K. (2021). LGBTQIA+ people and disasters. *DRR Dynamics*. Available at: https://www.preventionweb.net/publication/lgbtqia-people-and-disasters

Sphere Project. (Ed.). (2018). *The sphere handbook: Humanitarian charter and minimum standards in humanitarian response* (4th ed.). Sphere Association.

Stuart, E., & Woodroffe. J. (2016). 'Leaving no-one behind: can the Sustainable Development Goals succeed where the Millennium Development Goals lacked?'. *Gender and Development, 24*(1), 69–81.

Sweetman, C. (2012). Introduction. *Gender and Development, 20*(3), 389–403. https://doi.org/10.1080/13552074.2012.743266

The National LGBT Health Education Centre. (2016). *Emergency Preparedness and Lesbian, Gay, Bisexual and Transgender (LGBT) People: What Health Centres Need to Know*. Accessed June 4, 2021, from https://www.lgbthealtheducation.org/resources/

The Sphere Project (Ed.). (2011). *Sphere handbook: Humanitarian charter and minimum standards in disaster response, 2011* (3rd ed.). The Sphere Project.

UN. (2020). *Protection against violence and discrimination based on sexual orientation and gender identity. A/73/152*. United Nations General Assembly.

UNISDR. (2005). *Hyogo framework for action 2005–2015: Building the resilience of nations and communities to disasters*. United Nations International Strategy for Disaster Reduction Secretariat.

UNISDR. (2015). *The Sendai framework for disaster risk reduction 2015–2030*. United Nations International Strategy for Disaster Reduction Secretariat.

UNISDR. (2016). *UNISDR strategic framework 2016–2021*. United Nations International Strategy for Disaster Reduction Secretariat.

UNISDR. (2017). Consideration of marginalized and minority groups in a National Disaster Risk Assessment. In *Words into action guidelines: National Disaster Risk Assessment*. United Nations International Strategy for Disaster Reduction Secretariat.

UNISDR, UNDP and IUCN. (2009). *Making disaster risk reduction gender-sensitive. Policy and practical guidelines*. United Nations International Strategy for Disaster Reduction Secretariat.

Vickery, J. (2018). Using an intersectional approach to advance understanding of homeless persons' vulnerability to disaster. *Environmental Sociology, 4*(1), 136–147. https://doi.org/10.1080/23251042.2017.1408549

Woodroffe, J., & Daigle. M. (2017). Intersectionality: reflections from the Gender and Development Network. London: GADN, 1. https://gadnetwork.org/gadn-resources/2017/11/20/intersectionalityreflections-from-the-gender-development-network

Yamashita, A. (2012). *Beyond Invisibility: Great East Japan Disaster and LGBT in Northeast Japan*｜ヒューライツ大阪. FOCUS, 69. Accessed March 13, 2021, from https://www.hurights.or.jp/archives/focus/section2/2012/09/beyond-invisibility-great-east-japan-disaster-and-lgbt-in-northeast-japan.html

CHAPTER 12

Disasters, Intimate Partner Violence, and Sex and Gender-Diverse Communities

Clare E. B. Cannon and Katherine Kovach

Introduction

According to the United States of America (USA) Centers for Disease Control and Prevention, intimate partner violence (IPV) includes physical, emotional, psychological, and economic abuse, and aggression, stalking, or sexual harm, by a current or former partner or spouse. It has reached epidemic proportions, effecting almost 1 in 5 women and 1 and 7 men in the USA alone. Some research suggests IPV prevalence in the USA varies across different socio-demographic characteristics, such as race

C. E. B. Cannon (✉)
Department of Human Ecology, University of California, Davis, Davis, CA, USA

Department of Social Work, University of the Free State, Bloemfontein, South Africa
e-mail: ebcannon@ucdavis.edu

K. Kovach
Department of Social Work, University of the Free State, Bloemfontein, South Africa

D. Dominey-Howes et al. (eds.), *Queering Disasters, Climate Change and Humanitarian Crises*, Sustainable Development Goals Series, https://doi.org/10.1007/978-981-96-3857-4_12

and ethnicity (West, 2012), rural and urban regions, and sex and gender identity (Langhinrichsen-Rohling et al., 2012).[1]

Of the scant research available into IPV prevalence in sex and gender diverse minority (SGM) communities in the USA, studies suggest lesbian, gay, bisexual, transgender, and queer (LGBTQ) populations experience IPV at similar or greater rates than heterosexual populations (Breiding et al., 2014; Smith et al., 2018). SGM communities include, but are not limited to, individuals who identify as lesbian, gay, bisexual, transgender, queer, non-binary, asexual, and/or intersex (U.S. National Institutes of Health, 2023). IPV produces and/or exacerbates a host of psychological symptoms including depression, post-traumatic stress disorder (PTSD), anxiety, suicidal ideation, and substance abuse (Hattery & Smith, 2019), and adverse physical health outcomes, including cardiovascular, reproductive, musculoskeletal, and nervous system conditions.

Research suggests disasters drive an increase in prevalence and severity of IPV in the USA (Schumacher et al., 2010). Anthropogenic climate change increases the frequency and severity of disasters (e.g., hurricanes, flooding, wildfire, etc.) globally. Disasters disrupt millions of lives annually and are accompanied by long-term effects, exacerbating stressors, and worsening a host of physical, behavioural, cognitive, psychological, and emotional symptoms for survivors (Cannon & Kovach, 2023).

These more frequent and severe disasters result in widening inequalities, including health disparities related to IPV for marginalised communities, such as individuals with disabilities, new migrants, and people of colour and, though frequently understudied, SGM communities (Medzhitova et al., 2022). Marginalised SGM communities are chronically understudied and underserved in the USA particularly in relation to disasters and their disparate impacts (U.S. Federal Emergency Management (FEMA), 2022; Goldsmith et al., 2022), and to health disparities such as those resulting from IPV (Cannon & Buttell, 2020).

The increased risk of IPV experienced by SGM following disaster is dependent on context and varies within and between countries. However, the research shows that despite geographic and cultural differences, there are similarities in the kinds of heteronormative discrimination (e.g., homophobia, transphobia) experienced by SGM people and communities, globally (Dominey-Howes et al., 2016; Gorman-Murray et al., 2017).

[1]For in-depth information on socio-demographic variations in USAIPV prevalence rates see reviews by West (2012) and Langhinrichsen-Rohling et al. (2012).

Given the lack of research on the topic in the USA, it is important to draw from studies conducted in other countries that may help inform the development of US research and policy. For example, McKinnon et al. (2017a), in analyses of Australian and New Zealand disasters, argue that the possible exclusion or lack of queer disaster narratives may contribute to gender and sexual diverse people's experiences of marginality during disaster, a finding applicable to the media landscape in the USA.

Moreover, the impacts of systemic discrimination and violence against SGM in many countries, including the USA, include not only disparities in health outcomes compared to national averages but also reduced access to IPV-related supports and services. Systemic discrimination against SGM is likely to negatively impact the provision of services to SGM who have experienced violence in their intimate relationships, including psychological abuse and physical violence; appropriate supports for SGM survivors of IPV, including shelters and other types of supports; and treatments for abusers including group therapy and batterer intervention programmes (Renzetti et al., 2011; Guadalupe-Diaz & Jasinski, 2017). Furthermore, IPV rates are likely to be exacerbated by crises, such as natural hazards, public health emergencies, and economic recessions that produce additional stressors including loss of income and/or housing (Cannon, 2020).

Non-disaster Contexts in the USA

IPV and SGM Communities

Though it is difficult to ascertain IPV prevalence rates in SGM populations due to the lack of empirical research, recent estimates suggest that in the USA, IPV is experienced in SGM relationships at similar or greater rates as in heterosexual relationships (Walters et al., 2013). The National Intimate Partner and Sexual Violence Survey (NISVS), the largest survey of its kind in the USA, indicates that nearly half of all self-identified lesbians (43.8%) have reported instances of physical violence, stalking, or rape by an intimate partner in their lifetime, almost two-thirds of all bisexual women (61.1%), and a little more than a third of heterosexual women (35%) (Walters et al., 2013). Moreover, of those who have experienced IPV, 89.5% of bisexual women and 98.7% of heterosexual women indicated their abusers were male only. Of lesbian women, the majority (67.4%) indicated only female perpetrators (Walters et al., 2013). Approximately a quarter of gay men (26%), a third of bisexual men

(37.3%), and almost a third of heterosexual men (29%) reported they had experienced rape, physical violence, and/or stalking by an intimate partner sometime during their lifetime (Walters et al., 2013). Of men who reported IPV experiences, nine out of ten gay men (90.7%) identified only male abusers, while roughly three out of four bisexual men (78.5%) and almost all heterosexual men (99.5%) indicated only women abusers (Walters et al., 2013).

It is important to note that the NIPSVS is the most comprehensive survey of its kind in the USA, and yet it does not ask any questions on transgender, genderqueer, or non-binary experiences or identity—though the next survey instrument may have more comprehensive data when it is released. Consequently, there is little data from a US national representative sample on prevalence rates of IPV in transgender, non-binary, and genderqueer relationships.

Although the NIPSVS fails to collect data on trans-identified people, the US Transgender Survey, the largest US survey for transgender people, does collect some IPV data. The US Transgender Survey, deployed by the National Centre for Transgender Equality, has 27,715 trans-identified people in its sample from all 50 states of the USA. Approximately one out of every ten respondents reported an experience of sexual assault in the previous year, while almost half (47%) reported having experienced a sexual assault over their lifetime (James et al., 2016). Of those surveyed, 54% have experienced some form of IPV (James et al., 2016). Recent research has found that transgender individuals are at greater risk of exposure to multiple types of violence over their life course than cisgender people, that is, people whose gender identity corresponds with their sex assigned at birth (Messinger, Guadalupe-Diaz, & Kurdyla, 2022. See Cava, 2016 for a definition of cisgender). Transgender people are also likely to experience transphobia-related barriers, including discrimination, to receiving support (Guadalupe-Diaz, 2019; Messinger, Guadalupe-Diaz, & Kurdyla, 2022).

Several external relational stressors, and the interactions among them, are known to exacerbate the use and experiences of IPV (Capaldi et al., 2012). Some of these may be more prevalent among transgender people. For instance, transgender individuals tend to experience higher rates of poverty than the general US population (14%), with almost one-third (29%) of respondents reporting that they live in poverty (James et al., 2016). Transgender individuals tend to experience unemployment rates higher than the national average. At the time of the US Transgender

Survey, respondents reported an employment rate of 15%, which was three times the national average (James et al., 2016). Research demonstrates high poverty levels and high unemployment rates are risk factors underpinning relationship violence (Capaldi et al., 2012).

Transgender people are also at increased risk of homelessness and unemployment due to transphobic discrimination, both of which are risk factors for poorer health and increased risk of IPV. Many transgender people are forced out of their homes because of gender-identity discrimination from family and are at increased risk of unemployment due to workplace discrimination, abuse, and lack of protections (James et al., 2016). They are also more likely to suffer from multiple health disparities and limited or no access to healthcare compared to cisgendered individuals (Meyer et al., 2017). These compounding stressors increase transgender people's risk of IPV. Under disaster conditions these stressors and their interactions are magnified for many transgender people, further increasing their risk of IPV.

Such social and health disparities experienced by SGM communities are likely to be compounded even further by climate change-fuelled disasters, which frequently exacerbate already existing social inequalities (Gorman-Murray et al., 2016, 2017, 2018; McKinnon et al., 2017a). For instance, in their study of queer residents in Queensland during the 2011 floods, Gorman-Murray et al. (2017) found that SGM people and communities experienced discrimination and faced barriers in accessing assistance, which negatively impacted their sense of self-reliance and coping strategies. Such lack of access to assistance and support services for SGM communities is also prevalent in the USA (Goldsmith et al., 2022). Meyer et al. (2017) suggest that reduced access to healthcare may be linked to minority stress. More data (specifically) and more research (generally) are needed to advance our understanding of the health disparities stemming from IPV that SGM individuals experience as they are impacted by disasters in the USA.

Minority Stress Theory, IPV, and SGM Communities

Proponents of minority stress theory argue that stigmatisation of minority groups leads to specific forms of chronic social stress (based on, e.g., gender, sexuality, race, ethnicity, socioeconomic status), which can, in turn, lead to adverse physical and mental health outcomes (Meyer, 2003). Research into minority stress and IPV in SGM communities in times outside of disaster suggests that stressors, such as sexual orientation-related victimisation, sexual minority stigma, internalised homonegativity, and sexual identity concealment, are related to physical, sexual, and psychological IPV among SGM individuals (Edwards & Sylaska, 2013; Edwards et al., 2021).

In their systematic review of the literature, Longobardi and Badenes-Ribera (2017) found there was a relationship between such minority stressors as described above with IPV in same-sex relationships. Similarly, in their study of men who have sex with men in Atlanta, Georgia (USA), Stephenson and Finneran (2017) found that minority stress measures—specifically internalised homophobia, sexuality-based discrimination, and racism—were linked to experiences of IPV. Some research, such as that conducted by Swann et al. (2022), has found that intersectional minority stress—the interlocking axes of social location that leads to minority stress (such as female assigned at birth, sexual minority, person of colour)—led to both perpetration and victimisation of IPV in the USA. Minority stress theory can also help to identify potential protective factors, with research suggesting social and community support, identity, supporting alternative values, and coping may reduce minority stress for SGM individuals experiencing IPV (Lewis et al., 2012). However, more research is required.

While some research applies minority stress theory to IPV in SGM relationships in non-disaster settings, research conducted in Australia suggests that minority stressors can also lead to increased social, political, and economic marginalisation of SGM individuals during disaster with implications for disaster response and recovery (Gorman-Murray et al., 2017). Similarly, research suggests that SGM communities may experience a range of vulnerabilities specific to the marginalisation they experience *as* SGM individuals (McKinnon et al., 2017a). Furthermore, disaster policy settings have frequently excluded SGM diverse people in disaster risk reduction, response, and recovery efforts, including consideration of the impacts of disasters on the risk of IPV (Dominey-Howes et al., 2022).

Given the promising research in applying minority stress theory to IPV in SGM relationships and studies of SGM populations during disaster in other countries, these lines of research can be extended to consider IPV in SGM relationships impacted by disaster in the USA. Given complementary research that has found that disasters exacerbate social stressors, generally, including food and mortgage or rent stress, while reducing resilience and increasing perceived stress and depression (Cannon, Ferreira, Buttell, & Anderson, 2022a, 2022b; Cannon, Ferreira, Buttell, & Jarquin, 2022c), including for survivors of IPV (Buttell et al., 2021; Cannon et al., 2021; Cannon, Ferreira, Buttell, & O'Connor, 2022d), disasters likely exacerbate minority stressors (e.g., sexual orientation-related victimisation, sexual minority stigma, internalised homonegativity) increasing their vulnerability to disaster (Goldsmith et al., 2022).

Given the lack of research into the intersections of minority stress theory, disasters, IPV, and SGM communities, future research should consider developing and deploying measurements of minority stress specific to SGM populations, including those developed in the literature reviewed above, such as sexual orientation-related victimisation, sexual minority stigma, internalised homonegativity, and sexual identity concealment. Such research could take place across different sites in the USA after disasters, using both quantitative methods and case study approaches to tease out disaster impacts related to IPV for SGM communities. Work with an explicit focus on SGM communities could build on prior research such as Schumacher et al.'s (2010) study of the aftereffects of Hurricane Katrina that showed a significant rise in IPV rates for men and women. The Covid-19 pandemic has also provided an opportunity to study the impacts of public health emergency strategies, such as lockdowns, on IPV in SGM communities. Some research has found that IPV prevalence likely increased during the pandemic in the USA, generally (Peitzmeier et al., 2022), and globally. Other research has found such lockdowns had negative implications for SGM individuals' well-being (Grant et al., 2021), mental health (Kidd et al., 2021), and exacerbated stress and depression (Moore et al., 2021).

Disaster Planning and Response in the USA

The Absence of SGM and IPV Communities

Only recently are disaster planning and response experts beginning to acknowledge disaster's disparate impacts on SGM communities and with it the additional support necessary to recover from these impacts in the USA (e.g., USA FEMA, 2022; Frank, 2020). For example, SGM individuals in the USA may not feel comfortable sheltering during or after a disaster (USA FEMA, 2022), and the lack of being able to safely use such resources increases the length of time it takes to recover from disaster (Dominey-Howes et al., 2016; McKinnon et al., 2017a, 2017b; Dominey-Howes et al., 2022). Some research has found that the ways in which disaster response has been outsourced to organisations, for instance to Christian-based groups, has led to SGM people not being treated equally in some places (Dominey-Howes et al., 2016).

For the most part, in the USA, members from SGM communities have not been included in disaster planning or preparedness policy (Jones, 2019). Preparedness is understood as collecting a set of supplies and/or creating a plan that can be deployed when a disaster is either imminent or encountered (Kohn et al., 2012). Although there is relatively more research into the relationships between disaster and IPV, there has been little progress in the way of preparedness and policy that seeks to prioritise and address IPV during and after disaster in the USA. At the same time, much remains unknown about key dynamics between disaster, IPV, and SGM communities beyond which these communities are likely to be especially hard hit by disaster (Goldsmith et al., 2022). Because of the dearth of research on these topics in the USA, there are few recommendations for how to support these communities during and after disaster. To fill this gap, we draw on research reviewed above along with research done in other parts of the world described below (Dominey-Howes et al., 2016; McKinnon et al., 2017a) and by adapting disaster preparedness best practices for other socially vulnerable groups. Taken together, we offer four recommendations for addressing IPV in SGM communities during and after disaster in the USA.

Researchers, practitioners, and policymakers have been working extensively in other countries to study and produce guidelines and tools to support people affected by IPV during disaster. There may be lessons from this work that are useful for developing SGM-specific recommendations for the US context. For instance, Australia has a national coalition that has developed guidelines for researching and addressing gendered dimensions of disaster impacts (see https://www.genderanddisaster.com.au/). This coalition has created trainings for practitioners working in disaster areas and developed national Gender and Emergency guidelines to assist them in their work. These efforts could be a model for developing a partnership between local, state, and federal disaster management agencies and US-based non-profits, such as the Human Rights Campaign and local LGBT community centres, to cultivate a network of organisations that can respond to the specific needs of the LGBT community after disaster. This network could also develop similar guidelines for the USA to help inform local, state, and federal disaster preparedness, planning, response, and recovery efforts.

Recommendations

The following are recommendations aimed at improving the support provided to SGM IPV survivors during disasters by government and non-government agencies involved in disaster planning and preparedness in the USA. Research into the increased prevalence of IPV due to disaster suggests that supporting SGM individuals during and after disaster will likely reduce the likelihood and prevalence of IPV in these communities. Applying minority stress theory to IPV in SGM communities during and after disaster also suggests that there are unique and specific stressors that SGM individuals face as members of SGM communities and that, as such, there needs to be disaster planning and preparedness that addresses these stressors.

It is necessary to note that in the USA, the systems that pose a threat to SGM communities—legal, political, and cultural—are the same ones that contribute to minority stress in the first place. Threats to safety, well-being, and health vary across local areas, regions, and states in the USA. For example, there has been a slate of anti-LGBT legislation across the USA, including bills that seek to restrict medical care for trans youth and adults, impede bathroom access that corresponds with one's gender, and censorship of gender and sexual identity curriculum from K-12 and higher education settings (American Civil Liberties Union (ACLU), 2024). Moreover, there is no guaranteed protection from discrimination afforded to LGBT peoples in the US Constitution or in federal law. Such attacks and threats also constrain the effectiveness and implementation of the proposed recommendations. The below recommendations may work in some places in the USA, such as San Francisco or New York City, that have stronger protections for SGM individuals and their communities but may not be feasible in other places, such as Florida, where there has been a sustained and ongoing assault on SGM communities (ACLU, 2024). These recommendations are intended to move the needle on IPV in SGM communities during disaster and acknowledge the current context in which SGM communities are frequently and routinely discriminated against.

Given the sexuality-based discrimination many SGM individuals face, they may be uncomfortable or unsafe utilising shelter services during and after a disaster (U.S. FEMA, 2022). Because of such minority stressors, disaster assistance must implement safety guidelines and clearly communicate them to keep SGM individuals safe when utilising shelter services. A model the USA could follow is the Australian National Gender and Emergency Management Guidelines (Parkinson et al., 2018). These guidelines are organised into three primary areas of 'gender equity, gender-sensitive communication, and domestic violence' (Parkinson et al., 2018, p. 28). These guidelines also provide a checklist for emergency personnel to assess the inclusiveness of their work, which can improve the national understanding of gendered issues of disaster.

At the same time, there are some shared commonalities across vulnerable communities, which have conducted more extensive disaster planning and policy research (Renne, 2018; Nick et al., 2009), which can be adapted to SGM communities. One such lesson is to include members of a vulnerable group in disaster planning and policy development and implementation. Similarly, cities and counties, as well as federal agencies, should consider including SGM individuals in disaster planning and response (Dominey-Howes et al., 2018; Dominey-Howes et al., 2022; Haworth et al., 2022). For instance, Haworth et al. (2022) argue for increased conceptual and methodological focus on pathways for inclusive knowledge production as one way to improve disaster responses for SGM communities. Having a member or members of SGM communities included in developing and crafting disaster planning procedures and responses can help incorporate the specific perspectives, needs, and challenges they may face during and after disaster. This recommendation may only work in some places where there are greater protections for SGM residents and where SGM residents feel safe enough to participate.

Another recommendation is for disaster-focused agencies to connect to SGM communities through already existing infrastructure, including Lesbian Gay Bisexual Trans (LGBT) centres, and local healthcare centres that specialise in LGBT+ health. For instance, some research has found that SGM communities demonstrate a range of vulnerabilities along with a wide range of resilient strategies (Dominey-Howes et al., 2022). Involving inclusive organisations such as LGBT centres in disaster management both can help inform disaster planning and response with insights from population who are directly impacted and can help to communicate disaster planning and response efforts to SGM communities. Research into IPV in SGM communities has also suggested connecting with LGBT centres to help inform IPV services and treatment for survivors and perpetrators in non-disaster contexts (Cannon, 2019; Cannon & Buttell, 2020).

A final recommendation is for disaster planners and response agencies to reach out to local and national IPV centres and groups, including the

USA. National Domestic Violence Hotline and the US National Resource Centre on Domestic Violence to work together to incorporate IPV awareness into disaster response as well as to connect with IPV services and treatment programmes for survivors and perpetrators across the USA.

Conclusions

As the frequency and severity of disasters increase due to anthropogenic climate change so too will the prevalence and severity of IPV. Both disaster and IPV create widening social and health disparities, particularly for SGM communities. Here, we have summarised the nascent research into disasters, IPV, and SGM communities to identify key recommendations for disaster planning and response in the USA. To develop these recommendations, we reviewed scholarship at the intersections of IPV, SGM identity, and minority stress theory in the USA and abroad. We applied insights from the review to develop these recommendations given the dearth of research into IPV, SGM populations, and disaster. Doing so yields fruitful insights into the unique stressors SGM individuals face from disaster, such as sexual discrimination in shelters, in the USA (U.S. FEMA, 2022). Finally, we provide four recommendations summarised here to improve disaster planning and response for SGM communities in the USA.

These recommendations may not be feasible or possible in all places across the USA given the cultural, legal, and political threats SGM individuals face and the varying degrees of protection they are afforded in different places across the USA. Nonetheless, they are intended to build and support the conversation around SGM and IPV during and after disaster.

1. Disaster assistance must implement safety guidelines and clearly communicate them to keep SGM individuals safe when utilising shelter services and other disaster-related emergency services.
2. Disaster-focused agencies at local, state, and federal levels should include SGM individuals in their disaster planning and response efforts, where it is safe for SGM individuals to meaningfully participate.
3. Disaster-focused agencies should connect with SGM communities through already existing infrastructure, including Lesbian Gay Bisexual Trans (LGBT) centres and local healthcare centres that specialise in LGBT+ health.

4. Disaster planners and response agencies must reach out to local and national IPV centres and groups, including the National Domestic Violence Hotline and the National Resource Centre on Domestic Violence to work together to incorporate IPV awareness into disaster response and to connect with IPV services and treatment programmes for survivors and perpetrators.

By employing these recommendations and by conducting future research directions suggested in this chapter, we hope that the social and health disparities experienced by SGM communities exacerbated by disaster can be reduced and eliminated.

Conflict of Interest The authors have no conflict of interests.

References

American Civil Liberties Union ACLU. (2024). *Mapping attacks on LGBTQ rights in U.S. State Legislatures in 2024.* Retrieved March 1, 2024, from https://www.aclu.org/legislative-attacks-on-lgbtq-rights-2024

Breiding, M. J., Chen, J., & Black, M. C. (2014). *Intimate partner violence in the United States—2010.* National Centre for Injury Prevention and Control, Centres for Disease Control and Prevention.

Buttell, F., Cannon, C. E., Rose, K., & Ferreira, R. J. (2021). *COVID-19* and intimate partner violence: Prevalence of resilience and perceived stress during a pandemic. *Traumatology, 27*(1), 20–28. https://doi.org/10.1037/trm0000296

Cannon, C. E. B. (2019). What services exist for LGBTQ perpetrators of IPV in batterer intervention programs across North America? A qualitative study. *Partner Abuse, 10*(2), 222–242. https://doi.org/10.1891/1946-6560.10.2.222

Cannon, C. E. B. (2020). Learning what you need: Modifying treatment programs for LGBTQ perpetrators of IPV. In B. Russell (Ed.), *Intimate partner violence and the LGBT+ community* (pp. 177–193). Springer. https://www.springer.com/us/book/9783030447618

Cannon, C., & Buttell, F. (2020). Supported recommendations for treating LGBTQ perpetrators of IPV: Implications for policy and practice. *Partner Abuse, 11*(4), 485–504.

Cannon, C. E., & Kovach, K. (2023). Intimate partner violence and disasters: A review of the literature. *Traumatology.* Advance online publication. https://doi.org/10.1037/trm0000491

Cannon, C. E., Ferreira, R., Buttell, F., & First, J. (2021). *COVID-19*, intimate partner violence, and communication ecologies. *American Behavioral Scientist, 65*(7), 992–1013.

Cannon, C. E., Ferreira, R., Buttell, F., & Anderson, C. (2022a). Changes in depression levels for US rural communities before and after the start of the *COVID-19* pandemic. *Traumatology, 28*(3), 411–414. https://doi.org/10.1037/trm0000403

Cannon, C. E., Ferreira, R., Buttell, F., & Anderson, C. (2022b). Sociodemographic predictors of depression in US rural communities during *COVID-19*: Implications for improving mental healthcare access to increase disaster preparedness. *Disaster Medicine and Public Health Preparedness, 17*(e-208), 1–4. https://doi.org/10.1017/dmp.2022.203

Cannon, C. E., Ferreira, R., Buttell, F., & Jarquin, T. (2022c). Assessing explanatory variables of perceived stress to disaster: Implications for risk research. *Journal of Risk Research, 25*(11–12), 1259–1271.

Cannon, C. E., Ferreira, R., Buttell, F., & O'Connor, A. (2022d). Impacts of the *COVID-19* pandemic on rural survivors of intimate partner violence. *Psychological trauma, 15*(3), 511–514. https://doi.org/10.1037/tra0001244

Capaldi, D. M., Knoble, N. B., Shortt, J. W., & Kim, H. K. (2012). A systematic review of risk factors for intimate partner violence. *Partner Abuse, 3*(2), 231–280. https://doi.org/10.1891/1946-6560.3.2.231

Cava, P. (2016). Cisgender and cissexual. In N. A. Naples (Ed.), *The Wiley Blackwell encyclopedia of gender and sexuality studies* (pp. 1–4). https://doi.org/10.1002/9781118663219

Dominey-Howes, D., Gorman-Murray, A., McKinnon, S., Itaoui, R., & Keppel, J. (2016). Emergency management response and recovery plans in relation to gender and sexual minorities in New South Wales, Australia. *International Journal of Disaster Risk Reduction, 16*, 1–11.

Dominey-Howes, D., Gorman-Murray, A. & McKinnon, S. (2018). On the disaster experiences of sexual and gender (LGBTI) minorities: Insights to support inclusive disaster risk reduction policy and practice. *Australian Journal of Emergency Management, Monograph, 3*, 59–67.

Dominey-Howes, D., McKinnon, S., Gorman-Murry, A., & Eriksen, C. (2022). Gender and sexual minorities in disasters. In T. McGee & E. Penning-Rowsell (Eds.), *Routledge handbook of environment and society* (pp. 259–272). Routledge.

Edwards, K. M., & Sylaska, K. M. (2013). The perpetration of intimate partner violence among LGBTQ college youth: The role of minority stress. *Journal of Youth and Adolescence, 42*(11), 1721–1731.

Edwards, K. M., Siller, L., Littleton, H., Wheeler, L., Chen, D., Sall, K., & Lim, S. (2021). Minority stress and sexual partner violence victimization and perpetration among LGBQ+ college students: The moderating roles of hazardous drinking and social support. *Psychology of Violence, 11*(5), 445–454. https://doi.org/10.1037/vio0000394

Frank, T. (2020). LGBTQ people are at higher risk in disasters. *Scientific American*. https://www.scientificamerican.com/article/lgbtq-people-are-at-higher-risk-in-disasters/

Goldsmith, L., Raditz, V., & Méndez, M. (2022). Queer and present danger: Understanding the disparate impacts of disasters on LGBTQ+ communities. *Disasters, 46*(4), 946–973.

Gorman-Murray, A., McKinnon, S., & Dominey-Howes, D. (2016). Masculinity, sexuality and disaster: Unpacking gendered LGBT experiences in the 2011 Brisbane floods in Queensland, Australia. In E. Enarson & B. Pease (Eds.), *Men, masculinities and disaster* (pp. 128–139). Routledge.

Gorman-Murray, A., Morris, S., Keppel, J., McKinnon, S., & Dominey-Howes, D. (2017). Problems and possibilities on the margins: LGBT experiences in the 2011 Queensland floods. *Gender, Place and Culture, 24*(1), 37–51.

Gorman-Murray, A., McKinnon, S., Dominey-Howes, D., Nash, C., & Bolton, B. (2018). Listening and learning: Giving voice to Trans experiences in disasters. *Gender, Place and Culture, 25*(2), 166–187.

Grant, R., Gorman-Murray, A., & Briohny Walker, B. (2021). The spatial impacts of *COVID-19* restrictions on LGBTIQ wellbeing, visibility, and belonging in Tasmania, Australia. *Journal of Homosexuality, 68*(4), 647–662.

Guadalupe-Diaz, X. L. (2019). *Transgressed: Intimate partner violence in transgender lives*. NYU Press.

Guadalupe-Diaz, X. L., & Jasinski, J. (2017). "I wasn't a priority, I wasn't a victim:" Challenges in help seeking for transgender survivors of intimate partner violence. *Violence against Women, 23*(6), 772–792.

Hattery, A. J., & Smith, E. (2019). *Gender, power, and violence: Responding to sexual and intimate partner violence in society today*. Rowman and Littlefield.

Haworth, B. T., McKinnon, S., & Eriksen, C. (2022). Advancing disaster geographies: From marginalisation to inclusion of gender and sexual minorities. *Geography Compass, 16*(11), e12664.

James, S. E., Herman, J. L., Rankin, S., Keisling, M., Mottet, L., & Anafi, M. (2016). *The report of the 2015 U.S. Transgender Survey*. National Centre for Transgender Equality.

Jones, L. (2019). How to include the LGBT community in disaster preparedness. *Homeland Security Today. US*, 30 May. https://www.hstoday.us/subject-matter-areas/emergency-preparedness/how-to-include-the-lgbt-community-in-disaster-preparedness/

Kidd, J. D., et al. (2021). Understanding the impact of the *COVID-19* pandemic on the mental health of transgender and gender nonbinary individuals engaged in a longitudinal cohort study. *Journal of Homosexuality, 68*(4), 592–611.

Kohn, S., Eaton, J. L., Feroz, S., Bainbridge, A. A., Hoolachan, J., & Barnett, D. J. (2012). Personal disaster preparedness: An integrative review of the literature. *Disaster Medicine and Public Health Preparedness, 6*(3), 217–231.

Langhinrichsen-Rohling, J., Misra, T. A., Selwyn, C., & Rohling, M. L. (2012). Rates of bi-directional versus unidirectional intimate partner violence across samples, sexual orientations, and race/ethnicities: A comprehensive review. *Partner Abuse, 3*(2), 199–230. https://doi.org/10.1891/1946-6560.3.2.199

Lewis, R. J., Milletich, R. J., Kelley, M. L., & Woody, A. (2012). Minority stress, substance use, and intimate partner violence among sexual minority women. *Aggression and Violent Behavior, 17*(3), 247–256.

Longobardi, C., & Badenes-Ribera, L. (2017). Intimate partner violence in same-sex relationships and the role of sexual minority stressors: A systematic review of the past 10 years. *Journal of Child and Family Studies, 26*(8), 2039–2049.

McKinnon, S., Gorman-Murray, A., & Dominey-Howes, D. (2017a). Disasters, queer narratives and the news: How are LGBTI disaster experiences reported by the mainstream and LGBTI media? *The Journal of Homosexuality, 64*(1), 122–144.

McKinnon, S., Gorman-Murray, A., & Dominey-Howes, D. (2017b). Remembering an epidemic during a disaster: Memories of HIV/AIDS, gay identities and experiences of disaster. *Gender, Place and Culture, 24*(1), 52–63.

Medzhitova, Y., Lai, B. S., Killenberg, P., Riobueno-Naylor, A., & Goodman, L. A. (2022). Risk factors for intimate partner violence in the context of disasters: A systematic review. *Trauma, Violence, and Abuse, 24*(4), 2265–2281. https://doi.org/10.1177/15248380221093688

Messinger, A. M., Guadalupe-Diaz, X. L., & Kurdyla, V. (2022). Transgender polyvictimization in the U.S. transgender survey. *Journal of Interpersonal Violence, 37*(19–20), NP18810–NP18836. https://doi.org/10.1177/08862605211039250

Meyer, I. H. (2003). Prejudice, social stress, and mental health in lesbian, gay, and bisexual populations: Conceptual issues and research evidence. *Psychological Bulletin, 129*(5), 674–697. https://doi.org/10.1037/0033-2909.129.5.674

Meyer, I. H., Brown, T. N., Herman, J. L., Reisner, S. L., & Bockting, W. O. (2017). Demographic characteristics and health status of transgender adults in select US regions: Behavioral risk factor surveillance system, 2014. *American Journal of Public Health, 107*(4), 582–589. https://doi.org/10.2105/AJPH.2016.303648

Moore, S. E., Wierenga, K. L., Prince, D. M., Gillani, B., & Mintz, L. J. (2021). Disproportionate impact of the *COVID-19* pandemic on perceived social support, mental health and somatic symptoms in sexual and gender minority populations. *Journal of Homosexuality, 68*(4), 577–591.

Nick, G. A., et al. (2009). Emergency preparedness for vulnerable populations: People with special health-care needs. *Public Health Reports, 124*(2), 338.

Parkinson, D., Duncan, A., Archer, F., Dominey-Howes, D., Gorman-Murray, A., & McKinnon, S. (2018). Introducing new national Gender and Emergency Management (GEM) Guidelines to support more inclusive disaster risk reduction work. *Australian Journal of Emergency Management, Monograph #3*, 27–29.

Peitzmeier, S. M., Fedina, L., Ashwell, L., Herrenkohl, T. I., & Tolman, R. (2022). Increases in intimate partner violence during *COVID-19*: Prevalence and correlates. *Journal of interpersonal violence, 37*(21–22), NP20482–NP20512.

Renne, J. L. (2018). Emergency evacuation planning policy for carless and vulnerable populations in the United States and United Kingdom. *International Journal of Disaster Risk Reduction, 31*, 1254–1261.

Renzetti, C. M., Edleson, J. L., & Bergen, R. K. (Eds.). (2011). *Companion reader on violence against women*. Sage Publications.

Schumacher, J. A., Coffey, S. F., s, F. H., Tracy, M., Clements, K., & Galea, S. (2010). Intimate partner violence and Hurricane Katrina: Predictors and associated mental health outcomes. *Violence and Victims, 25*(5), 588–603.

Smith, S. G., Zhang, X., Basile, K. C., Merrick, M. T., Wang, J., Kresnow, M., & Chen, J. (2018). *The National Intimate Partner and Sexual Violence Survey (NISVS): 2015 data brief—Updated release*. National Centre for Injury Prevention and Control, Centres for Disease Control and Prevention.

Stephenson, R., & Finneran, C. (2017). Minority stress and intimate partner violence among gay and bisexual men in Atlanta. *American Journal of Men's Health, 11*(4), 952–961.

Swann, G., Dyar, C., Baidoo, L., Crosby, S., Newcomb, M. E., & Whitton, S. W. (2022). Intersectional minority stress and intimate partner violence: The effects of enacted stigma on racial minority youth assigned female at birth. *Archives of Sexual Behavior, 51*(2), 1031–1043.

U.S. Federal Emergency Management Agency. (2022). *National preparedness report*. https://www.fema.gov/sites/default/files/documents/fema_2022-npr.pdf

U.S. National Institutes of Health. (2023). *Sexual and Gender Minority Research Office*. https://dpcpsi.nih.gov/sgmro

Walters, M. L., Chen, J., & Breiding, M. J. (2013). *The National Intimate Partner and Sexual Violence Survey (NISVS): 2010 findings on victimization by sexual orientation*. National Centre for Injury Prevention and Control, Centres for Disease Control and Prevention.

West, C. M. (2012). Partner abuse in ethnic minority and gay, lesbian, bisexual, and transgender populations. *Partner Abuse, 3*(3), 336–357.

CHAPTER 13

Nowhere to Turn: Everyday Emergency Response Failures

Marcilyn Cianfarani

I had been working as a paramedic in one of Canada's largest urban centres for close to ten years. During those years, the systemic issues of sexism, misogyny and homophobia I experienced weighed heavily on me and my emotional well-being. I knew I needed a new career. In 2011 I applied to grad school and in 2013 obtained a Master of Arts (MA) in Disaster and Emergency Management (DEM). As part of my MA I investigated the vulnerabilities, needs and capacities of gender and sexual diverse people during disasters and emergencies. I wanted to understand the experiences of gender and sexual diverse people during different disruptions that had occurred such as pandemics (SARS and HIV), power disruptions (2003 northeastern blackout), weather emergencies (heat waves and ice storms) and mass shootings/police action (G20 demonstration), to name a few.

The author now resides in Ontario.

M. Cianfarani (✉)
Disaster and Emergency Management Practitioner, Ontario, Canada

D. Dominey-Howes et al. (eds.), *Queering Disasters, Climate Change and Humanitarian Crises*, Sustainable Development Goals Series, https://doi.org/10.1007/978-981-96-3857-4_13

In total, I have spent 20 years working alongside police, fire and other emergency response services on various provincial and federal disaster response teams, on specialised Emergency Medical Service (EMS) units and finally as a disaster and emergency management (DEM) consultant. As a DEM consultant I was responsible for leading numerous preparedness, response and recovery initiatives for mass casualty incidents, CBRNE (chemical, biological, radiological, nuclear and explosive) events, infectious diseases and mass gatherings for the province of Ontario's health system.

Earlier in my career when I was working as a paramedic, I witnessed verbal homophobic and transphobic attacks on queer and trans patients by my co-workers. I remember waiting in triage at a fairly busy hospital listening to a co-worker humiliate a patient—a trans woman—by referring to her as 'it' and joking about her gender identity while she lay semi-conscious on an ambulance stretcher. My attempt to advocate for trans rights in that moment was laughed away by my co-workers.

Years later while conducting research for my MA, a trans respondent shared their anxieties about needing an ambulance fearing that 'if they discover I'm not cisgender, they'll mistreat me or neglect me while I'm unconscious or otherwise unable to help myself' (Cianfarani, 2013). I came out four years before I was hired by that paramedic service and felt like I needed to crawl back into the closet. It took me about three years on the job before I felt comfortable telling my co-workers I was queer despite my perceived non-binary identity. While I do not identify as non-binary, my gender identity is often misread as non-binary or trans and consequently I am frequently treated poorly.

My experiences working for first responder agencies were rarely positive—systems rife with sexism, misogyny, racism, ableism and homo/transphobia—cultures fuelled by abuses of power. I had taken the position as the women's representative at paramedic service I worked for—a newly formed and much-needed position—with the hope of advocating for ongoing workplace gender equality at the workplace. During my first week in the role, I was bombarded with email messages by male colleagues, who could not understand the reason for affinity groups and felt they needed a 'men's group' for fear of 'reverse sexism'. Being a woman in a male-dominated, militant organisation was challenging. When intersecting identities (i.e. gender identity, race, disability) are added to the mix, these environments become increasingly unsafe. The glass ceiling was thick.

Despite other queer co-workers outing themselves to me over the years, many of them never came out to others, and some left to find careers that felt less threatening. It made sense that years into my career I would gravitate towards a research study that focussed on understanding the needs, vulnerabilities and capacities of gender and sexual diverse people and the systems that support or fail them during disasters and emergencies.

From my research I learned that gay and lesbian people express lower levels of satisfaction with police performance (i.e. treating people fairly) than do their heterosexual counterparts (Beauchamp, 2004). As part of my research, I asked the question, 'What services (police, fire, EMS, etc.) would be safe and welcoming for you as a 2SLGBTQIA[1] person during a disaster or emergency?' The responses, while not surprising, were heartbreaking.

Most respondents mentioned fear of police stigma as a barrier to accessing emergency services during a disaster or emergency. One respondent reported, "I'm a gay sex worker, Police Services are totally off my list,"while another said, "I avoid interactions with the police," and another suggested,"'I'm not actually comfortable dealing with them [the police]." Past negative experiences with the local police service also influenced respondents negatively. A genderqueer individual reported having a "profound distrust of police" because of their "historical treatment of LGBTQ peopl"' and reflected on a time when they called for help and were "harassed by the police". Another respondent reported, "I do not trust the police not to single out queers for harsher treatment. During the G20 they isolated queers in detention and treated us much worse than most other people." Protests actually evolved in 2010 following the unlawful arrest and detainment of queer people during the G20 Summit in Toronto, Canada (McCann, 2010). Lisa Walters, a lesbian journalist reporting on the G20 and one of many people unlawfully arrested, was the victim of homophobic insults, verbal abuse and public humiliation by a male police officer (McCann, 2010; McNeilly, 2012).

It's now been ten years since I completed my Master's and very little, if anything, has changed. In fact, according to recent media headlines, it appears that hate crimes against gender and sexual diverse people, people

[1]At the time of my research the acronym 2SLGBTQIA was commonly used in Canada to represent sexual and gender diverse populations and stands for Two-spirit, Lesbian, Gay, Bisexual, Trans, Queer, Intersex and Asexual. Other variations of this acronym exist including LGBT, LGBTQ and LGBTQIA+.

of colour, and other oppressed populations have been increasing in Canada and that marginalised people continue to face discrimination and harassment (Jain, 2022; Egale Canada, 2023; Statistics Canada, 2023). Fae Johnstone, a trans rights activist, states, "We're seeing more hate messages on social media. We're seeing more demonization of our communities and is *[sic]* a genuine fear, and it's a reasonable fear, that we're going to regress in Canada" (MacSweeney, 2023). Here, in Atlantic Canada, my partner and I are already feeling this regression.

In 2019, my partner and I purchased a home in rural Cape Breton, Nova Scotia. In early 2020 we finally made it to our new home after a gruelling journey during the early stages of Covid-19 when the country was in lockdown. It was a big move and one I was so hopeful and excited about. I grew up in a small rural hamlet in southwestern Ontario and was eager to return to rural living. What was new for me was moving to a rural community at the height of what became the biggest right-wing extremist uprising in the country stemming from Covid-19 vaccine mandates in 2020 and burgeoning into the far-right Freedom Convoy Occupation that immobilised the city of Ottawa, Ontario, in 2022. The threats to the security of Canada during the Freedom Convoy Occupation were so serious that the federal government declared a public order emergency on February 14, 2022 (GOC, 2022). According to the report from the Ottawa People's Commission (OPC) on the Convoy Occupation, the people of downtown Ottawa were abandoned by police and that abdicating responsibility was nothing less than a human rights failure (OPC, 2023).

The far-right extremist group Diagolon, formed in Nova Scotia, became increasingly prevalent during the occupation. My partner and I found ourselves living among a local community of what seemed to be right-wing extremists who flew their 'Save Canada' and 'F*CK Trudeau'[2] flags from the back of their trucks and off-road vehicles while driving by our home. We flew our rainbow flag with pride despite our nervousness. Media articles about hate crimes against gender and sexual diverse people in Nova Scotia were becoming more frequent and in 2022/2023 it was not uncommon to read about these crimes taking place in different Nova

[2] The 'Save Canada' and 'F*CK Trudeau' flags were used initially to symbolise far/alt right extremist anger towards the Covid-19 vaccine mandates in Canada. These flags replaced swastika and confederate flags that demonstrators were using in the early days of the Freedom Convoy Occupation. Following the occupation the flags remained a far-right symbol of hate.

Scotia communities on a regular basis (Munoz, 2022; Armstrong, 2023; Roach, 2023; Sampson, 2023).

Shortly after we moved to Cape Breton, we started being harassed. We do not have direct neighbours and so could not understand the occasional drive-by when people we had never met would scream obscenities out the window or make obscene gestures at us. When the harassment increased to daily occurrences, we started to understand that we had become the target of extremists. Stephanie Carvin, a professor of International Relations at Carleton University, describes these acts of spontaneous violence as 'berserking'—a practice where far-right extremists get inebriated and target 2SLGBTQ people or their property (Smith, 2023).

We started collecting evidence—photos and videos that we would eventually submit to our local police service. We invested in a security system because of instances of trespassing that were occurring on our property, and we photographed people hiding in the bushes outside our home filming us while we did random tasks around the yard. We have videos of our harassers making obscene gestures while driving by on their off-road vehicles, and others of them doing donuts near the top of our driveway, causing damage to our property including uprooting vegetation and removing gravel from our driveway. We are constantly picking up cigarette butts that they throw intentionally on our property—a major risk in a province that is susceptible to wildfires. Through writing this vignette, I learned that 29 right-wing extremist groups were active in Atlantic Canada between 2000 and 2019 with the majority of extremist activity (property crime, harassment, hate speech and violent crimes) taking place in Nova Scotia and that "rural communities tend to be natural environments for the growth of far-right extremist ideas"(Hofmann et al., 2021).

We stopped flying our pride flag. We knew eventually it would also become a target.

It did not seem to matter that my partner was born and raised in Cape Breton. We thought we might have some small-town protections seeing as her family and some friends still live here. What we have found instead is a culture of fear—a culture where no one feels comfortable speaking out for fear of retribution or abuse. A culture where, in our experience, the police protect right-wing extremists instead of the people the extremists are terrorising.

In the summer of 2022, we were assaulted in front of our property. We called the police to report the incident and for protection. We provided video and photo evidence of the assailant. The officers that arrived on scene barely spoke with us and they did not seem to take any notes (which I questioned them about while on scene). The officers clearly knew the assailant and we watched while they talked buddy-buddy with him. We were told by one officer that the assailant fixes her car and that he has a "bad tempe"' but "isn't going to do anything else". I promptly reminded the officer that he had just assaulted us. When the assailant's friends surprisingly showed up on scene, the officers prioritised speaking with them over taking our statements. We would later find out these friends were part of the group of locals that had been harassing us.

We learned from the police that our assaulter and his friends were upset that there were rocks on the road and that they were concerned about their old run-down trucks being damaged because of them. We live on a dirt road. It's not uncommon for provincial work trucks to scatter rocks along the road when performing regular road maintenance like grading and ditching. Instead of prioritising the assault, the officers told us we should pick the rocks up so that"'your neighbours think you take care of your property".

We have not picked up those rocks.

These remarks made by the police officers were dehumanising, belittling and tried to minimise our attempts at being good neighbours and to ultimately try to turn attention away from the assault. Instead of addressing the fact that we had been assaulted, they acted to destabilise our sense of self, home and safety.

Initially, I was baffled at the response by the police. Why weren't they taking notes or documenting the situation? Why were they ignoring us? Why were they telling us to pick up rocks on the road? Why were they not doing their job? And then I remembered my research.

The police refused to press charges.

Because the police actively decided against charging the person who assaulted us, the only avenue for protection left to my partner and I was to go through the provincial court system to obtain a court order against the assailant. This court order requires police enforcement by the same officers who refused to lay charges in the first place. We did not feel much farther ahead and the harassment continued. We have asked the police to investigate the criminal harassment and they have refused. We have written letters to provincial and federal Members of the Legislative Assembly (MLAs) and our local council asking for support and to address these issues at meetings. We have spoken with our local health care provider as well as a retired police officer for advice. We've had to ask friends to act as witnesses for us in court, and if we can temporarily stay at their homes for safety.

We are terrified every single day. We call a friend who lives around the corner instead of the police when we need protection and we have contacted the local rainbow alliance for support. Hesitatingly, I have applied to be on the police services board for our local police detachment so I can advocate for community safety. We have shared our story with the media who will not publish it because we have asked for anonymity, because we are afraid. And now, we feel we are being forced to leave our beautiful home—a move that isn't economical or financially ideal at all—but we have little choice—if we stay we might very well be killed.

It did not matter that the Mass Casualty Commission was in the middle of reviewing the police response to the deadliest killing spree in Canadian history and deemed the police response to the attacks in Portapique, Nova Scotia, a failure. The Commission has subsequently called for an overhaul of policing in Canada (MCC, 2023). It did not seem to matter to these small-town police officers that the umbrella organisation they work for was under the microscope. It seems they felt protected by their uniform and a culture of impunity that has created a society where cops feel like they will not face consequences for their actions (Hickey, 2023).

As a medic I supported and cared for patients who were struggling with everyday disasters like bullying, harassment and violence, often because of their gender/gender identity, class, race or religion. When I started experiencing daily occurrences of harassment, it reminded me that, for many marginalised people, there is no one to call for help during everyday disasters. I wanted to understand how these instances of 'everydayisms' (risks from everyday hazards and vulnerabilities) fit into disaster and emergency management planning. It is the very same institutions, responders and systems that are tasked with supporting gender and sexual diverse people on a daily basis that will be needed during disasters and emergencies.

Queer communities have historically had to devise their own support systems following crises, like the AIDS pandemic, when mainstream services failed to step up. One participant from my research study recalled:

> education, outreach, homecare, and hospices, free condoms and needle exchanges were all community initiatives from the get-go. The mainstream media/public called it a gay disease, what choice did we have?

In many rural locations, queer people and communities exist quietly and sometimes secretively, and my partner and I have had a hard time finding queer allies. We remain thankful to our heterosexual feminist friends for taking us under their wing—these new friendships are a nod to the silver linings of disaster research that Dr. Shelley Pacholok and I started investigating years ago (UBC Media Release, 2013).

I have learned that the *University College London (UCL) Institute for Risk and Disaster Reduction's Everyday Disasters and Violences Research Group* investigates risks from everyday hazards and vulnerabilities, including gender-based and structural violences and systems of marginalisation. According to the Center's website, this knowledge will contribute to improving the way that policy, practice, governance, social and legal systems improve their responses to marginalisation, gendered and intersectional injustice, and violence, which lead to perpetual everyday disaster risks (UCL, 2023). I would love to see one of these centres in every country.

From my research, I provided a list of recommendations for use by emergency managers, first responder organisations and aid workers. While one of my recommendations to Canadian policing services remains the same—provide sensitivity training and education about homophobia, transphobia, heterosexism and anti-oppression—I am not certain it is enough. My experiences of needing the police for protection were everything my ten-year-old research study suggested they might be. I was just another queer person in daily distress that the police did not protect. We have felt not only ignored and unsafe but actually silenced by the institutions and services tasked to protect us. In 1987, during the AIDS crisis in America, the Silence = Death project rallied queer communities to act rather than remain silent (Finkestein, 2017). Thirty-six years later the project remains just as important to our queer communities—our voices will not be silenced.Conflict of InterestThe author has no conflict of interests.

References

Armstrong, L. (2023). Homophobic vandalism reminder of risks of being out in rural Nova Scotia. *Global News*, 7 February. Retrieved July 22, 2023, from https://globalnews.ca/news/9468099/homophobic-vandalism-rural-nova-scotia/

Beauchamp, D. L. (2004). *Sexual orientation and victimization*. Canadian Centre for Justice Statistics Profile Series. Retrieved September 1, 2023, from www.statcan.gc.ca/pub/85f0033m/85f0033m2008016-eng.pdf

Cianfarani, M. A. (2013). *Supporting the resilience of LGBTQI2SA people and households in Canadian disaster and emergency management*. A Major Research Project in fulfillment of a Master of Arts for Royal Roads University.

Egale Canada. (2023). Brief on police-reported crime statistics in Canada, 2021: Hate crimes targeting sexual orientation and sex or gender. (Updated, March 2023), *Egale Canada*. Retrieved August 30, 2023, from https://egale.ca/awareness/brief-on-police-reported-hate-crimes-2021/

Finkestein, A. (2017). Silence = death. How an iconic protest poster came into being. *Literary Hub*, 1 December. Retrieved August 28, 2023, from https://lithub.com/silence-death-how-an-iconic-protest-poster-came-into-being/

Government of Canada. (2022). February 14, Declaration of Public Order Emergency. *Department of Justice*. Retrieved September 5, 2023, from https://www.justice.gc.ca/eng/csj-sjc/section58.html

Hickey, A. (2023). Why are we still trusting the police? *The American Agora*, 27 February. Retrieved August 28, 2023, from https://www.americanagora.org/single-post/why-are-we-still-trusting-the-police

Hofmann, D. C., et al. (2021). An exploration of right-wing extremist incidents in Atlantic Canada. *Dynamics of Asymmetric Conflict*. https://doi.org/10.1080/17467586.2021.1876900

Jain, S. (2022). Hate crime surge in Canada during pandemic. *Reuters*, August 5. Retrieved July 7, 2023, from https://www.reuters.com/world/americas/hate-crimes-surge-canada-during-pandemic-2022-08-05/

MacSweeney, C. (2023). 2SLGBTQ+ advocates sound alarm over federal conservative proposals. *CityNews*, August 24. Retrieved August 28, 2023, from https://vancouver.citynews.ca/2023/08/24/2slgbtq-federal-conservative-proposals/

Mass Casualty Commission (MCC). (2023). *Mass casualty commission—Turning the tide together*. Retrieved June 29, 2023, from https://masscasualtycommission.ca/

McCann, M. (2010). Pride Toronto: Toronto Police cocktail party turns ugly. *Xtra Magazine*, 29 June. Retrieved September 1, 2023, from https://xtramagazine.com/power/pride-toronto-toronto-police-cocktail-party-turns-ugly-30132

McNeilly, G. (2012). *Policing the right to protest, G20 Systemic Review Report*. Retrieved September 1, 2023, from https://www.oiprd.on.ca/wp-content/uploads/G20-Systemic-Review-2012_E.pdf

Munoz, U. (2022). Nova Scotia town holds queer prom despite threats of violence against event. *Xtra Magazine,* June 20. Retrieved August 5, 2023, from https://xtramagazine.com/power/nova-scotia-queer-prom-threat-225337

Ottawa People's Commission (OPC). (2023). *Making our voices heard*. Retrieved August 15, 2023, from https://www.opc-cpo.ca

Roach, C. (2023). Some people in CBRM's LGBTQ community say bullying is on the rise. *CBC News Nova Scotia*, 24 April. Retrieved July 22, 2023, from https://www.cbc.ca/news/canada/nova-scotia/lgtbq-bullying-cape-breton-1.6818760

Sampson, A. (2023). Former student, school officials condemn destruction of Pride flag at N.S. school. *CBC News Nova Scotia*, 22 April. Retrieved July 22, 2023, from https://www.cbc.ca/news/canada/nova-scotia/desecration-pride-flag-nova-scotia-school-1.6819521

Smith, D. (2023). Canadian conservatives can't cherry-pick the 'culture war' they want to fight. *Xtra Magazine*, 17 February. Retrieved September 1, 2023, from https://xtramagazine.com/power/canada-conservatives-culture-war-245693

Statistics Canada. (2023). *Police-reported hate crime, 2021*. Retrieved August 28, 2023, from https://www150.statcan.gc.ca/n1/daily-quotidien/230322/dq230322a-eng.htm

UBC Okanagan News. (2013). *Looking for the silver lining in the wake of hurricane sandy*. Retrieved September 1, 2023, from https://news.ok.ubc.ca/2013/01/29/looking-for-the-silver-lining-in-the-wake-of-hurricane-sandy/

University College London Institute for Risk and Disaster Reduction. (2023). *Everyday disasters and violences research group*. Retrieved July 22, 2023, from https://www.ucl.ac.uk/risk-disaster-reduction/research/research-themes/everyday-disasters-and-violences-research-group-edv-rg

PART IV

Doing It for Themselves: Queering Disaster Scholarship, Media Representations and Allyship

CHAPTER 14

Can't Stop Change: Queer Climate Stories from the Florida Frontlines

Vanessa Raditz, Yarrow Koning, Natalia Villarán-Quiñones, and Jess Martínez

Discussion

Vanessa: Thank you all for agreeing to this conversation. To start us off, Yarrow, since you are now the lead organiser on the documentary's impact campaign, could you briefly introduce the documentary and what we have been trying to do through its creation?

Yarrow: Absolutely. Over the past several years, Florida's legislative violence against marginalised communities, including queer and

V. Raditz (✉) • J. Martínez
University of Georgia, Athens, GA, USA
e-mail: vanessa.raditz@uga.edu;

Y. Koning
Queers for Climate Justice, West Palm Beach, FL, USA

N. Villarán-Quiñones
Queers for Climate Justice, Rio Grande, PR, USA

D. Dominey-Howes et al. (eds.), *Queering Disasters, Climate Change and Humanitarian Crises*, Sustainable Development Goals Series, https://doi.org/10.1007/978-981-96-3857-4_14

> trans communities, has appeared frequently in the headlines. For this project, we wanted to go beneath the headlines to look more deeply at how LGBTQ2S+ communities are *also* on the frontlines of accelerating climate change and build a narrative that could support intersectional organising. To do that, *Can't Stop Change* weaves interviews with 15 LGBTQ2S+ artists, organisers, and educators across Florida (and the new Florida diaspora) into an intersectional climate justice narrative.

Our impact campaign strategy for *Can't Stop Change* furthers the work of disaster justice by supporting queer and trans communities responding to climate crises and political disasters. We see *Can't Stop Change* as an organising tool that can both uplift and support LGBTQ+ resistance movements across Florida. For instance, we have been materially resourcing folks through a mutual aid fund, which has helped our film collaborators in Florida through yet another devastating hurricane season this year.

We also see *Can't Stop Change* as an educational tool, especially for those outside of Florida. The film shares stories of 'the real Florida' that push back against negative mainstream media depictions of the state. We provide screening hosts with a detailed resource guide that includes readings and suggestions for curricula around the film.

Lastly, we see *Can't Stop Change* as a way to build bridges between LGBTQ+ and environmental organisations that too often exist siloed from each other.

Vanessa: Bringing environmental and LGBTQ+ communities together in a shared audience is also the longer-term goal of our work through Queers for Climate Justice. Natalia, could you share how we started working together?

Natalia: You know I can! I live in Borikén, mostly known by its colonial name, Puerto Rico. Vanessa had been working on a documentary here about queer and trans experiences since Hurricane Maria.

Vanessa: Yes, *Fire & Flood: Queer Resilience in the era of Climate Change*. I am still working on that film. That project started with my experience of the Tubbs Fire in Northern California in 2017, which was just after Hurricane Maria hit Puerto Rico.

Natalia: Exactly. You showed how Puerto Rico was impacted much more and received less support. The 'disaster' of Maria was not just because of the hurricane, but because of our colonial status and the U.S. imposed 'Fiscal Oversight Board.' People talk about Puerto Rico as part of the U.S. but fail to recognise the colonial status. We have to fight against the U.S. imperial interests in order to care for our land and each other. Your work documented our reality of how climate crisis, gender, and class intersect, so I reached out to you.

Through Queers for Climate Justice, I have been bringing these ideas to new spaces, while adding race, disability, and political status to the conversation as a Black, disabled, queer, femme Boricua.

Vanessa: I am grateful for all you've taught me! I have been thinking about LGBTQ+ vulnerability to disasters for close to 10 years, and I am concerned about how much of this work points towards one-dimensional solutions, such as updating the policies of the Federal Emergency Management Agency (FEMA) to include non-discrimination protections for Sexual Orientation and Gender Identity. One-dimensional solutions like these tend to address the needs of predominantly affluent LGBTQ+ people for whom their gender or sexuality are their only experience of marginalisation. Often, these narratives ignore the disproportionate poverty among LGBTQ+ people and how systems of impoverishment create vulnerability among people of all genders and sexualities.

Puerto Rico is a fitting example of that. Getting a non-discrimination protection through FEMA for LGBTQ+ people does not change the fact that FEMA has routinely failed communities in the U.S. colonial territories in the wake of disasters. Queer Boricuas are vulnerable, with or without these non-discrimination 'rights.'

Natalia: Multi-dimensional problems can't be solved long term with one-dimensional solutions. To restore the balance and live in tune with the planet, we must understand that our experience is not the only one that matters. Creating multi-dimensional solutions will require learning from others and our more-than-human kin. We must think holistically, *with* the Earth, about our problem-solving approaches. That's when community-building comes in, because connecting with each other helps us exist in solidarity.

Jess: Hearing Natalia's answer, I think of so many connections to my experience growing up in Georgia, which is in the former Confederacy and deeply shaped by racial capitalism. My embodied experiences as a disabled, queer, gender non-binary, working-class Latine Southerner, combined with what I have seen as a community organiser for housing and immigrant justice, have emphasised the complex contours of 'vulnerability.'

For instance, relief mobilisations after hurricanes often don't attend to the needs of un(der)documented Spanish-speaking immigrants who are ineligible for government aid or are distrustful of organisations at a time of escalating anti-immigrant violence in the U.S. That includes queer and trans immigrants. People who are multiply marginalised in different ways must be at the forefront of climate change responses, decision-making, and also crafting the narrative, because our experiences embody that intersectionality.

Vanessa: I agree. I want to shift us to thinking about how this project started back in April 2023. Natalia and Yarrow, could you share more about that pivotal meeting we had?

Natalia: It was serendipitous. Vanessa and I had been invited to speak at Florida Atlantic University. Vanessa was sharing the media analysis of Hurricane Ian, and my presentation was about environmental crises in Puerto Rico linked to our colonial status and how the LGBTQ+ community has been organising to resist and thrive. The U.S. military is the biggest polluter on Earth. To this day, Vieques is waiting for the U.S. military to clean up its toxic wastes, which contribute to high cancer rates and perpetuate hurricane vulnerabilities.

Yarrow: At the time Vanessa and Natalia visited Florida Atlantic University (FAU), I had been a graduate student there for close to two years and experienced the rise in anti-trans and anti-queer legislation across Florida. These policies significantly shifted the field of education across Florida, including my Master's in Education in Environmental Education. These policies negatively affected my life as a queer, trans, and non-binary educator, and disrupted my research as a student studying queer ecopedagogies.

Vanessa and I already knew each other through a screening of *Fire & Flood*, and they reached out to ask if I could introduce them to folks involved in disaster response after Hurricane Ian. I immediately thought of my friend Jimmy Dunson with Mutual Aid Disaster Relief (MADR). In 2019, I collaborated with Jimmy on a disaster preparedness training in Atlanta that centred queer and trans folks, and we have maintained a friendship ever since. I reached out to Jimmy and made plans for myself, Natalia, and Vanessa to travel to Orlando to meet with him and Robin Harris, a Black, queer, mutual aid organiser with Central Florida Mutual Aid.

Vanessa: I still have the recording of that conversation. It was powerful. Yarrow, I remember us driving back after that meeting. We were both already over-committed and exhausted, but we didn't need to talk ourselves into doing it. It was more like resigning ourselves to the fact that we were always already going to do it.

Yarrow: And we did it quickly. Within the week after you left Florida, you found a fiscal sponsor, Southern Documentary Fund, and a week or two later we had our first rapid grant.

Vanessa: Urgent Action Fund! It's interesting to note that we never received any traditional documentary grants. It was social justice movement rapid grants that made this project possible. With that rapid grant and funding support from Out for Sustainability and the Centre for Cultural Power's 'Climate Woke' programme, we had enough to start planning.

Yarrow: A little more than a month from that conversation with Robin and Jimmy, you were back down in Florida and we were recording. The timeline was wild.

Vanessa: It really was. Once we received funding, that is when Jess and Shoog got involved.

Jess: Yes! Vanessa and I are partners and live together. When they got home from their FAU trip and recounted the idea of the film, I immediately offered to join the team as an audio technician. I had already done some sound work with Vanessa on interviews for *Fire & Flood*. As a person with difficulties in verbal communication, I prefer writing and listening, which are also central to storytelling.

The production road trip began on an early Saturday in May; Vanessa and I rented a van in Athens, Georgia, and travelled south to pick up Shoog outside of Gainesville, Florida. Shoog is a Southern, queer, non-binary, fat photographer and artist, who joined as our cinematographer. The three of us continued to West Palm Beach on Sunday to pick up Yarrow and then to Miami, where we joined Natalia. Upon checking in at the hotel, we gathered in a room to plan. That was the first time all five of us talked about the technical details of what we were about to do.

Yarrow: That next day we had two interviews. Libre met us bright and early in one of our hotel suites, and then that same day, we also drove out to the Everglades for the interview with Lee and Houston.

Jess: Like you said, the timeline was wild. It took a toll on our bodies.

Yarrow: As a group with a number of disabilities, that was the biggest challenge.

Jess: But somehow, it was really fluid, like the water that accompanied us throughout the road trip. The next day, we had a walking interview with Barbara along the polluted canal behind the hotel, then that afternoon, learnt from Rebecca and Nasim about the creation of those canals and the importance of water to Florida's ecology. That evening, we ended the day at Miami Beach for a photo shoot celebrating queer and trans joy. Water was everywhere.

Wednesday morning, the five of us re-packed 'Roxanne the Van' to meet with Jimmy in Tampa and then Robin and Osvaldo in Orlando on Thursday. Water kept moving with us. On Friday, we returned to Shoog's house for a day of rest and swimming at the nearby Springs. Before saying goodbye on Saturday, we stopped at a popular birding spot by a sinkhole and watched the water flow back down into the Earth.

Yarrow: Again, this was all one week.

Jess: It took all five of us to make it happen.

Vanessa: The whole process was collaborative: Instead of thinking of ourselves in specific roles like 'audio technician' or 'cinematographer,' we created a Co-Director structure where we shared multiple responsibilities as a team. We contracted with others for specific roles, like Hamza Acevedo's work on editing, but Co-Director collaborations continued in post-production, as we shared roles in transcribing, illustrating, editing, translating, captioning, and beyond.

Yarrow: This queerly collaborative process also included the people we interviewed, giving them some agency over how their stories were told. We sent collaborators a draft transcript and a loose storytelling structure, then hosted a series of focus groups (with choice of English or Spanish). These focus groups provided a space for collaborator input early in post-production, which was critical to the story's development, prompting us to conduct further interviews beyond our one-week road trip.

After completing the rough-cut, we kept working with collaborators to organise 'first look' screenings in Miami, Orlando, Tallahassee, and Lake Worth Beach, to see how the film was received by the communities it represents. Unfortunately, these 'first look' screenings did not meet our goals due to ongoing political and climate disasters. For instance, Hurricane Idalia hit Florida during this period, with some film collaborators experiencing personal upheaval and property damage.

However, that was also a moment where I felt the queer climate justice vision was really coming to life: Co-Director Shoog and collaborator Jimmy, who had met through *Can't Stop Change*, worked together to distribute mutual aid supplies. It emphasised for me that we weren't just creating a film product but also crafting a process that connects people and supports new organising.

Vanessa: Absolutely. Despite these early set-backs, this process also helped us create a product that is now reaching and resonating with people far outside of Florida. I would love it if we could each share something about the story and what makes it a climate justice narrative.

Coming from my media analysis of Hurricane Ian, I was interested in building a narrative that situated Ian within other ways that environmental harm pervades queer and trans lives. The first third of the film is structured around the road trip, covering sea-level rise in Miami and biodiversity loss in the Everglades before we journey to Fort Myers, when Jimmy begins to narrate the impact of Hurricane Ian. The story of Ian continues to unfold as the road trip moves inland to Orlando. Robin takes us on a walk through the predominantly immigrant neighbourhood of Orlo Vista. Although far from the coast, this neighbourhood flooded due to the state's failure to maintain pumps in a nearby pond. Robin insists that the 'disaster' was not the hurricane: It was political, economic, and social systems that made Ian disastrous for marginalised communities.

One of my favourite sections to use in teaching is when our road trip approaches Tallahassee, and the multi-dimensional story of state violence emerges through narration from Valencia Gunder, founder of the Freedom Lab. Valencia weaves together attacks on education, critical race theory, abortion, and transgender people into what she calls a 'policy disaster.' My students really click with this idea of 'policy disaster.' It is a simple but powerful concept.

Yarrow: For me, what's most compelling about *Can't Stop Change* is that it is both deeply place-based and transcends colonial borders. *Can't Stop Change* uplifts the social and ecological diversity of Florida, combining personal narratives with archival footage and current events media to create a film that feels thoroughly Floridian. Yet the themes of the film are relevant to communities across the South and even across the globe in a time of rising global fascism, intensifying climate crisis, and increased attacks on LGBTQ+ and other marginalised groups.

For instance, the middle chapter of the film, titled *What Lies Beneath*, connects these many issues by highlighting the role of fossil fuels and extractive industries. We start with Miccosukee Two-Spirit artist and organiser Houston Cypress, who describes the impacts of fossil-fuel infrastructure on the Everglades and efforts to stop their expansion. Environmental educator Rebecca then brings in a historical analysis, highlighting how the colonial settlement of Florida modified the landscape to extract wealth, particularly under the leadership of Henry Flagler, founder of Standard Oil. This development harmed natural systems that now make people more vulnerable, especially when the various pumps and canals imposed on the swamp to accommodate dense housing are then not maintained in marginalised communities—which connects back to that story from Robin about the pond flooding during Ian.

Climate justice narratives often centre colonialism, capitalism, and fossil fuels. *Can't Stop Change* adds a gender analysis to this. In our interview with graduate student Barbara Perez, she describes her experience teaching the Gender and Climate Change class at Florida Atlantic University that has been specifically derided by Florida lawmakers. She provides an ecofeminist layer to this story, highlighting the patriarchal logics of the colonial mindset: 'Marginalised genders, women, people of colour,' said Perez, 'they have all been connected or associated to the environment through the Western construction of knowledge. And that association has justified the exploitation of the environment, but at the same time justified the exploitation of all of these marginalised groups.'

Moving from the past to the present, we wrap up that section with Lakey Love of the Florida Coalition for Transgender Liberation, who describes the power of fossil-fuel money in anti-trans legislation, as channelled by Koch Industries and the American Legislative Exchange Council.

Vanessa: That interview was particularly moving for me. I want to share the quote from Lakey that we used to close that chapter on *What Lies Beneath*: 'If you're separating the battle, that the fight for LGBTQ and trans liberation, the fight for racial justice, and the liberation of black and brown people and immigrant people, and the fight for the land for the air and the water ... If you're separating them ... it's because you're too close to the mosquito netting. If you pull it away from your face, then you can see that they're woven together ... for one paradigm and one purpose, which is white, patriarchal corporate poVwer.'

Yarrow: It's a powerful analysis. I often say that 'What happens in Florida doesn't stay in Florida,' since Florida is a test state for legislation that has spread to other conservative states and which is now reflected in federal policy.

The fight for climate justice is simultaneously local and global. Climate disasters are not bound by colonial borders; neither are the responses to them.

Jess: I agree. One way that we demonstrate a queer climate justice analysis is how we situate Florida 'as a construct' which is a quote from Houston Cypress. We show the many Floridas in between the Caribbean and the U.S. South, queering the colonial borders and reintegrating the land into the continuous ecosystems beyond state lines.

This is important because these violent systems also cross state lines. It was really illuminating when Jimmy discussed disaster militarisation and the way that the National Guard is an increasingly important part of disaster response in the U.S. We fund our military and police much more than social services, including disaster relief.

My favourite part was adding the additional interview about this with Terra in the garden in Tallahassee.

Vanessa: That was an interview that only made it into the film because our collaborators told us during the focus groups that we needed a more balanced depiction of North Florida, since so many interviews were in South Florida. We listened to our collaborators, and that led us to Terra, whose story crossed Florida's border into Georgia.

Jess: Terra was a close friend of Manuel Teran, also known as Tortuguita, who travelled to Atlanta, Georgia, to be part of the Stop Cop City movement, which was defending the Weelaunee forest from being cut down for the building of a police training facility. They were killed by Atlanta police officers while defending the forest. This is really the climax of the film. As this section draws to a close, the audience is left with the dystopian image of rising authoritarianism in response to climate instability.

Natalia: We don't shy away from this. State violence is not only in legislation, it is also through the police. That is something that climate justice movements have been talking about a lot. We have seen the way the police respond to protestors. It is clear who they are there to defend: the billionaires and the extractive system enriching them. Then we're supposed to believe the police are there to 'save' us after a hurricane?

If I'm honest, while these parts are important, it's always the ending that makes me cry. Alongside the trauma this violent system inflicts on us individually and collectively, we also profile LGBTQ2S+ artists, ecologists, and visionaries dreaming for a liberated future. This section weaves stories of queer ecologies, community care, and cultivating queer and trans belonging in nature. We want audiences to experience the many ways queer and trans communities maintain resistance by creating spaces of joy. These efforts are blossoming all across the state.

It's a narrative that inspires audiences to craft creative coalitions that work with the changing planet to shape more liberatory futures. In the conclusion, Valencia and Jimmy describe how disasters—whether hurricanes or policy disasters—are an important time to bring people into our movements. Disasters create cracks in the power structures, and we can plant our seeds in those cracks to break it further from the ground up.

Vanessa: Like Shoog wrote to us after reviewing the first look edit: 'the sad is so sad, but the joy is so joy.' We hope this representation can inspire audiences to hope not for technological quick-fixes that prop up the violent status quo but for the possibility that ecologically rooted, community-based strategies can world a habitable future.

Yarrow: *Can't Stop Change* screenings are the most essential component of honouring the stories in the film and doing the work of 'shaping change'—an idea from science fiction novelist Octavia Butler that frames the conclusion. These screenings are places for organising, building relationships, disaster preparedness strategising, and even resource distribution.

Natalia: The *Can't Stop Change* screenings so far have been all that and more. Despite the disappointing way the weather impacted the 'first look' screenings, the film has now made it into festivals and Universities all over the U.S. and even Europe! In all of these spaces, audiences have received it eagerly, and we have already seen it lead to new connections and more opportunities to organise.

Jess: It has been meaningful to speak with audiences outside of the South. The U.S. South is home; I've lived in the region all my life. To me, the film represented an opportunity to 'restory' the South. It disrupts representations of this region and its peoples as damaged, oppressed, and 'backward,' and honours the long histories and ongoing work of Black and brown resistance. It is healing to share the South that I know and love and to rally support for our communities.

Yarrow: I agree. I've had the opportunity to attend *Can't Stop Change* screenings across the country and am consistently moved by audience reflections. People without Florida roots are regularly surprised to learn about the incredible diversity of Florida and its legacies of queer and trans resistance. Those with roots in Florida frequently share how powerful it is to see their communities represented and celebrated.

Moving forward, I want to focus on our organising within Florida. I am excited to see how we can more fully utilise screenings as spaces for disaster preparedness, like distributing resources to vulnerable communities.

Vanessa: Thank you all for this conversation. I started this whole conversation with you by talking about the idea of wedges and bridges. To me, this film is a bridge. We are making sure our climate justice narratives recognise intersectionality and don't leave out the queer and trans voices needed for collective liberation. And we are letting our LGBTQ+ community see themselves *in* nature and *as* nature, and to imagine futures outside of colonial capitalism and outside of homonormativity.

We hope this vignette may help others imagine different ways of telling stories of queer and trans disaster experiences to disrupt normalised violence towards bodies and lands, and offer glimpses of collective liberation for people and the planet. As we say at the end of the credits: May this work be in service of healing and transformation.

For information on how to book a screening of *Can't Stop Change*, please visit queerecoproject.org/cant-stop-change. Follow @queers4climatejustice Instagram page for recent updates.

Conflict of Interest The four authors are all describing a documentary that they created. The documentary is for educational use, fiscally sponsored by a non-profit organisation, and does not generate profit for the directors.

Vanessa Raditz is the author of another chapter in this volume, which discusses research that is thematically connected to the content of the film described in this vignette.

CHAPTER 15

Calling Disaster Scholarship *in*: Towards Gender and Sexual Diverse Inclusive Research Practices

Ashleigh Rushton

Introduction

People with diverse sexual orientations, gender identities, gender expressions and sex characteristics (SOGIESC)—queer[1] for short—lead complex and multifaceted lives (Henrickson et al., 2020; Haworth et al., 2023). However, the complexity of queer experiences and ways of knowing are frequently ignored, stigmatised or devalued. This extends to current disaster research methodologies and methods that determine whose voice is

[1] Language is temporally and spatially situated, signifying the ever-evolving and changing nature of terminology. The evolution and progression of language is always a celebratory feat. That said, this chapter presents current uses of terms and expressions that may become dated and fade into disuse.

A. Rushton (✉)
Disaster and Emergency Management Researcher and Planner, Cleveland, UK

D. Dominey-Howes et al. (eds.), *Queering Disasters, Climate Change and Humanitarian Crises*, Sustainable Development Goals Series, https://doi.org/10.1007/978-981-96-3857-4_15

heard and whose is not. The silencing of minority voices and the privileging of majority perspectives in disaster research become especially problematic when considering how these biases impactn disaster planning, policy and practice (Meagher et al., 2008).

Disaster research captures the complex and adverse experiences that accompany disruptions to people's social and physical environments. Research methods are continually being adapted, expanded and refined to ensure they are appropriate to the diversity of disasters and people's disaster experiences. For example, the development of trauma-informed research provides supports such as counselling for participants who may have experienced trauma. While research and practice require flexibility, disaster research methodologies are deeply heterocisnormative and have excluded non-binary sexualities and genders (Rushton et al., 2019).

The concept of disaster justice can assist in understanding prejudice and biases in the field of disasters. According to Lukasiewicz (2020, p. 9) disaster justice involves 'an inclusive decision-making process, free of discrimination, that pays attention to risks disproportionately imposed on socially disadvantaged populations.' Research design is an important component of disaster justice. Aligning research practices with disaster justice, where justice is considered a process rather than a fixed aftereffect (Lukasiewicz, 2020), can extend theoretical understandings of disaster inequities and injustices. To attain justice in planning, policy and practice, disaster research must be inclusive and free of discrimination, judgements, prejudices and pre-conceived ideas.

Rather than calling disaster researchers and practitioners *out*, I seek to call them *in* (Minalga et al., 2022), to consider how their work can be more inclusive and contribute to the growing application of nonnormative practices (Gaillard et al., 2017; Haworth et al., 2022). I introduce the concept of *epistemologies of ignorance* to showcase how research methodologies and methods, and researcher biases, can ignore marginalised voices and lead to deficits in disaster risk reduction (DRR) policies and practices. Injustices to diverse populations in disaster policy and practice are tied to, or are a result of, ignorance. Ultimately, there has been a silencing of diverse voices in disaster research and emergency management.

This chapter is a guide to understanding some of the nuances and ways of engaging with queer populations. While the chapter focuses on conducting inclusive research, its recommendations have implications for disaster practitioners and policy makers.

Blending Disaster and Queer Research Methodologies

Disaster research has historically employed qualitative methods that provide the social and contextual information required to showcase the complexity of human experience (Aitsi-Selmi et al., 2016). Akin to many disaster scholars, queer researchers favour qualitative methods as they support flexible, subjective and meaningful data collection and analysis, enabling a detailed understanding of identity, place and power. However, qualitative research remains on the margins in discussions on disasters and DRR policies and practices (Donovan et al., 2012; Pickering et al., 2020). The prioritisation of quantitative approaches in DRR initiatives and plans is exemplified in the 2015 *Sendai Framework for Disaster Risk Reduction* (Aitsi-Selmi et al., 2016; Pickering et al., 2020). The exclusion of qualitative data in the Sendai Framework contributes to the silencing and absence of minority voices and experiences in DRR guidelines and represents a form of epistemological ignorance that is at odds with the principles of disaster justice, equity and inclusion.

Queer scholarship has focused on the ethical dilemmas that research bias poses. While quantitative research commonly views bias as something to be reduced or eliminated (Pickering et al., 2020), qualitative research acknowledges the impossibility of removing bias. Qualitative researchers embed methods and practices within research designs to understand how unconscious and conscious perspectives shape research development, implementation and data interpretation (Pickering et al., 2020). As many queer and feminist scholars argue, attempting to remove rather than acknowledge research biases is concerning for marginalised populations. For example, the failure to include diverse gender options in questionnaires is indicative of cisnormative biases in research. Paradoxically, assuming biases are eliminated becomes part of how these biases are perpetuated.

Assumptions about objectivity and the truth of dualisms underpin much research. As Doan (2016) points out, the inclusion of only two options for sex/gender in data collection assumes that this binary or dualism is natural and not to be questioned. This *preemptively* excludes people who identify as non-binary, for example, increasing the invisibility of gender diverse individuals within demographics. van Anders et al. (2023) contend that if a participant's gender does not fit within 'standard' gender categories, researchers will group them with one of the set categories or exclude them altogether. Furthermore, surveys and questionnaires designed for qualitative and quantitative disaster research commonly exclude sexuality and/or genders beyond the binary. In those rare cases where they are present, it is not uncommon for the terms 'sex' and 'gender' to be used incorrectly and interchangeably (Rushton et al., 2019).

Employing queer approaches to disaster research destabilises heteronormative and cisnormative assumptions. It showcases the fluidity of sexualities and genders and promotes diverse, inclusive research and data collection (Das, 2020). As Phillips (2014) contends, the value and importance of disaster scholarship is that it has real-life outcomes. In order to maximise these outcomes by providing stronger disaster policies and protocols, research design needs to be inclusive and disrupt unquestioned assumptions and power imbalances.

Practices to Ensure Inclusive Research

Diverse voices and experiences have been deliberately silenced, through violence, laws, heteronormative practices and assumptions and the policing of the media. Although data is scarce, studies show that queer people account for approximately 4–20% of a population (IPSOS, 2021; Office for National Statistics, 2022; Spizzirri et al., 2022; Statistics Canada, 2022; Stats NZ Tatauranga Aotearoa, 2022). It is therefore probable that any study sample will include at least one SOGIESC diverse person (Henrickson et al., 2020).

Heterocisnormativity is embedded in disaster research and practice and in the media's ability to control whose voices are and are not heard in disaster narratives (McKinnon et al., 2017). The conditioning of the production of knowledge and the privileging of the majority at the expense of minorities can be understood as epistemologies of ignorance (Bowleg et al., 2017). Practices of ignorance can be recognised in biases, exclusions, and oppressions, showcasing how ignorance is socially constructed (Tuana & Sullivan, 2006). The social construction of ignorance has a dependency on reinforcing social and institutional systems (Alcoff, 2007; El Kassar, 2018). In a community survey following the 2011 floods in Brisbane, Australia, over 50% of queer respondents reported avoiding support services because they felt uncertain or anxious (Gorman-Murray et al., 2014). This highlights how oppression, ignorance and discrimination cultivate environments that exclude certain groups. Synergies between epistemologies of ignorance and disaster (in)justice can be drawn on as injustices are underpinned by the negative beliefs towards a person or group. A paradox lies in the virtue of justice being socially expected (Lukasiewicz, 2020), even as injustices are formed through the social manifestation of ignorance.

Disaster research and practice are shaped by epistemologies of ignorance. Who is considered, included and (intentionally and/or unintentionally) excluded matters. Ultimately, all-inclusive research strives towards disaster justice. Queering the academy has lent itself to disrupting normative ways of thinking about lived experience and creating questions on ways forward in conducting inclusive research. This section provides practices that can be applied within the study processes to support inclusion.

Inclusive Terminology

Embedding inclusive terminology in all stages of a research project, from conceptualisation to the dissemination of findings, sets practical and ethical pathways for communication over the life of the project. Consideration of the geographic and cultural context in which research is situated is critical. Western languages and terminologies have been prioritised and assumed as universal, such as the acronym 'LGBT' (lesbian, gay, bisexual and transgender). Recognising the geographic and cultural diversity of identities and terminologies undermines the use of absolute terms. It demands that researchers engage with local people and communities as active agents to ensure their gender and sexual diverse frameworks, identities and practices are reflected in research design and methodologies. In Canada, for example, *two-spirit* is adopted by Indigenous People who identify with a gender beyond the woman/man dichotomy (Robinson, 2019). '[T]wo spirit identity,' writes Robinson, 'asserts that the meaning of sexual or gender difference among Indigenous people is to be found in Indigenous cultural frameworks rather than Settler categories' (p. 7).

Engaging meaningfully with community groups or people who are sexual or gender diverse can minimise the risks of cultural misappropriation and offence. This can involve asking for support in reviewing the language, terminology and frameworks being used and community consultations to develop recruitment strategies that maximise participation in part by not using poorly chosen words or making inappropriate assumptions. For example, Lowik et al. (2022) provide a research scenario where advertisements for a study on menstruation and sport that solely focuses on women are likely to exclude people who are transgender or non-binary who menstruate. Small considerations such as language can have significant impacts on the recruitment process and participant's sense of safety.

An important practice is the use of correct pronouns. This creates safe spaces for people who are gender diverse through normalising the display of individual pronouns. Making pronouns visible in research, while not assuming gender identity, removes a major barrier to recruiting gender diverse participants. A sound practice that ensures people are not misgendered is to make 'no assumptions' and ask participants as early as possible what their pronouns are (Edward & Greenough, 2019). Acknowledging a persons pronouns extends dignity and respect to gender diverse and gender non-binary people (Henrickson et al., 2020). Additionally, offering participants the opportunity to choose a pseudonym provides further autonomy. Principally, it is important not to ask personal or inappropriate questions relating to gender and sexuality unless directly relevant to the research and participant's have been informed about the question topics (Haworth et al., 2022). As researchers and practitioners use and employ more inclusive terminology, methods, research designs and practices, the prospect of more all-inclusive disaster and emergency policies and plans increases.

Disrupting Categories

Disaggregated data can determine elevated needs in and across differing populations and support the allocation of emergency funding and disaster planning (Benelli et al., 2012). An issue, however, is that disaggregated disaster data is insufficient (Akerkar & Fordham, 2017; WHO, 2020). While research advances knowledge it also shapes future DRR resources and protocols. It is therefore imperative to collect and collate data accurately, ensuring the data is diverse and inclusivee. Inclusive terminology also extends to disrupting normative categories to enable the collection of detailed data that includes identities previously overlooked or excluded. Removing set categories, such as 'female' or 'male' only, may result in extra work for researchers including additional and more complex analysis (Henrickson et al., 2020). However, including options for participants to self-identify as sexual or gender diverse will lead to more accurate reporting whilst giving participants agency. Nonetheless, caution is warranted. Using a third term like 'other' as the 'inclusive' option has the effect of othering participants and grouping a diversity of identities under one category (Henrickson et al., 2020).

Qualitative and, more notably, quantitative data on gender and sexually diverse people and communities is scarce (Office of the High Commissioner for Human Rights, 2019). Even for research that does not have a specific focus on gender or sexuality, cultivating pathways for people with diverse identities to be involved (as opposed to excluding them) ensures that their voices are included, improving understandings of disasters and working towards disaster justice.

The Research Environment

For qualitative researchers, participant comfort during the research process is paramount. This helps build trust, safety and a relaxed setting between the researcher and participant (Liamputtong, 2007). Agreeing on a safe and comfortable place to conduct research is important for SOGIESC diverse people. There are a multitude of unique risks that sexual or gender diverse people may face when participating in research. Participation may expose them to harm including concerns about privacy and confidentiality and someone overhearing sensitive and private information. This could happen in both a public place and their home. People know what is best for their safety and comfort levels and need to be consulted on where their interview or focus group is conducted. Safeguards should be put in place, and if at any point you feel that the participant is at risk, steps should be followed to protect them. This may mean cancelling the interview or focus group or removing their contributions from the research. All actions must be considered in consultation with the participant. In addition, pay attention to *where* the research is being carried out. Ask questions such as, *What are the laws in place around gender and sexuality? What is the political environment like at the time of the research?* and *What are the media discourses around sexuality and gender?* These are important considerations given that even within countries where being queer is no longer criminalised and same-sex marriage is legal, there are increasing tensions, resulting in homophobic and transphobic hate speech, and diminishing political support for gender and sexually diverse populations (Rushton & Scarlett, 2023; Stonewall, 2023).

Consideration should also be given to the research team and the organisation or university where the research is being conducted. For example, consider what the institution's history is regarding Pride or their commitment to creating safe spaces for sexually or gender diverse students and staff. It is imperative that sensitivities are not ignored or overlooked. It is possible that participants may feel uncomfortable or unsafe if the research is being conducted at a university that has a history of treating gender and sexual diverse people unfairly. Making meaningful and genuine connections with relevant community organisations and stakeholders would be a positive step towards building positive bridges between researchers, institutions and community members. Furthermore, SOGIESC diverse individuals may be reluctant to volunteer to participate in research that is being conducted primarily by privileged researchers, such as white, able-bodied heterosexual cisgender men (Lowik et al., 2022). A diverse team can be part of an inclusive recruitment strategy as well as fostering a safe space for diverse voices to be heard.

Social media is an important research tool, however it is critical to consider the risks associated with using social media for communication purposes with sexually or gender diverse communities. Unfortunately, studies concerning sexuality and gender that are shared on social media have been subject to homophobic and transphobic abuse (Littler & Joy, 2021). Dominey-Howes and colleagues (this collection, Chap. 19) experienced considerable hate speech and homophobic abuse online when their research on queer disaster experiences was publicly shared. It is recommended that a safeguarding plan is implemented, including actively monitoring social media comments or turning them off. These protocols can help minimise unwanted attention and/or harm to participants and the research team, but they are not 100% successful (Lowik et al., 2022).

Disaster research is inherently sensitive and can pose a diversity of risks, including re-traumatisation (Drozdzewski & Dominey-Howes, 2015; Haworth et al., 2022). Consequently, researchers and practitioners have a responsibility to provide safe and inclusive environments as best they can in which to conduct research.

Strength-Based Approaches

Studies concerning sexuality and gender have routinely been undertaken through a deficit-lens, focusing on negativities. Ignoring strengths, resilience and the adaptive capacities of people and communities produces incomplete understandings of their experiences and gaps in knowledge (Vaughan & Rodriguez, 2014). Gender and disaster scholarship has relied on a problem-focused approach, but there has been a recent shift towards exploring women's agency in DRR initiatives to 're-balance' knowledge disparities (Rushton et al., 2020). The current literature on gender and sexual diversity and disasters has steadily incorporated SOGIESC diverse peoples and communities' challenges ***and*** strengths (see Balgos et al., 2012; Gorman-Murray et al., 2017). While disaster research needs to highlight the adversity and problems experienced by marginalised populations, it also needs to acknowledge their individual and collective capacities to develop the most effective and appropriate disaster planning, response, and recovery.

Gender and sexual diverse people and communities have extraordinary capacities (Gorman-Murray et al., 2014; Larkin, 2019) which should not be ignored. SOGIESC diverse people have developed strategies to navigate and overcome the systemic discrimination and abuse they experience in their everyday lives (Henrickson et al., 2020). For some, these experiences include stress, adversity and hardship, and yet, amongst accounts of difficultly, there are stories of surviving and thriving—for example, the strength and capacity required to mobilise a grass roots, gender and sexual diverse community response to the HIV/AIDS pandemic in the 1980s and 1990s (This volume, Chap. 3). There are clear lessons to be learnt from populations and communities routinely marginalised in DRR policies and practices (Dominey-Howes et al., 2014).

The Praxis of Reflexivity

Feminist and queer studies have similar theoretical foundations and the shared goal of inclusivity (van Anders et al., 2023) (for thorough reviews see Fordham & Meyreles, 2014; Rushton et al., 2020). Reflexivity is a research practice championed by feminist, and later by queer scholars that promotes inclusivity (Haraway, 1988). The practice encourages reflection on one's own biases, interpretations and background. To be reflexive means to engage with the assumption that science is never bias-free or objective, but rather, research is fundamentally shaped by our own positions, beliefs and assumptions (McDowell, 1992; Das, 2020; van Anders et al., 2023). Acknowledgement is given to how situated knowledges are dynamic, fluid and socially constructed (Haraway, 1988; Hesse-Biber, 2014). Moreover, how researchers conduct research and interpret knowledge is underpinned by background, training, discipline and worldviews which are often taken for granted or not interrogated (Bourdieu, 1988). Knowledge is a constructed reality (Bourdieu, 1988), and researchers hold responsibilities to be diligent, fair and ethical in research endeavours which should include considerations of how to acquire and understand knowledge.

To address problems of research bias and subjectivity, tools and practices are available to guide researchers become more self-reflexive. These tools help researchers recognise how their positionality may construct a form of reality and interpretation that are dissimilar to those of their research participants (Gorman-Murray et al., 2010). Attention must be given to similarities and differences between participants and researchers and how these may be (re)negotiated throughout the research process. This may include how identities are interpreted by both the researcher and participant, with scrutiny over how these identities may intersect (Hesse-Biber, 2014; Valentine, 2002).

Sou (2021), in her ethnographic work in an evangelical Christian community in Bolivia, discusses the tensions between conducting research in a highly conservative locale and her queer, cisgender woman identity. For reasons relating to safety and research success, Sou concealed her sexuality. However, she continuously questioned how her concealment and her initial perceived beliefs that the community she was working with may be homophobic—later confirmed to be true—shaped the research as well as the relationships she formed. In her reflexive analysis, Sou (2021) recognised that by constructing a false heterosexual identity in the field, she had possibly silenced queer voices that may have connected with her. Sou's account is an insightful reflection on the importance of self-reflexivity and the complexities of personal and perceived interpretations of identities in research.

Reflexive practices map the ways research is influenced by the entanglement of multiple, overlapping social identities, beliefs and experiences. This includes how our backgrounds and beliefs underpin our chosen methodologies, methods and which funding to apply for. It also includes team composition, which may encompass differences in gender, age, sexuality, ethnicity, religion, class, income, immigration status etc. van Anders et al. (2023) explain the importance of research engaging with the subjective self and discuss how research concerning transgender people has been uncritically dominated by cisgender scholars. The authors argue that cisgender people experience gender differently to people who are transgender, and this includes a limited of understanding about transphobic discrimination. A lack of critical interrogation of the positionalities of cisgender scholars conducting research on the lived experiences of transgender individuals not only results in 'reinforcing the dominance of cisgender perspectives on gender' (van Anders et al., 2023, p. 7), but also distorts and limits research outcomes.

Member checking is a practice that can reduce bias and increase the accuracy of interpretation. Member checking provides participants with opportunities to review their contributions (such as their interview transcript) and to address any inconsistencies, errors or misinterpretations. It also enables participants to remove or add information as they see fit. Fundamentally, member checking ensures more accurate interpretation of data. An extension of member checking is sharing research outputs (such as journal articles or conference documents) with participants before publication to further check that information has been accurately interpreted—extending the principle that participants are collaborators or co-researchers.

Reflexivity holds an important place in research concerning gender and sexuality precisely because it places social constructions of identity at the core of knowledge and demands reflection on power and perceived truth. It can be argued that disaster research is inherently underscored by biases and distorted understandings of the self (Rushton et al., 2020), particularly as it is a discipline derived from military discourses (Coppola, 2020) that traditionally encourage hyper-masculinity and disciplined behaviours. DRR policies and practices, for example, lend themselves to the privileging of heterosexual and cisgender beliefs recognised in emergency management, including the military. DRR research, policy and practice are tied to what the powerful believe to be just or unjust (Lukasiewicz, 2020). They reflect the beliefs, values and cultural norms of the powerful and of those who undertake and then translate the research into DRR policy and practice. The issues lie in the evidence that decisions about who gets included or excluded from DRR research, policy and practice go uncritically unexamined.

Ethical Considerations

Ethics is a fundamental aspect of research practice. Ethics is particularly salient in studies concerning sensitive topics, such as disasters and the lives of marginalised or vulnerable populations (Liamputtong, 2007; Dubois et al., 2020). This must include SOGIESC diverse people in disasters.

A core research principle is *do no harm.* One challenge for disaster scholars when developing sensitive research projects is forming questions that do not cause distress or lead to re-traumatisation. There is increased awareness that people who have experienced disasters or work in post-disaster environments may experience trauma (Dominey-Howes, 2015; Rushton et al., 2021). Consequently, researchers have a responsibility to reduce the emotional impacts on participants as much as possible. There are standard practices to minimise distress and re-trauma which have been discussed in detail (see Kavanaugh & Ayres, 1998; Liamputtong, 2007; Dempsey et al., 2016). However, threading through these practices needs to be a consideration of the unique needs and lived experiences of SOGIESC diverse people. For example, a practice in trauma-related research is making referrals to mental health resources and/or services, such as counselling, when needed (Liamputtong, 2007; Drozdzewski & Dominey-Howes, 2021) and should include services/resources that are appropriate for gender and sexual diverse persons. A service that publicly supports and works with SOGIESC diverse individuals may be more appropriate given comfort levels accessing visibly inclusive spaces.

Although human research often requires compliance with institutional ethics processes, approval to proceed does not guarantee all ethical dilemmas have been addressed. This is not to say that institutional ethics processes are inadequate, but rather, acknowledgement is given to the complexities of research projects and gaps in knowledge. How sexuality and gender diversity relate to ethics, for example, is often uncritically assumed in discussions or ignored. This is based on heterocisnormative assumptions and epistemologies of ignorance embedded in the academy. For example, a human research ethics committee considers the ethical dilemmas of researchers speaking with gender and/or sexual diverse youth, focusing on obtaining parental consent (Henrickson et al., 2020; Flores et al., 2018). However, consideration of the risks relating to obtaining parental consent posed to the gender and/or sexual diverse participant is devalued or not reviewed appropriately. It cannot be assumed that the parent(s)/guardian(s) is aware that their child is gender or sexual diverse or that they are supportive. The parental consent requirement for youth participation potentially creates conflict at home and effectively silences queer youth's voices (Flores et al., 2018; Henrickson et al., 2020). As Henrickson et al. (2020) propose, adhering to ethics protocols in an age-appropriate manner, such as informed consent and confidentiality, ensures that the research meets the conditions of ethics processes and that parental consent for participation is not needed.

Researchers, in concert with human ethics boards, must recognise potential concerns and respond accordingly. Additional considerations may be required when working with SOGIESC diverse people—for example, identifying a participant as gay within a project on emergency managers in a small town. The likelihood of there being many queer emergency managers in one small town is likely to be low, thus this participant would be easily identifiable. Using large sample sizes ensures not only a diverse sample but also maintains confidentiality in circumstances where there is a risk that a participant might be identifiable (Haworth et al., 2022).

To consider and address all possible ethical pitfalls, researchers should work to build genuine partnerships with relevant stakeholders or gender or sexual diverse people with knowledge of research processes. Establishing meaningful consultations with advisors supports not only the research design inception, but also ensures the interests of the communities are prioritised (Dubois et al., 2020). These partnerships with advisors will also facilitate respectful and more accurate interpretations and presentations of the data, and reduce misunderstandings (Henrickson et al., 2020). Moreover, this approach may lead to greater recruitment (Haworth et al., 2022).

Disaster researchers manage continually evolving situations (Phillips, 2014), which create unique ethical considerations. As Donner and Diaz (2018) and van Zijll de Jong et al. (2011) ask, is it ethical to request interviews or survey responses from services that are potentially already overstretched? Since disasters often occur suddenly, it is common for there to be a rapid influx of international researchers who descend on the affected population to collect time-sensitive data. This can lead to affected communities feeling overwhelmed with sudden attention, and 'participant research fatigue,' whereby people who experienced/are experiencing the disaster get repeatedly asked to participate in primary studies (van Zijll de Jong et al., 2011). Fatigue is salient in disaster research and researchers must be sensitive to how overburdening participants, particularly minority populations, can cause increased stress and reduced willingness to participate (Patel et al., 2020). At some point, people will just want to 'move on' and stop talking about a traumatic event that affected their lives (van Zijll de Jong et al., 2011). Research must benefit the group(s) participating. Consequently, their needs, time and priorities must be respected. This includes determining *when* it is most appropriate and feasible to conduct the research (van Zijll de Jong et al., 2011; Dubois et al., 2020).

Upholding ethics begins with terminology. There is a distinct difference between identifying people who are part of the research as 'participants,' 'collaborators,' 'co-authors,' 'co-researchers' or 'subjects'

(Pickering et al., 2020). The first four terms are considered more appropriate (Dubois et al., 2020; Pickering et al., 2020) because they acknowledge participants as active agents in the research whose contributions (which include their time and knowledge) add to the quality and value of the research. Inclusive terminology works to minimise power differentials while highlighting the researcher's appreciation of participant contributions. This is especially relevant when working with marginalised populations, where past research was often driven by exploitation and filtered through the lenses of deficit, problems and blame.

To summarise, when conducting research on or with gender and/or sexual diverse populations, researchers should include the following: (1) are there limitations to the research? (2) What ethical considerations have been or need to be applied? (3) What potential harms to queer individuals and/or communities might occur? (4) How might the study further stigmatise queer people? (Minalga et al., 2022) and (5) is the research meaningful and does it benefit queer populations? Participant recruitment and collaboration must be fair, equitable and genuine and must not be determined by bias, access and privilege (Dubois et al., 2020). As Dubois et al. (2020) contend, any exclusions in research must 'be based on valid scientific justification' (p. 188). Therefore, it is unethical to exclude SOGIESC diverse people.

Considerations for DRR Practitioners

The considerations presented below build on the proceeding discussion of research practices and are intended to be used by DRR managers and practitioners.

- Adopt inclusive terminology in registration documents. This encourages gender and/or sexually diverse individuals to register for support and statistically acknowledges diversity in disaster-affected environments.
- Be consistent in the use of terminology (e.g., see Rushton et al., 2019).
- Where possible, liaise with community leaders who can offer guidance in relation to supporting gender and/or sexually diverse individuals and communities.
- Recognise unique risks and/or barriers to SOGIESC diverse people in disasters (e.g., see Haskell, 2014; Dunn, 2016; Goldsmith et al.,

2022; Rushton & Scarlett, 2023) while also understanding strengths. This can aid in understanding how best to support SOGIESC diverse people.

- Acknowledge the cultural, political and historical context concerning gender and sexual diverse populations in which you are working. For example, it may be unsafe for a SOGIESC diverse person to ask for support in disaster, thus ways should be made (appropriately and safely) to reach this population.
- Recognise that some spaces may be unsafe for gender and sexual diverse people. For example, evacuation centres can often be in religious premises, which are spaces which have historically excluded queer people (Goldsmith et al., 2022). Establishing evacuation centres in areas that do not have a history of exclusion is recommended.
- Overall, diversity of response personnel has the potential to build rapport and trust with diverse communities.

Concluding Thoughts

Through calling scholars and practitioners *in*, this chapter has showcased methods and practices that support inclusivity. The inclusion of SOGIESC diverse people and communities will enhance the vigour and strength of informed research and practices. Consideration of diverse communities must begin in the research design and followed throughout the life of the project. As Haworth et al. (2022, p. 7) remind us, while 'heteronormativity persists, so too will heterosexism,' which ultimately leads to the ignorance of diverse identities that is endemic to disaster scholarship and practice. Failures to prevent harm or assist the needs of people in disaster cultivate a social environment of injustice. It can be said that epistemologies of ignorance are at work in DRR research and practice, showcased in the absence of local, national and international guidelines, policies and practices as well as across disaster research endeavours, especially in relation to SOGIESC diverse individuals.Conflict of InterestThe author has no conflict of interests.

References

Aitsi-Selmi, A., Blanchard, K., & Murray, V. (2016). Ensuring science is useful, usable and used in global disaster risk reduction and sustainable development: A view through the Sendai framework lens. *Palgrave Communications, 2*(1), 1–9.

Akerkar, S., & Fordham, M. (2017). Gender, place and mental health recovery in disasters: Addressing issues of equality and difference. *International Journal of Disaster Risk Reduction, 23*, 218–230. https://doi.org/10.1016/j.ijdrr.2017.03.014

Alcoff, L. (2007). Epistemologies of ignorance. three types. In S. Sullivan & N. Tuana (Eds.), *Race and epistemologies of ignorance* (pp. 39–57). State University of New York Press.

Balgos, B., Gaillard, J. C., & Sanz, K. (2012). The warias of Indonesia in disaster risk reduction: The case of the 2010 Mt Merapi eruption in Indonesia. *Gender and Development, 20*(2), 337–348.

Benelli, P., Mazurana, D., & Walker, P. (2012). Using sex and age disaggregated data to improve humanitarian response in emergencies. *Gender and Development, 20*(2), 219–232.

Bourdieu, P. (1988). Vive la crise! For heterodoxy in social science. *Theory and Society, 17*, 773–787.

Bowleg, L., et al. (2017). Intersectional epistemologies of ignorance: How behavioral and social science research shapes what we know, think we know, and don't know about US Black men's sexualities. *The Journal of Sex Research, 54*(4-5), 577–603.

Coppola, D. (2020). *Introduction to international disaster management* (4th ed.). Elsevier.

Das, S. (2020). Queer methodologies and social science. In N. A. Naples (Ed.), *Companion to sexuality studies* (pp. 95–121). John Wiley and Sons.

Dempsey, L., Dowling, M., Larkin, P., & Murphy, K. (2016). Sensitive interviewing in qualitative research. *Research in Nursing and Health, 39*(6), 480–490.

Doan, P. L. (2016). To count or not to count: Queering measurement and the transgender community. *Women's Studies Quarterly, 44*(3/4), 89–110.

Dominey-Howes, D. (2015). Seeing 'the dark passenger'—Reflections on the emotional trauma of conducting post-disaster research. *Emotion, Space and Society, 17*, 55–62.

Dominey-Howes, D., Gorman-Murray, A., & McKinnon, S. (2014). Queering disasters: On the need to account for LGBTI experiences in natural disaster contexts. *Gender, Place and Culture, 21*(7), 905–918.

Donner, W., & Diaz, W. (2018). Methodological issues in disaster research. In H. Rodríguez, W. Donner, & J. Trainor (Eds.), *Handbook of disaster research* (2nd ed., pp. 289–309). Springer.

Donovan, A., Oppenheimer, C., & Bravo, M. (2012). Social studies of volcanology: Knowledge generation and expert advice on active volcanoes. *Bulletin of Volcanology, 74*, 677–689.

Drozdzewski, D., & Dominey-Howes, D. (2015). Research and trauma: Understanding the impact of traumatic content and places on the researcher. *Emotion, Space and Society, 17*, 17–21.

Drozdzewski, D., & Dominey-Howes, D. (2021). Researcher trauma: considering the ethics, impacts and outcomes of research on researchers. In *Research Ethics in Human Geography* (pp. 168–181). Routledge.

Dubois, C., Wright, K., & Parker, M. (2020). Ethics in research. In R. Kayano, V. Murray, M. Clarke, & E. Chan (Eds.), *WHO guidance on research methods for health emergency and disaster risk management* (pp. 185–199). World Health Organisation.

Dunn, L. (2016). Integrating men and masculinities in Caribbean disaster risk management. In E. Enarson & B. Peace (Eds.), *Men, masculinities and disaster* (pp. 210–218). Routledge.

Edward, M., & Greenough, C. (2019). Queer literacy. In R. Iphofen (Ed.), *Handbook of research ethics and scientific integrity* (pp. 1–12). Springer. https://doi.org/10.1007/978-3-319-76040-7_57-1

El Kassar, N. (2018). What ignorance really is. Examining the foundations of epistemology of ignorance. *Social Epistemology, 32*(5), 300–310.

Flores, D., McKinney, R., Jr., Arscott, J., & Barroso, J. (2018). Obtaining waivers of parental consent: A strategy endorsed by gay, bisexual, and queer adolescent males for health prevention research. *Nursing Outlook, 66*(2), 138–148.

Fordham, M., & Meyreles, L. (2014). Gender aspects of disaster management. In A. López-Carresi, M. Fordham, B. Wisner, I. Kelman, & J. C. Gaillard (Eds.), *Disaster management: International lessons in risk reduction, response and recovery* (pp. 23–40). Routledge.

Gaillard, J. C., Sanz, K., Balgos, B. C., Dalisay, S. N. M., Gorman-Murray, A., Smith, F., & Toelupe, V. A. (2017). Beyond men and women: A critical perspective on gender and disaster. *Disasters, 41*(3), 429–447.

Goldsmith, L., Raditz, V., & Méndez, M. (2022). Queer and present danger: Understanding the disparate impacts of disasters on LGBTQ+ communities. *Disasters, 46*(4), 946–973.

Gorman-Murray, A., Johnston, L., & Waitt, G. (2010). Queer(ing) communication in research relationships: A conversation about subjectivities, methodologies and ethics. In K. Browne & C. J. Nash (Eds.), *Queer methods and methodologies: Intersecting queer theories and social science research* (pp. 97–112). Routledge.

Gorman-Murray, A. W., Morris, S., Keppel, J., McKinnon, S. J., & Dominey-Howes, D. (2014). The LGBTI community in the 2011 Queensland floods: Marginality, vulnerability and resilience. *LES Online, 6*(1), 4–20.

Gorman-Murray, A., Morris, S., Keppel, J., McKinnon, S., & Dominey-Howes, D. (2017). Problems and possibilities on the margins: LGBT experiences in the 2011 Queensland floods. *Gender, Place and Culture, 24*(1), 37–51.

Haraway, D. (1988). Situated knowledges: The science question in feminism and the privilege of partial perspective. *Feminist Studies, 14*(3), 575–599.

Haskell, B. (2014). Sexuality and natural disaster: Challenges of LGBT communities facing Hurricane Katrina. Retrieved December 20, 2023, from https://ssrn.com/abstract=2513650 or https://doi.org/10.2139/ssrn.2513650

Haworth, B. T., McKinnon, S., & Eriksen, C. (2022). Advancing disaster geographies: From marginalisation to inclusion of gender and sexual minorities. *Geography Compass, 16*(11), e12664.

Haworth, B. T., Cassal, L. C. B., & Muniz, T. (2023). "No-one knows how to care for LGBT community like LGBT do": LGBTIQA+ experiences of *COVID-19* in the UK and Brazil. *Disasters, 47*(3), 484–507. https://doi.org/10.1111/disa.12565

Henrickson, M., Giwa, S., Hafford-Letchfield, T., Cocker, C., Mulé, N. J., Schaub, J., & Baril, A. (2020). Research ethics with gender and sexually diverse persons. *International Journal of Environmental Research and Public Health, 17*(18), 6615.

Hesse-Biber, S. (Ed.). (2014). *Feminist research practice: A primer* (2nd ed.). Sage Publications.

IPSOS. (2021). *LGBT+ pride 2021 global survey. A 27-country Ipsos survey.* Retrieved June 15, 2023, from https://www.ipsos.com/sites/default/files/ct/news/documents/2021-06/LGBT%20Pride%202021%20Global%20Survey%20Report_3.pdf

Kavanaugh, K., & Ayres, L. (1998). "Not as bad as it could have been": Assessing and mitigating harm during research interviews on sensitive topics. *Research in Nursing and Health, 21*(1), 91–97.

Larkin, B. (2019). Pride and prejudice: LGBTIQ community responses to disaster events worldwide. *Australian Journal of Emergency Management, 34*(4), 60–66.

Liamputtong, P. (2007). *Researching the vulnerable: A guide to sensitive research methods.* Sage Publications.

Littler, C., & Joy, P. (2021). Queer considerations: Exploring the use of social media for research recruitment within LGBTQ communities. *Research Ethics, 17*(3), 267–274.

Lowik, A., et al. (2022). *Gender and sex in methods and measurement—Research equity toolkit. "Tool #2: Effective recruitment strategies."* Centre for Gender and Sexual Health Equity, University of British Columbia.

Lukasiewicz, A. (2020). The emerging imperative of disaster justice. In A. Lukasiewicz & C. Baldwin (Eds.), *Natural hazards and disaster justice: Challenges for Australia and its neighbours* (pp. 3–23). Palgrave Macmillan.

McDowell, L. (1992). Doing gender: Feminism, feminists and research methods in human geography. *Transactions of the Institute of British Geographers, 17*(4), 399–416.

McKinnon, S., Gorman-Murray, A., & Dominey-Howes, D. (2017). Disasters, queer narratives, and the news: How are LGBTI disaster experiences reported by the mainstream and LGBTI media? *Journal of Homosexuality, 64*(1), 122–144.

Meagher, L., Lyall, C., & Nutley, S. (2008). Flows of knowledge, expertise and influence: A method for assessing policy and practice impacts from social science research. *Research Evaluation, 17*(3), 163–173.

Minalga, B., Chung, C., Davids, J. D., Martin, A., Perry, N. L., & Shook, A. (2022). Research on transgender people must benefit transgender people. *The Lancet, 399*(10325), 628.

Office for National Statistics. (2022). *Sexual orientation, UK: 2020*. Retrieved June 15, 2023, from https://www.ons.gov.uk/peoplepopulationandcommunity/culturalidentity/sexuality/bulletins/sexualidentityuk/2020

Office of the High Commissioner for Human Rights. (2019). *Data collection and management: An essential component in creating awareness and providing effective measures to address violence and discrimination based on SOGI*. Retrieved June 15, 2023, from https://www.ohchr.org/sites/default/files/2022-01/Report_on_data_summary.pdf

Patel, S. S., Webster, R. K., Greenberg, N., Weston, D., & Brooks, S. K. (2020). Research fatigue in *COVID-19* pandemic and post-disaster research: Causes, consequences and recommendations. *Disaster Prevention and Management: An International Journal, 29*(4), 445–455.

Phillips, B. (2014). *Qualitative disaster research*. Oxford University Press.

Pickering, C., Phibbs, S., Kenney, C., & O'Sullivan, T. (2020). Qualitative research. In R. Kayano, V. Murray, M. Clarke, & E. Chan (Eds.), *WHO guidance on research methods for health emergency and disaster risk management* (pp. 383–401). World Health Organisation. Retrieved June 15, 2023, from https://extranet.who.int/kobe_centre/sites/default/files/pdf/WHO%20Guidance_Research%20Methods_Health-EDRM_4.12.pdf

Robinson, M. (2019). Two-spirit identity in a time of gender fluidity. *Journal of Homosexuality, 67*(12), 1675–1690. https://doi.org/10.1080/00918369.2019.1613853

Rushton, A., & Scarlett, J. (2023). The blame game: Disaster, queerness and prejudice. *Disaster Prevention and Management: An International Journal, 32*(2), 352–363.

Rushton, A., Gray, L., Canty, J., & Blanchard, K. (2019). Beyond binary: (Re) defining 'gender' for 21st century disaster risk reduction research, policy, and practice. *International Journal of Environmental Research and Public Health, 16*(20), 3984. https://doi.org/10.3390/ijerph16203984

Rushton, A., Phibbs, S., Kenney, C., & Anderson, C. (2020). The gendered body politic in disaster policy and practice. *International Journal of Disaster Risk Reduction, 47*, 101648. https://doi.org/10.1016/j.ijdrr.2020.101648

Rushton, A., Phibbs, S., Kenney, C., & Anderson, C. (2021). "She'll be right": The place of gendered emotions in disasters. *Gender, Place and Culture, 29*(6), 793–815. https://doi.org/10.1080/0966369X.2021.1892595

Sou, G. (2021). Concealing researcher identity in fieldwork and social media: Sexuality and speaking for participants. *Area, 53*(3), 473–480.

Spizzirri, G., Eufrásio, R. Á., Abdo, C. H. N., & Lima, M. C. P. (2022). Proportion of ALGBT adult Brazilians, sociodemographic characteristics, and self-reported violence. *Scientific Reports, 12*(1), 11176.

Statistics Canada. (2022). *A statistical portrait of Canada's diverse LGBTQ2+ communities.* Retrieved June 15, 2023, from https://www150.statcan.gc.ca/n1/daily-quotidien/210615/dq210615a-eng.htm

Stats NZ Tatauranga Aotearoa. (2022). *LGBT+ population of Aotearoa: Year ended June 2021.* Retrieved June 15, 2023, from https://www.stats.govt.nz/information-releases/lgbt-plus-population-of-aotearoa-year-ended-june-2021/

Stonewall. (2023). New data: Rise in hate crime against LGBTQ+ people continues, Stonewall slams UK Gov "inaction." *Stonewall*, 5 October. https://www.stonewall.org.uk/about-us/news/new-data-rise-hate-crime-against-lgbtq-people-continues-stonewall-slams-uk-gov-

Tuana, N., & Sullivan, S. (2006). Introduction: Feminist epistemologies of ignorance. *Hypatia, 21*(3), vii–ix.

Valentine, G. (2002). People like us: Negotiating sameness and difference in the research process. In P. Moss (Ed.), *Feminist geography in practice: Research and methods* (pp. 116–126). Blackwell.

Vaughan, M. D., & Rodriguez, E. M. (2014). LGBT strengths: Incorporating positive psychology into theory, research, training, and practice. *Psychology of Sexual Orientation and Gender Diversity, 1*(4), 325–334.

van Anders, S. M., Schudson, Z. C., Beischel, W. J., & Chadwick, S. B. (2023). Feminist and queer science: Principles for research with gender, sex, and sexuality in psychology and beyond. In *Psychology of sexual orientation and gender diversity*, advance online publication. https://doi.org/10.1037/SGD0000646

van Zijll de Jong, S., et al. (2011). Process, practice, and priorities—Reflections on, and suggestions for, undertaking sensitive social reconnaissance research as part of an (UNESCO-IOC) International Tsunami Survey Team. *Earth Science Reviews, 107*, 174–192.

World Health Organisation (WHO). (2020). *Gender and Covid-19.* Advocacy Brief. Retrieved June 15, 2023, from https://www.who.int/publications/i/item/gender-and-*Covid--19*

CHAPTER 16

Recognising Queer Climate Justice: A Queer Analysis of Media Representations of LGBTQ+ Communities Impacted by Hurricane Ian in Florida

Vanessa Raditz

Introduction

> In our movement, we claim no more or less than our human place among the creatures on the planet … This re-energized movement will be, in Suzanne Pharr's eloquent terms, 'not a wedge, but a bridge'; not a point of division, but of expansion and connection. To those who insist on denying us our full humanity, we will insist on the sacred humanity of all people. (Segrest, 1994, pp. 245–46)

At the heart of this chapter is a concern that at a time of accelerating changes in our earth-body, one-dimensional, assimilationist politics are driving wedges between the most vulnerable of our human family on the

V. Raditz (✉)
University of Georgia, Athens, GA, USA
e-mail: vanessa.raditz@uga.edu

D. Dominey-Howes et al. (eds.), *Queering Disasters, Climate Change and Humanitarian Crises*, Sustainable Development Goals Series, https://doi.org/10.1007/978-981-96-3857-4_16

frontlines of climate disaster. This chapter expresses a desperate desire for a coalitional movement that works against the extractive systems at the root of harm to bodies and land. Only such a coalition can foster a queer climate justice movement that rises to Mab Segrest's vision, delivered at the National LGBTQ Task Force's 1993 Creating Change conference, that our queerness be *a bridge, and not a wedge.*

The chapter opens with a discussion of the impacts of Hurricane Ian on LGBTQ+ communities in Florida and how these were compounded by the worsening political, economic, and ecological context. It presents a theoretical framework to guide the analysis, including a distinction between studying *disaster impacts on LGBTQ+ people* compared to a *queer climate justice approach to disasters.* The chapter then provides an analysis of media accounts, using a queer theoretical approach to deconstruct these representations. The chapter emphasises that a 'one-dimensional' approach to LGBTQ+ politics has contributed to 'homonormative' representations that drive wedges between the LGBTQ+ community and other politically targeted communities, which makes coalitional organising towards root-cause solutions harder to achieve.

The chapter accompanies a vignette (this collection, G. Vignette) on *Can't Stop Change: Queer Climate Stories from the Florida Frontlines.* The vignette is a discussion with three of my Co-Directors, on our roles in the creation of this new documentary project. Together, the chapter and vignette propose possibilities of storying LGBTQ+ people in disasters using a liberationist queer lens that aligns with the broader climate justice movement.

Background

Multi-dimensional Disasters in Florida

On the morning of September 24, 2022, as Tropical Storm Ian intensified into a Hurricane in the Caribbean, Florida's Governor, Ron DeSantis, prepared to announce a State of Emergency. Meanwhile, the director of the Pride Community Centre of North Central Florida awoke to the news that someone had thrown a brick through the centre's window (Wiggins, 2022). The incident was part of an uptick in LGBTQ+ hate crimes that has accompanied the rise of hundreds of state legislature bills in the United States of America (USA) that deny civil rights and medical care to LGBTQ+ communities, particularly transgender communities (HRC, 2023b). Six months later, the Human Rights Campaign (HRC) declared their own 'State of Emergency' in response to over 75 anti-LGBTQ+ bills signed into law in 2023 alone (HRC, 2023a). According to journalist Imara Jones, these anti-LGBTQ+ bills are emerging from a nexus of 'dark money, right wing organizations, radical figures and extreme ideology' (Translash, 2021). Florida has led this political movement, passing model bills from right-wing think tanks that have since been adopted in other states (American Civil Liberties Union, 2024; McMahon & Moore-O'Neal, 2018). The brick through the window highlights the local consequences of this worsening political context—emboldened violence against LGBTQ+ people and organisations.

As anti-trans legislation increases, so too is the severity and frequency of Hurricanes, creating a precarious political and ecological landscape. A few days after someone threw a brick through the Pride Centre's window, Category 4 Hurricane Ian made landfall in Southwest Florida devastating coastal communities. While hurricane severity is traditionally measured by wind speed, hurricane cyclones now carry more water than before due to rising sea temperatures, contributing to catastrophic flooding and destruction (IPCC, 2022). Naples, a LGBTQ+ tourist destination known for its annual Pride event (Covello, 2021), was severely impacted by flood waters, including their Pride Centre. Hurricane Ian impacted an estimated 83,000 LGBTQ+ residents of this South Florida region, with some facing water cut-offs, power outages, and economic insecurity weeks after (Queer News Tonight, 2022).

Beneath these headlines of violence and devastation lurks an unspoken story: the 'dark money' networks fuelling anti-trans sentiment and legislation which are deeply entwined with the extractive, fossil-fuel-based economy driving climate change. Alongside Governor Ron DeSantis's attacks on social issues, such as transgender health care, teaching racial and gender justice in schools, and access to abortion and reproductive health care, he has also pioneered legislative attacks on 'Environmental, Social, and Governance' (ESG) investments. The legislation bars state officials from choosing investments with environmental or social benefit if they would reduce fiduciary returns. As the environmental journalism outlet Grist reports:

> At the heart of the issue is the perceived threat ESG poses to the oil and gas industries. It seems that the more companies and investors recognize the need to transition away from fossil fuels, the more ESG opponents have

ramped up efforts to protect the oil and gas sector—a huge source of lobbying money for the Republican party. (Hu, 2023).[1]

Paradoxically, this anti-ESG position comes at the same time as DeSantis champions climate adaptation measures like hardening points to the increasing severity of storms, and the need to increase the state budget for disaster response, recovery, and mitigation (Staff Report, 2023). While DeSantis continues to deny anthropogenic causes of climate change and vehemently opposes actions that would target fossil fuel industries, he simultaneously recognises the threat of climate change for the state's infrastructure and businesses and is taking steps to protect these interests (Nehemas & Mazzei, 2023). What connects these paradoxical political decisions are groups such as the American Legislative Exchange Council (ALEC), which represents the interests of member corporations by writing model bills and connecting lobbyists with politicians. It is perhaps no coincidence that it is also playing a key role in attacks on transgender communities, particularly through their Parents Bill of Rights model legislation (Kotch, 2021a, 2021b; ALEC Action, 2023; Hoban, 2023).

Here we are full circle: a brick through a window, a flooded Pride centre, states of emergency, and the enrichment of the oil and gas industry. Altogether, the multi-dimensionality of compounding social, political, economic, and ecological disasters in Florida provides an interesting case study for thinking about disaster justice.

Theoretical Framework

Queering Disaster Justice Through 'the Justice of Recognition'

The complexity of social and environmental legislation in Florida draws attention to the undergirding politico-economic logic driving increased disaster vulnerability. As Lukasiewicz (2020, p. 44) writes, 'talking about justice immediately involves talking about authorities and institutions which are meant to administer or ensure justice.' Quoting Robert Vershick, Lukasiewicz (p. 46) argues that 'disaster justice is a moral claim on

[1] Grist also reports that progressive environmentalists have long been critical of these investments, but for the opposite reason, believing that they are a form of neoliberal greenwashing that distracts from the urgent need for government regulation on fossil fuel extraction and emissions.

governance, which arises from anthropogenic interventions in nature that incubate environmental crises and magnify their socially and spatially uneven impacts.' In the case of socially vulnerable communities impacted by Hurricane Ian, this is a 'moral claim on governance' of a specific sort borne of the political economy of neoliberalism. This moral claim must simultaneously address the environmental, social, and economic legislation that is precipitating the intensifying hurricanes, exacerbating risks for people in already precarious positions, and siphoning public funds towards wealthy corporate executives.

This chapter argues that attending to justice claims of LGBTQ+ people as a stable identity group with one-dimensional political needs fails to address the complex set of power relations that simultaneously drive climate change, the harm to gender and sexual minorities, and the extinguishing of diversity across all its ecological and social forms.

Queer theory offers useful tools for the emerging field of disaster justice to address this nexus of socioecological power relations. It provides a way of questioning and deconstructing discourses and representations that make socially constructed categories appear natural, including those employed in normative notions of justice. Lukasiewicz (2020, p. 49) suggests that the two most discussed forms of 'justice' are *distributive justice*, which is concerned with the equitable allocation of resources, and *procedural justice*, which is concerned with the process of decision-making and who is considered in those decisions. However, there are other forms of justice, including 'justice as recognition.' Quoting Shreshta and colleagues, Lukasiewicz (p. 50) argues that 'recognition' is not limited to 'different stakeholder groups' but is also about the 'recognition of knowledge systems, justice conceptions and value structures.' This chapter pursues this idea further. While the chapter addresses sexual and gender minority peoples' experiences, its emphasis is not on recognising them as a unique and stable 'stakeholder group.' Rather, it explores the benefits to disaster justice of thinking with the lineages of queer 'knowledge systems' and analytical approaches.

From LGBTQ+ Disaster Studies Towards Queer Climate Justice

There is an emerging field of disaster research on gender and sexual minorities, including a global body of empirical evidence of disparities in disaster preparedness, response, and recovery among LGBTQ+ communities. A review of the literature on the impacts of disasters on LGBTQ+

people by Goldsmith et al. (2022) describes how LGBTQ+ people are disproportionately affected by disasters because of discriminatory response policies, and because of 'contextual vulnerabilities,' such as high rates of poverty, homelessness, disability, chronic illness, and incarceration. In other words, vulnerability is constructed through the everyday disasters of our current political economy.

There is also a legacy of research into the lack of representation of queer vulnerabilities in the mainstream media and an inadequate representation of intersectional vulnerability in LGBTQ+ media. McKinnon et al. (2017) addressed the absence of LGBTQ+ stories in media about disasters in Australia and New Zealand. They raised 'the possibility that the exclusion or absence of queer disaster narratives may contribute to marginality through the media's construction of disasters as experienced exclusively by heterosexual family groups' (p. 122). Their study also addresses how this invisibility within mainstream media is accompanied by inequitable reporting within LGBTQ+ media. They found that these reports were 'heavily weighted toward stories involving gay men,' with little or no reporting of 'the specific vulnerabilities experienced by lesbians and transgender individuals' (p. 139). While this research did not examine differences in stories based on race or class, it provided a starting point for questioning whether 'this inequitable representation potentially creates deeper layers of marginalization and vulnerability within LGBTI communities' (p. 141).

Indeed, media representation of wealthy LGBTQ+ people proliferated through the 1990s and 2000s with the popularity of shows such as *Will and Grace*, *The L Word*, and *Queer as Folk*, creating a 'myth of gay affluence' that obscured the widespread economic injustices experienced by LGBTQ+ people (Hollibaugh & Weiss, 2015; DeFilippis, 2016). This kind of representation is one aspect of what Lisa Duggan (2002) named 'homonormativity,' a political strategy that prioritises assimilation into mainstream society, including consumerism, domesticity, and nationalism. Duggan argues that this strategy, epitomised by the campaigns for gay marriage and inclusion in the military, represents a neoliberal privatisation of public issues.

Homonormativity aligns with Nancy Fraser's framework of *progressive neoliberalism*, a political economy that combines neoliberalism's extraction and enclosure of wealth with progressive politics of diversity, equity, and inclusion (Fraser, 2019). This combination produces an identity politics with meritocratic principles that focuses on the tokenistic representation of select numbers of the most elite members of a minoritised community, while preserving the systems committing violence against most of that community (Fraser, 2019). These representations give the optics of social justice while keeping us locked in the extractive economy driving climate change. The progressive neoliberalism of homonormativity furthers the inequities among Black trans women, who are most targeted by state legislation, interpersonal violence, and climate disasters (Stanley & Smith, 2015; Gossett et al., 2017; Goldsmith et al., 2022).

The social, political, and economic construction of disaster vulnerability among multi-marginalised LGBTQ+ communities can be better addressed through a climate justice lens. Climate justice is the global movement fighting against injustice in the causes and consequences of climate change, and advocating solutions that address the underlying political and economic drivers of social and ecological injustice. These solutions include

> dismantling fossil fuel dependency, defetishization of endless capitalist growth on a finite planet, challenging non-participatory democracy, resisting extractive exploitation of natural resources, confronting racial capitalism and Indigenous erasure. (Sultana, 2022)

They also encompass other approaches that question normalised aspects of our political economy. In recent years, there have been many calls for intersectional approaches to climate justice that incorporate gender and sexuality when theorising the root causes and solutions to the harm to bodies and land (Kaijser & Kronsell, 2014; Mikulewicz et al., 2023). In contrast to assimilationist homonormativity, there is also a long tradition of queer liberationist politics that aligns with climate justice. The politics of queer liberation addresses how the current marginality of LGBTQ+ communities has been constructed through centuries of entanglements between cisheteropatriarchy and white supremacy, racial capitalism, and settler colonialism (Somerville, 2000; Ferguson, 2018; Hong and Ferguson, 2011).

Thinking with Queer Theory

A queer theoretical lens provides a way of crafting a queer climate justice as both a liberationist politic and a representational strategy. This theoretical lens developed in the academy in alignment with queer liberationist social movements that pushed against assimilationist agendas

(Ferguson, 2012). Multiply marginalised and radically politicised activists have worked in coalition with other oppressed groups to make multi-dimensional demands for social change, such as occurred during the Stonewall riots that launched the Gay Liberation movement (Ferguson, 2018). As Michael Warner (1991) argued during the height of the AIDS crisis, the immediate struggles for community survival crafted a broader political consciousness. 'The more lesbians and gay men elaborate our positions in this political environment,' wrote Warner, 'the more we are called upon to consider our resistance to normalized sexuality in terms that are not always initially evident as sex-specific' (p. 7). This included a growing understanding of sexuality's 'embeddedness in modernity, colonialism, structures of civil society, ideologies of liberalism, and the like' (p. 7). This was increasingly in contrast to the normalising political agenda of organisations like the Human Rights Campaign that sought civil rights within the political structure of the United States, rather than questioning the power of the settler colonial state.

Within the academy, queer theory emerged as an alternative to the limitations of gay and lesbian studies, informed by the highly productive analytical approaches of post-structuralism, including the attention to deconstruction and discourse analysis. As Jagose (1997, p. 3) argued, queer theory represented a 'reworking of the post-structuralist figuring of identity as a constellation of multiple and unstable positions.' Rather than researching people with 'deviant' genders and sexualities, queer theorists asked broader questions about the processes by which heterosexuality came to be understood as natural and endowed with the social power that has justified violence to those who do not conform. This shift in attention allowed queer theorists to ask questions about intersecting processes that co-produce the violences of racialisation, impoverishment, disablement, and beyond.

While early queer theorists brought post-structural tools to the study of sex, gender, and sexuality, the next generation of theorists brought this analytical lens to the political economy of neoliberalism. In 2005 Eng, Halberstam, and Muñoz brought together a special double issue of *Social Text* documenting queer's radical political utility. Addressing the compounding crises of neoliberalism and 'corporate domination,' they adopted Duggan's concept of homonormativity to argue that this gay and lesbian discourse is 'rhetorically remapping and recoding freedom and liberation in narrow terms of privacy, domesticity, and the unfettered ability to consume in the "free" market.' In so doing, they argued, it 'abandons a more global critique of capitalist exploitation and domination, state violence and expansion, and religious fundamentalisms and hate' (p. 11). By contrast, they asserted that a queer way of knowing 'insists that we embark on expanded investigations of normalizations and intersectionality' and uplift perverse modernities, 'those queer bodies and knowledges that exist outside the boundaries of sanctioned time and space, legal status, citizen-subjecthood, and liberal humanism' (p. 13). Doing so highlights the many interlocking violences that exceed the one-dimensional concept of 'homophobia' to examine the multi-dimensional processes that create social othering and condemn these others to social and literal death (Cacho, 2012).

Moving from a focus on articulating the disaster vulnerability of LGBTQ+ people to a queer climate justice approach to disasters is useful for enriching the concept of disaster justice. This approach calls for critical attention to strategies for justice that normalise gender and sexually diverse people within systems that contribute to the violence against them. Instead of focusing on LGBTQ+ identity categories, I argue that recognising queer climate justice requires that we attend to 'people who live outside the logics of capital accumulation and bourgeois reproduction' (Cacho, 2012). Doing so can contribute to the climate justice movement's critiques of the political economy driving climate change and inform root-cause solutions to interlocking socioecological crises.

Media Critique

This section uses this queer climate justice lens to analyse the multi-dimensional disasters of Hurricane Ian in Florida. The analysis relies on a Google News search conducted in the Spring of 2023, six months after Hurricane Ian's landfall. The search used 'Hurricane Ian' in combination with a range of identity terms for gender and sexually diverse people in the United States, including Lesbian, Gay, Bisexual, Trans*, Queer, and LGBT*. While the results of the Google search are not exhaustive, they point to some of the limitations of using these 'standard identity terms' on generic platforms like Google, including what these searches fail to uncover. As such, this critique highlights the usefulness of applying a queer theoretical lens in disaster justice research.

Temporal Connection Without Analysis

Most results from the Google News search came from online news website pages with roundups of headlines describing the temporal overlap between the devastation caused by the hurricane and the proposing, passing, and resistance to anti-LGBTQ+ bills, but without analysis linking them. This includes articles cited in this chapter's introduction that describe separately the impacts to LGBTQ+ centres from hate crime violence and hurricane flooding.

One notable example of the timing of these parallel violences is a headline from Queerty, an LGBTQ+ media outlet, declaring that Florida State House Representative Spencer Roach, one of the authors of the 'Don't Say Gay' bill,[2] lost his home in the hurricane (Hudson, 2022). The article first discusses Roach's political stances, then details damages from the hurricane as a separate section. What remained unsaid is that Roach is part of a privileged segment of Florida society that has access to resources to recover, such as home insurance. Meanwhile, preliminary data analysis of the US Census Household Pulse Study (Geiger et al., 2023) suggests 'LGBTQ+ disaster displacement is nearly 2 times higher than cisgender, heterosexual displacement,' and this disparity is greatest for LGBTQ+ people of colour. Furthermore, 'states with more anti-LGBTQ+ policies also generally have higher LGBTQ+ displacement during disasters.'

The article's (lack of) framing seemed to provide readers with a vindictive pleasure at the destruction of Roach's home, as evident in the comments section, where readers left snide remarks saying that this was 'God's wrath' for his bigotry. As one commenter explained, 'it does feel kinda good to throw it back in their faces when they use God to bash us,' referring to the history of conservative religious leaders blaming disasters on the sins of homosexuals (Blumenfeld, 2012). While it may be satisfying to flip the script of 'divine justice' onto a conservative politician, it leads the LGBTQ+ media to focus on Roach as an individual. In so doing, it ignores the links among Roach's politics, the relative protection of the affluent against the worst effects of disasters, and the increasing vulnerability of

[2] HB 1557 in Florida, the 'Parental Rights in Education' act, is colloquially known as the 'Don't Say Gay' bill because it prohibits classroom instruction on sexual orientation or gender identity in K-3 (additional bills have expanded this age range). Other activists have tried to rebrand it 'Don't Say or Be LGBTQ+' to emphasize that it not only restricts education on these topics, it also harms LGBTQ+ youth by forcing schools to non-consensually 'out' LGBTQ+ students to parents who may not be supportive.

gender and sexually diverse communities under Florida's new legislation. It also ignores how these structural issues are linked to the increasing vulnerability of those marginally positioned under climate change.

Homonormativity in LGBTQ+ Media Accounts of Hurricane Ian

Other reports provided very clear examples of homonormativity within LGBTQ+ media accounts of the hurricane's impacts. A key example is the reporting on Hurricane Ian from the LGBTQ+ media outlet Queer News Tonight (2022). The video broadcast, days after Ian made landfall, features interviews with the owner of a gay bar, the publisher of a gay tourism magazine, a drag queen entertainer, and the directors of impacted LGBTQ+ centres. The coverage directly addresses the income precarity among drag performers in the era of climate change. However, this report focused on a white, gay performer who is cis-male presenting in his daily life, and who reported that his savings enabled him to weather the storm financially, a privilege not available to many drag performers (Berkowitz & Liska Belgrave, 2010). The show's focus on LGBTQ+ communities with economic privilege was made more apparent as the anchor read an opening statement from a gay member of the Lee County Tourism Bureau who flew to Chicago to escape the incoming hurricane. His statement expresses his unbearable shock at seeing the devastation of beaches and islands that he has spent years marketing to visitors.

Taken together, this news coverage represents the neoliberal cultures of homonormativity by narrowly addressing the economic impact to LGBTQ+ businesses and tourism. Importantly, previous research from Hurricane Katrina has shown that these forms of gay tourism destinations are quicker to reopen than the centres of support for the most-marginalised in queer communities, which can distract from other experiences and needs. As D'Ooge (2008, p. 23) writes, the focus on reassuring gay tourists that they can return 'has resulted in focusing on the unaffected traditionally gay male neighbourhoods such as the French Quarter while rendering invisible the suffering of LGBT New Orleanians living elsewhere, consisting predominantly of lesbians and African Americans.' These marginalised stories include the spike in convictions of sex workers in the wake of Katrina, who were being charged with 'Solicitation of a Crime Against Nature' (SCAN), a law specifically criminalising sex work involving oral or anal sex that requires the convicted to be registered as a sex offender (Piano & Haywood, 2011). While the law was initially written to target gay men and had fallen out of use over time, it was applied with fervour after the hurricane, predominantly to women of colour and trans women. This included many who were engaging in sex work out of desperation in the post-Katrina economy (Piano & Haywood, 2011).

LGBTQ+ Issues Portrayed as 'Wedge' in Local Media Coverage of Response Efforts

The homonormative representation of LGBTQ+ experiences in disasters is not limited to LGBTQ+ media; it also appears in mainstream media. This was evident in an article from the *Naples Daily News* on the community disaster relief work organised by the Naples Pride Centre, where the journalist's depiction of this relief work differed from the description by the centre's staff and volunteers.

The Naples Pride centre is an example of queer community resilience and care in action. While dealing with their own recovery from flooding, Naples Pride volunteers used their centre to collect supplies. One recipient of these supplies was a neighbouring mobile home park that also flooded during Ian. This form of neighbourhood-level action differs from the economically privileged myopia and one-dimensional queerness described in the previous example, and aligns with many other 'mutual aid' approaches to disaster relief from marginalized communities (Kinane, 2022). As the director of the Naples Pride Centre, Cori Cracium, said in the *Queer News Tonight* broadcast:

> We don't ask them if they're LGBT or not. We know everybody needs help … and because when we ask for help, also everybody helps us … This is the only way we're going to get through this, as a united community. It doesn't matter who you love or what your background is, everyone needs help at this time. (Queer News Tonight, 2022, 30:50–31:30)

Following this humanist response from Cracium, the *Queer News Tonight* anchor moves back to the full panel to ask again about the unique impacts of the hurricane on the LGBTQ+ community.

It is the impact of this one-dimensional approach that is of concern. The article in the *Naples Daily News* emphasises a binary opposition between the seemingly privileged LGBTQ+ benefactors and the impoverished, racialised neighbours in the mobile home community who are presupposed to be uniformly straight and cisgender. According to the journalist, Heithus (2022):

> Pride is used to fighting for *itself*. The LGBTQ+ support organization has dealt with antagonistic Florida laws, attempts to shut down its annual celebration at Cambier Park, even purging of school library books that deal with its issues. But it has turned to fighting for *others* after Hurricane Ian, beginning with the hard-hit Shadowlawn neighbourhood surrounding its quarters. [Italics added]

This exemplifies how framing LGBTQ+ issues as special rights 'others' LGBTQ+ people and communities. This othering drives a wedge between LGBTQ+ communities and those groups portrayed as being at the forefront of harm. Cracium's comments bring us back to Segrest's (1994) opening quote at the beginning of this chapter and her desire that queerness be a bridge, not a wedge, 'not a point of division but an expression of connection' (pp. 245–246).

Absent Coverage of Any Uniquely Targeted Discrimination Towards LGBTQ+ People

One of the limitations of the Google search conducted for this project is that media reports tend to focus on the dramatic impacts of disasters on people and communities. One of the consequences of this focus is, once again, to background the broader structural forces and how they lead to increased risks and impacts of disasters on multiply marginalised gender and sexually diverse people. This could explain a key finding in this media search: While Goldsmith et al.'s review (2022) listed many examples of targeted discrimination towards LGBTQ+ people, particularly within shelters and encounters with faith-based disaster relief organisations, the Google search failed to find similar reports following Hurricane Ian. This is despite public reports documenting extensive hurricane damage, with estimates of upwards of 10,000 people housed in emergency shelters (Robertson et al., 2022; Reed, 2022). In part, this absence may reflect a lack of understanding of how systemic discrimination impacts access to relief services within the media. I suggest that it also may reflect the limitations of using LGBTQ+ identity terms when attempting research that foregrounds these systemic processes.

Personal reflections written after this media search help highlight these limitations and emphasise the difference between an LGBTQ+ disaster studies approach as compared to a queer climate justice lens:

> As I scrolled through page after page of stories, I found myself longing for a story with a clear and compelling example of homophobic discrimination, as if finally finding that one story would be proof that this is an issue worth attending to. Hours into the search, not finding any, I kept trying to make sense of this absence: *How could there be no stories of targeted discrimination? With the general statewide political climate, were LGBTQ+ people simply avoiding these places? Were they closeting themselves while in them? Were they embarrassed about speaking to the media? Was the media disinterested in these stories?* Catching myself in this stream of speculation, I put back on my queer theoretical lens to ask myself: *Why am I desiring this one-dimensional story? What difference would it make to finally find* one *story of homophobia along my hunt, amidst thousands of stories on the extensive loss of life and massive dispossession?* When I reoriented my gaze from the search for one-dimensional violence against LGBTQ+ people to the broader socioecological processes at work, a thousand stories came to me, carrying in them the thousands of different ways that people are constructed as perverse, pathological, unworthy, and disposable. (Raditz, 2023)

My notes call attention to the different perspective provided by a queer theoretical lens, where 'queer' is not being used as a synonym for LGBTQ+ identities. Instead, queer represents a critical analytical framework that deconstructs taken-for-granted social, political, and economic processes that render some communities 'normal' and legible, and others as deviant or 'abnormal.' This has been an important distinction within critical queer studies, such as Eng et al.'s insistence that queer remains a 'political metaphor without a fixed referent' (2005, p. 1). Recognising this queer way of knowing opens different ways of doing this media search, leading to multi-dimensional insights that lend themselves to coalition-building possibilities.

For example, a media headline reading 'Lee County Skipped Evacuation of Jail Located in Evacuation Zone' (DeLuca, 2022), despite containing no LGBTQ+ identity terms, can be read through a queer climate justice lens. As I wrote in my initial notes:

> I connect the dots to the disproportionate number of gender and sexually diverse people who are criminalized and incarcerated. Rates are three times

> higher than the general population for LGB people (Meyer et al., 2017) and even higher among transgender communities. They are higher still among Black trans people, nearly half of whom have been incarcerated at some point in their life (Grant et al., 2011, p. 163). I think about the reasons for these rates, including the explicit discrimination by police officers, but also the broader criminalization of poverty, homelessness, neurodiversity, and survival economies like sex work (Stanley & Smith, 2015*), and how incarceration creates disaster vulnerability* amongst all these many marginalized communities in our country. And then I remember what I have learned about the prison industrial complex and the ways carceral systems support the neoliberal extraction of wealth (Gilmore, 2007; Story, 2019). This then reminds me of ALEC's role in proliferating private prisons through model bills (Cooper et al., 2016), similar to their model bills enabling anti-trans policies. All of a sudden, the pieces click into place … What this change in perspective provided for me was the opportunity to release the obsession with finding that one story of one-dimensional, homophobic discrimination against LGBTQ+ people, and to instead look at the broader story that was omnipresent in my search: the marginalization and criminalization of 'others' as a tool of maintaining the extractive economy. (Raditz, 2023)

A narrow focus on the immediate impacts of disaster on LGBTQ+ people can mask the ways in which they are implicated in broader, interacting systems. My argument is not that homophobia and transphobia do not exist or that the individual, personal harms to LGBTQ+ people during a disaster don't matter. Rather, this chapter suggests that a singular focus on finding *that* story can keep us from looking at the multi-dimensional processes of how certain people are constructed as disposable within the economic systems driving climate change. This creates epistemological impediments that prevent researchers from questioning how cisheteronormativity benefits an extractive, fossil-fuel-based economy, why corporations would seek to invest in maintaining this system of social control, and what social movements can do about it. It is these questions that are of greatest concern for queer climate justice.

Conclusion

This chapter challenges scholars of disaster justice who focus on people with LGBTQ+ identities to consider the insights that can be gained from engaging a queer climate justice lens. It argues that the goal of LGBTQ+ disaster studies has largely been to highlight the vulnerability of LGBTQ+

identified people, whereas the concern of queer climate justice is to identify the interlocking social, political, and economic systems that simultaneously drive climate change and social death. The chapter demonstrates that this queer way of knowing provides tools for better understanding systems that perpetuate violence to people and the planet, and that accessing this lens requires that we learn to deconstruct discourses of homonormativity.

Recognising queer climate justice may foster an understanding of queerness as a 'bridge, not a wedge.' Instead of seeking unique experiences of LGBTQ+ people in disasters, this approach attempts to bridge the experiences of gender and sexually diverse people with systems that connect all communities who are positioned marginally, working within a politics of climate justice. Fostering this understanding may open pathways to new coalitions. As Farhana Sultana (2022, p. 122) describes in her articulation of Critical Climate Justice, when 'confronting capitalist extractivism and climate colonialism … collaboration becomes vital.' She emphasises that this coalitional collaboration is a necessarily non-linear process that requires multi-dimensional thinking:

> Ultimately, there is no singular or ideal climate justice for all, no singular arrival point, but a becoming, of doing more and better, unlearning to relearn, and continually taking stock in various collectivities for equitable and transformative praxis. (p. 122)

There are many examples of more liberatory climate justice efforts led by gender and sexually diverse people that do not show up when you use LGBTQ+ search terms. Instead of focusing on 'identity' as a stable and distinct category for justice claims, these initiatives emphasise 'queer' as a term of becoming-together. This is the focus of the vignette (this collection, Chap. 17) that accompanies this chapter.

This chapter expresses the hope that by recognising the emerging perspective of the queer climate justice movement, we can better engage with visionary strategies. Together, we can challenge human supremacy above the earth and devise new futures of collective liberation. In honour of the cycles of our earth-body, I would like to close where we began, with the dream of a queer movement where 'we claim no more or less than our human place among the creatures on the planet.'

Conflict of Interest The author has no conflict of interests.

References

ALEC Action. (2023). Support fundamental parental rights. *ALEC Action*, 7 March. https://www.alecaction.org/update/support-fundamental-parental-rights/

American Civil Liberties Union. (2024). Mapping attacks on LGBTQ rights in U.S. State Legislatures in 2024. https://www.aclu.org/legislative-attacks-on-lgbtq-rights-2024

Berkowitz, D., & Liska Belgrave, L. (2010). "She works hard for the money": Drag queens and the management of their contradictory status of celebrity and marginality. *Journal of Contemporary Ethnography, 39*(2), 159–186.

Blumenfeld, W. J. (2012). God and natural disasters: It's the gays' fault? *The Huffington Post*, 5 November. https://www.huffpost.com/entry/god-and-natural-disasters-its-the-gays-fault_b_2068817

Cacho, L. M. (2012). *Social death: Racialized rightlessness and the criminalization of the unprotected*. NYU Press.

Cooper, R., Heldman, C., Ackerman, A. R., & Farrar-Meyers, V. A. (2016). Hidden corporate profits in the US prison system: The unorthodox policy-making of the American Legislative Exchange Council. *Contemporary Justice Review, 19*(3), 380–400.

Covello, R. (2021). Exploring gay Naples: Everything you need to know. *OutCoast*. https://outcoast.com/gay-florida-travel-exploring-naples-everything-you-need-to-know/#:~:text=Bambusa's%20Bar%20%26%20Grill%20is%20the,alone%20is%20worth%20the%20visit!

D'ooge, C. (2008). Queer Katrina: Gender and sexual orientation matters in the aftermath of the disaster. In B. Willinger (Ed.), *Katrina and the women of New Orleans* (pp. 22–24). Tulane University.

DeFilippis, J. N. (2016). 'What about the rest of us?' An overview of LGBT poverty issues and a call to action. *Journal of Progressive Human Services, 27*(3), 143–174.

DeLuca, A. (2022). Lee County skipped evacuation of jail located in evacuation zone. *Miami New Times*, 28 September. https://www.miaminewtimes.com/news/hurricane-ian-lee-county-refused-to-evacuate-jail-located-in-evacuation-zone-15366221

Duggan, L. (2002). The new homonormativity: The sexual politics of neoliberalism. In R. Castronovo, D. D. Nelson, & D. E. Pease (Eds.), *Materializing democracy: Toward a revitalized cultural politics* (pp. 175–194). Duke University Press. https://doi.org/10.1515/9780822383901-008

Eng, D. L., Halberstam, J., & Muñoz, J. E. (2005). What's queer about queer studies now? *Social Text, 23*(3–4), 1–17.

Ferguson, R. A. (2012). *The reorder of things: The university and its pedagogies of minority difference*. University of Minnesota Press.

Ferguson, R. A. (2018). *One-Dimensional Queer*. John Wiley and Sons.

Fraser, N. (2019). *The old is dying and the new cannot be born: From progressive neoliberalism to Trump and beyond.* Verso Books.

Geiger, J., Méndez, M., & Goldsmith. (2023). Amplified harm: LGBTQ+ disaster displacement. *University of California Irvine School of Social Ecology*, 28 November. https://socialecology.uci.edu/news/amplified-harm-lgbtq-disaster-displacement

Gilmore, R. W. (2007). *Golden Gulag: Prisons, surplus, crisis, and opposition in globalizing California.* University of California Press.

Goldsmith, L., Raditz, V., & Mendez, M. (2022). Queer and present danger: Understanding the disparate impacts of disasters on LGBTQ+ communities. *Disasters, 46*(4), 946–973.

Gossett, R., Stanley, E. A., & Burton, J. (Eds.). (2017). *Trap door: Trans cultural production and the politics of visibility.* The MIT Press.

Grant, J. M., Mottet, L. A., Tanis, J., Harrison, J., Herman, J. L., & Keisling, M. (2011). *Injustice at every turn: A report of the national transgender discrimination survey.* National Centre for Transgender Equality and National Gay and Lesbian Task Force.

Heithus, H. H. (2022). Naples Pride's Butterfly Express was on the streets, helping Hurricane Ian victims early. *Naples Daily News*, 13 October. https://www.naplesnews.com/story/weather/hurricanes/2022/10/13/hurricane-ian-naples-pride-reaches-out-victims-supplies-help/10471048002/

Hoban, R. (2023). Legislature revives stalled NC Parents Bill of Rights targeting trans kids. *NC Health News*, 28 June. https://www.northcarolinahealthnews.org/2023/06/28/legislature-revives-stalled-bill-targeting-schools-gender-issues/

Hollibaugh, A., & Weiss, M. (2015). Queer precarity and the myth of gay affluence. *New Labor Forum, 24*(3), 18–27. https://doi.org/10.1177/1095796015599414

Hong, G. K., & Ferguson, R. (Eds.). (2011). *Strange affinities: The gender and sexual politics of comparative racialization.* Duke University Press.

HRC. (2023a). *LGBTQ+ Americans under attack: A report and reflection on the 2023 State Legislative Session.* Human Rights Campaign. https://hrc-prod-requests.s3-us-west-2.amazonaws.com/Anti-LGBTQ-Legislation-Impact-Report.pdf

HRC. (2023b). *The epidemic of violence against the transgender and gender non-conforming community in the United States: The 2023 report.* Human Rights Campaign. https://reports.hrc.org/an-epidemic-of-violence-2023#

Hu, A. (2023). What is ESG, the investment strategy under attack by republicans? *Grist.* https://grist.org/economics/what-is-esg-the-investment-strategy-under-attack-by-republicans/

Hudson, D. (2022). Florida GOP Lawmaker Who Took Aim at "Woke" Disney Sees Home Destroyed by Hurricane. *Queerty*, 30 September. https://www.

queerty.com/florida-gop-lawmaker-took-aim-woke-disney-sees-home-destroyed-hurricane-20220930

IPCC. (2022). *Climate change 2022: Impacts, adaptation and vulnerability.* Working Group II Contribution to the IPCC Sixth Assessment Report. Cambridge University Press.

Jagose, A. (1997). *Queer theory: An introduction.* NYU Press.

Kaijser, A., & Kronsell, A. (2014). Climate change through the lens of intersectionality. *Environmental Politics, 23*(3), 417–433.

Kinane, S. (2022). Mutual aid disaster relief helps Florida victims of Hurricane Ian by "sharing power with each other". *WMNF 88.5FM Community Radio.* https://www.wmnf.org/mutual-aid-disaster-relief-helps-florida-victims-of-hurricane-ian-sharing-power-with-each-other/

Kotch, A. (2021a). ALEC leaders boast about anti-abortion, anti-trans bills. *Exposed by Centre for Media and Democracy* (blog), 4 October. https://www.exposedbycmd.org/2021a/10/04/alec-leaders-boast-about-anti-abortion-anti-trans-bills/

Kotch, A. (2021b). Leading free-market policy network enabling anti-LGBTQ hate. *Southern Poverty Law Centre,* 7 December 7. https://www.splcentre.org/hatewatch/2021b/12/07/leading-free-market-policy-network-enabling-anti-lgbtq-hate

Lukasiewicz, A. (2020). The emerging imperative of disaster justice. In A. Lukasiewicz & C. Baldwin (Eds.), *Natural hazards and disaster justice: Challenges for Australia and its neighbours* (pp. 3–24). Palgrave Macmillan.

McKinnon, S., Gorman-Murray, A., & Dominey-Howes, D. (2017). Disasters, queer narratives, and the news: How are LGBTI disaster experiences reported by the mainstream and LGBTI media? *Journal of Homosexuality, 64*(1), 122–144.

McMahon, A. and Moore-O'Neal, W. (dirs.) (2018) *This little light: A Black feminist worker's story.* Jaliyah Consulting. https://www.thislittlelightfilm.com/

Meyer, I. H., Flores, A. R., Stemple, L., Romero, A. P., Wilson, B. D. M., & Herman, J. L. (2017). Incarceration rates and traits of sexual minorities in the United States: National Inmate Survey, 2011–2012. *American Journal of Public Health, 107*(2), 267–273.

Mikulewicz, M., Caretta, M. A., Sultana, F., & Crawford, N. J. W. (2023). Intersectionality and climate justice: A call for synergy in climate change scholarship. *Environmental Politics, 32*(7), 1275–1286.

Nehemas, N., & Mazzei, P. (2023). DeSantis, leading a state menaced by climate change, shrugs off the threat. *The New York Times,* 8 September. https://www.nytimes.com/2023/09/08/us/politics/desantis-florida-storm-climate-change.html

Piano, D., & Haywood, D. (2011). Working the streets of post-Katrina New Orleans: An interview with Deon Haywood, Executive Director, Women with a Vision, Inc. *Women's Studies Quarterly, 39*(3/4), 201–218.

Queer News Tonight. (2022). Hurricane IAN—Southwest Florida Prides | Gay Bars | Pride Centre | Devastation In Our LGBTQ Community [Video]', 5 October. https://www.youtube.com/watch?v=5hcLEYLZpYU

Raditz, V. (2023). *Homonormativity to queer climate justice*. Speaker's notes for presentation at National Women's Studies Association Conference. Unpublished.

Reed, J. (2022). Weeks after Hurricane Ian, hundreds of Floridians remain in shelters. *The New York Times*, 19 October. https://www.nytimes.com/2022/10/19/us/hurricane-ian-shelters.html

Robertson, L., Flechas, J., & Padró Ocasio, B. (2022). About 10,000 people still in Florida shelters after Hurricane Ian, Officials Say. *Miami Herald*, 3 October. https://www.miamiherald.com/news/weather/hurricane/article266659011.html

Segrest, M. (1994). A bridge, not a wedge. In *Memoir of a race traitor* (pp. 229–246). South End Press.

Somerville, S. B. (2000). *Queering the colour line*. Duke University Press.

Stanley, E., & Smith, N. (Eds.). (2015). *Captive genders: Trans embodiment and the prison industrial complex*. AK Press.

Story, B. (2019). *Prison land: Mapping carceral power across neoliberal America*. University of Minnesota Press.

Sultana, F. (2022). Critical climate justice. *The Geographical Journal*, *188*(1), 118–124.

Translash. (2021). *The anti-trans hate machine: A plot against equality—Trailer*. The TransLash Podcast with Imara Jones. https://translash.org/podcasts/the-anti-trans-hate-machine/the-anti-trans-hate-machine-a-plot-against-equality-trailer/

Warner, M. (1991). Introduction: Fear of a queer planet. *Social Text*, *29*, 3–17.

Wiggins, C. (2022). Florida police call LGBTQ+ Pride Centre vandalism a hate crime. *Advocate*. https://www.advocate.com/news/2022/9/26/florida-police-call-lgbtq-pride-center-vandalism-hate-crime

CHAPTER 17

Suck It up Princess: The Queer Back Lash

Dale Dominey-Howes

Activism and research have focused on the experiences, needs, and capacities of gender and sexual diverse people impacted by disasters, climate change, and humanitarian crises. However, there has been limited interest in understanding and addressing the impacts on those professional activists and researchers who undertake this work, many of whom identify as gender or sexually diverse. In this vignette I discuss my experiences as a researcher, practising disaster risk manager, and cis-gender gay man—experiences that have, largely, been ignored.

I have worked internationally in various disaster-impacted places involving a combination of disaster response and/or conducting interviews with survivors. Previously, I realised that the collective and compounding effects of repeated post-disaster work had triggered vicarious trauma. It is hard dealing with death, body recovery and identification, injury, loss, trauma, and suffering. Trauma is nasty and leaves its marks on our minds and bodies. Following counselling, I wrote about my trauma to learn about and live with it, and take back control (Dominey-Howes, 2015). An

D. Dominey-Howes (✉)
The Sydney Environment Institute and School of Geosciences,
The University of Sydney, Sydney, NSW, Australia
e-mail: dale.dominey-howes@sydney.edu.au

D. Dominey-Howes et al. (eds.), *Queering Disasters, Climate Change and Humanitarian Crises*, Sustainable Development Goals Series, https://doi.org/10.1007/978-981-96-3857-4_17

unexpected consequence was that in sharing my experiences with trusted colleagues, we became aware that professionals in varied disciplines were similarly suffering (e.g. anthropology, ecology, environmental studies, geography, history, law, and sociology).

Together with extraordinary colleagues including my friend Associate Professor Danielle Drozdzewski, many have contributed to advancing a branch of scholarship examining the 'secondary or vicarious trauma' professionals experience when hearing, documenting, reporting, and reliving traumatising events (e.g., Drozdzewski and Dominey-Howes, 2015, Markowitz, 2021; Pihkala, 2020). In November 2015, Danielle and I co-edited a Special Issue of *Emotion, Space and Society* focusing on *researcher vicarious trauma*. My experience with and of vicarious trauma continues, in this case, in relation to my work with gender and sexual diverse disaster survivors.

It may seem indulgent to focus on professionals rather than gender and sexual diverse survivors but like other workers in disaster studies and relief and recovery, I am not devoid of or separated from my emotions including compassion, love, anger, grief, and frustration (Plutchik, 2001). Like others, my work following disasters with gender and sexual diverse people aggravated my vicarious trauma. However, I have used those emotions and my position as a queer *AND* disaster management 'insider' to galvanise my research, policy development, and activist agenda. I use these experiences to drive change in government policy and professional practice, reflecting the enabling, doubled position of 'insider-outsider' described by black feminists (Hill Collins, 1986: S18). I seek to ensure gender and sexual diverse people, and their families are included in disaster risk management policy and planning and that the well-being of queer practitioners is included in disaster risk reduction professional practice. This reflects a feminist tradition of scholarship that pushed back on white, heterosexist and cis-normative oppression to protect women from discrimination and marginalisation (Lorde, 1984; Hill Collins, 1986). This tradition identified the need for professional self-care and practical solutions for the women who did the difficult work of 'pushing back'.

There is a resurgence in this approach to self-care, led by feminist human geographers, arguing for an 'emotional turn in the social sciences' (Caretta et al., 2018, p. 261), particularly in response to the masculinist pressures of working in neoliberal sectors (Drozdzewski, 2015; Mountz et al., 2015; Caretta et al., 2018; Jones & Whittle, 2021).

In 2013, Andrew Gorman-Murray and I applied for a prestigious Australian Research Council (ARC) Discovery Grant, 'Queering disasters in the Antipodes: investigating the experiences of LGBTI people in disasters'. The ARC is an independent (federal government funded body) employing rigorous, independent and expert peer-review where applicants provide a detailed response to assessor comments. The ARC College of Experts then makes funding recommendations to the Minister for their approval. Our successful application was awarded $325,000+.

At the time, the federal government was the Australian Labor Party. The opposition was a conservative coalition of the Liberal and National parties. Immediately, we became the focus of intense and at times highly vitriolic personal attacks including from Shadow Ministers of the opposition in the Federal House of Representatives (Parliament). A media release dated 5th September by Jamie Briggs MP announced:

> HOW will all we ignoramuses cope without research into queering disasters? A COALITION government, if elected, will crack down on Labor's addiction to waste by auditing increasingly ridiculous research grants.

Briggs' attack was picked up, reported, and added to by two right-wing commentators—Andrew Bolt and Gerard Henderson—in blog posts and by journalist Rebecca Urban in *The Australian*, one of the nation's leading, right-of-centre, newspapers.

All had negative things to say about our grant which they appeared to know little about and didn't seem to have read. Nor did any of the commentators take the time to contact us for comment, except Rebecca Urban who my university advised I ignore so as not to fan the flames. One of Henderson's most gratuitous insults was to ridicule the waste of money on academics who both happen to have double barrel surnames (Henderson, 2014). Henderson went on to encourage his readership to respond to our surveys and research so we may or may not be able to write a paper about a non-issue in an unread, insignificant journal. Bolt (2013) simply denigrated the value of the grant awarded to us.

Urban (2016) at *The Australian*, following publication of policy analysis I had led, questioned my academic qualifications, global expertise, and ability to lead ground-breaking research.

Urban (2016) claimed that we had 'caused an outcry from charities' by pointing out that faith-based charities who had received government grants to provide relief and recovery services in Australia had also been granted faith-based legal exemptions that allowed them to withhold services to the queer community. No, it had not caused an outcry. Initially, I was deeply hurt and offended, but with time, I took it as a mark of success that I got under their collective skins, irritating them so much they needed to vent.

Insults ignored, we continued our research and published the data. Attacks escalated. In December 2019, *The Australian* came at me hard. One article was accompanied by a photo of me lifted from a private Facebook post. *The Australian* selected an especially unflattering photo, changed it to black and white, dramatically darkening the image. To my eye it made me look sinister and monstrous. The paper has continued to target gender and sexual diverse communities, with the trans community their latest.

As we continued and published, I was invited by media outlets to write for newspapers, appear on radio and TV, and specifically write for the Australian edition of *The Conversation* (Dominey-Howes, 2021). The relentless trolling continued. Comments included:

> For Fucks Sake!!!
> More left-wing gender bullshit
> Oh for fucks sake, identity politics off the planet
> Were left-handed people treated with respect?

Even gender and sexual diverse individuals had hurtful commentary:

> I just HATE the way LGB is on the front of this. We're just homos who have nothing to do with the current madness.

Like many other out gay men, I have often experienced public abuse including verbal insults, threats to my physical safety, actual physical violence as a teenager that left me in hospital, and online attacks. However, as a professional, no one had forewarned me that I might experience such abuse. I was NOT trained to handle this. At the time there was no training for researchers who deal with subject matter that provokes personal attacks from the media and public. I thought of myself as a public civil servant doing his job that involved action-oriented research and policy. No one warned me, and more worryingly no one provided help. One less than sympathetic colleague simply said when I was discussing the personal attacks, 'Suck it up princess! You got the grant so just deal with it'.

Globally, approximately 80% of all disasters are weather- and climate-related and the future is bleak (Sewell et al., 2016). Since many Australian disasters are weather- and climate-related, and I am a valued expert and excellent communicator, I am regularly approached by the media. In Australia we have the Australian Science Media Centre (ausSMC) that connects journalists with scientific experts—in my case, to discuss and present information and data on extreme weather- and climate-related events and/or disasters as they occur. I have been privileged to make hundreds of media contributions. I love this work and feel a moral and ethical duty to communicate my knowledge, since both my salary and grants are publicly funded.

During the Covid-19 pandemic as various Australian states and territories and then the whole nation were locked down, daily media conferences were held by political leaders, accompanied by commentary from the nation's leading infectious disease experts—mostly women. As widely reported, female academics, lawyers, politicians, and journalists—in particular, women of colour who speak publicly on feminism, domestic violence, race, marginality—experience disproportionate levels of online abuse, hate, trolling, and threats of violence, rape, and death. This abuse has many and various negative impacts and consequences (Gardiner, 2018; Southern & Harmer, 2019; Trott, 2021).

Perhaps it is no surprise, then, that those expert and courageous women became the focus of insidious and awful trolling. Consequently, ausSMC conducted a national survey of professionals who comment in the media and the consequent trolling, and the impacts on their emotional and mental well-being. I was one contributor (ausSMC, 2021). This led to a publication on the impact professionals face in doing the job asked of them by their employers, and this formed part of a paper published in *Nature* (Nogrady, 2021). Following the ausSMC and *Nature* articles, I was interviewed by Nicolle White of the Australian Broadcasting Corporation (ABC).

I reproduce here in full White's writing (with permission):

> One such scientist is Dale Dominey-Howes, a professor of hazard and disaster risk sciences at Sydney University who has spoken publicly about natural disasters and climate change.
>
> Speaking with the media is usually an enjoyable part of his role, he said, including because it makes him feel he's giving back to the community. But increasingly it's exposing him and his colleagues to threats of harm.

> His worst experiences of online abuse followed a campaign to tackle climate change myths.
>
> 'I was getting trolled. I was getting insults,' Professor Dominey-Howes said. 'Somehow, I don't know how, individuals got a hold of my mobile and home address, so I was getting hate mail and death threats and poison letters where people had cut bits out of newspapers to create [messages like], "I'm going to kill you", ",I know where you live", "You're an idiot", "Watch yourself". And that was being sent to my home ... It did affect my emotional and mental wellbeing and I spent much of the year away from doing that work because it was just so overwhelming'. (White, 2022)

Yes, death threats sent to my home! My employer did step in, and the police became involved. Legal action was taken, and I was given space to 'relax'.

These experiences led to an increase in my underlying anxiety, poor mental well-being, and further vicarious trauma. As noted by Conti (2021), repeated exposure to trauma compounds trauma's impacts leading to a 'destabilising of self'—something I experienced. Consequently, I deleted all my social media accounts and reluctantly withdrew from all public engagements and outreach activities.

As a professional, no one tells you, nor is there any training about the harm you might face working on specific topics such as domestic violence, climate change, disasters, and gender and sexual diverse people. THIS MUST STOP! In addition to training in their disciplinary fields such as research techniques, statistical analysis, and so on, future professionals require training on how to deal with the emotional and mental burdens and impacts of researching potentially traumatising topics and events. There isn't space here to provide examples that activists and researchers might use to anticipate and manage vicarious trauma. However, as a start, others might read about self-care and compassion in Calgaro (2015) and Eriksen and Ditrich (2015).

I will not stop researching and caring for gender and sexual diverse people impacted by disasters, climate change, and humanitarian crises. However, I have and am suffering from the abuse I have received. I hope this vignette prompts others to develop appropriate training, care, and support that prepare researchers and practitioners for the media and public backlash that might accompany their research and minimises its potentially traumatic impacts on their mental health and well-being.

Conflict of Interest The research that this vignette is based on was supported by a Discovery grant from the Australian Research Council, DP130102658. The author has no other conflicts of interest.

References

ausSMC. (2021). Pandemic levels of abuse directed at *COVID-19* experts. Available here: EXPERT REACTION: Pandemic levels of abuse directed at*COVID-19*experts—Scimex.

Bolt, A. (2013). I mean. Seriously? *The Sun Herald*. Retrieved September 11, 2023, from https://www.heraldsun.com.au/blogs/andrew-bolt/i-mean-seriously/news-story/68464707c322834e2d2ed78276f555ac

Calgaro, E. (2015). If you're vulnerable and you know it raise your hand: Experiences from working in post-tsunami Thailand. *Emotion, Space and Society, 17*, 45–54.

Caretta, M., et al. (2018). "Who can play *this* game?" The lived experiences of doctoral candidates and early career women in the neoliberal university. *Journal of Geography in Higher Education, 42*(2), 261–275.

Conti, P. (2021). *Trauma: The invisible epidemic* (1st ed.). Vermilion, 175pp.

Dominey-Howes, D. (2015). Seeing 'the dark passenger'—Reflections on the emotional trauma of conducting post-disaster research. *Emotion, Space and Society, 17*, 55–62.

Dominey-Howes, D. (2021). "You never know if you'll be treated properly and with respect": voices of LGBTIQA+ people who lived through disasters. *The Conversation*, 15th February 2021.

Drozdzewski, D. (2015). Retrospective reflexivity: The residua and subliminal repercussions of researching war. *Emotion, Space and Society, 17*, 30–36.

Drozdzewski, D., & Dominey-Howes, D. (2015). Research and trauma: Understanding the impact of traumatic content and places on the researcher. *Emotion, Space and Society, 17*, 17–21.

Eriksen, C., & Ditrich, T. (2015). The relevance of mindfulness practice for trauma-exposed disaster researchers. *Emotion, Space and Society, 17*, 63–69.

Gardiner, B. (2018). "It's a terrible way to work:" What 70 million readers' comments on the Guardian revealed about hostility to women and minorities online. *Feminist Media Studies, 18*(4), 592–608.

Henderson, G. (2014). Gerard Henderson's Media Watch Dog—Issue No. 245. 3 October 2014. Retrieved September 11, 2023, from https://thesydneyinstitute.com.au/blog/issue-245/

Hill Collins, P. (1986). Learning from the outsider within: The sociological significance of black feminist thought. *Social Problems, 33*(6), S14–S32.

Jones, C., & Whittle, R. (2021). Researcher self-care and caring in the research community. *Area, 53*(2), 381–388.

Lorde, A. (1984). *Sister outsider: Essays and speeches*. Crossing Press.
Markowitz, A. (2021). The better to break and bleed with: Research, violence and trauma. *Geopolitics, 26*(1), 94–117.
Mountz, A., et al. (2015). For slow scholarship: A feminist politics of resistance through collective action in the neoliberal university. *An International Journal for Critical Geographies, 14*(4), 1235–1259.
Nogrady, B. (2021). Scientists under attack. *Nature, 598*, 250–253.
Pihkala, P. (2020). The cost of bearing witness to the environmental crisis: Vicarious traumatization and dealing with secondary traumatic stress among environmental researchers. *Social Epistemology, 34*(1), 86–100.
Plutchik, R. (2001). The nature of emotions. *American Scientist, 89*, 344–350.
Sewell, T., et al. (2016). Natural disaster declarations in New South Wales, Australia between 2004 and 2014. *Nature Scientific Reports, 6*, 36369. https://doi.org/10.1038/srep36369
Southern, R., & Harmer, E. (2019). Othering political women: Online misogyny, racism and ableism towards women in public life. In K. Lumsden & K. Harmer (Eds.), *Othering online: Exploring digital violence and discrimination on the web* (pp. 187–210). Palgrave.
Trott, V. (2021). Networked feminism: Counterpublics and the intersectional issues of #MeToo. *Feminist Media Studies, 21*(7), 1125–1142.
Urban, R. (2016). Retrieved September 11, 2023, from Outrage over gay protocols for crises in Australia | The Wentworth Report
White, N. (2022). How online abuse in Australia is shutting down crucial debates and damaging democracy. *The Australian Broadcasting Corporation*, 13th February 2022.

CHAPTER 18

Becoming Professional Allies: Opportunities to Advance SOGIESC Inclusion in UK Emergency Management

Louise Elstow and Matthew Hogan

Introduction

In 2009, I (Louise) was a volunteer registering evacuees at a rest centre from a tower block fire in London. Our forms recorded who was together and any special needs. I spoke to a male evacuee, who was there with his male partner but who did not explicitly state his sexual orientation, gender identity, gender expression and sex characteristics (SOGIESC) status. Nor did I specifically inquire about it. He asked if it was possible for the fire service to return to their home to retrieve antiretroviral medication, as it had been left behind during evacuation. The standard method for accessing medication in a rest centre is for replacements to be dispensed from a local pharmacy. However, this medication had been dispensed from a

L. Elstow (✉)
Fynbos Consulting Limited and University of Cambridge, Cambridge, UK
e-mail: lfe21@cam.ac.uk

M. Hogan
London Resilience Unit, London, UK

D. Dominey-Howes et al. (eds.), *Queering Disasters, Climate Change and Humanitarian Crises*, Sustainable Development Goals Series, https://doi.org/10.1007/978-981-96-3857-4_18

specialist pharmacy and could not be replaced. After speaking with responders and the man, we determined the best course of action would be to guide firefighters to the location of the medication by drawing a map of his home. They were able to re-enter the property and reunite the man with his medication promptly.

This example illustrates that it's not always necessary during disaster to know someone's sexual orientation, gender identity, gender expression and sex characteristics (SOGIESC). However, it can be helpful to both responders and the individual concerned, to determine whether specific needs arise for SOGIESC-diverse individuals that might not be addressed by standardised policies, documentation and professional practices. My volunteer training taught me to consider the ways in which people might be vulnerable in emergencies but did not include any specific awareness of SOGIESC-diverse characteristics. Despite this incident taking place over 15 years ago, the authors believe that there have been no significant changes to this aspect of emergency management (EM) in the United Kingdom (UK) since.

UK EM plans and practices tend to be generic and reductive (Easthope, 2018, 2022; Elstow, 2023), and understandably so. Generic emergency plans are designed to manage the consequences of multiple hazards (an *all-hazards approach*), so they can be applied to different incident scenarios and operate in different geographic areas (HM Government, 2022). Standardised response structures and principles of interoperability between responders enable broad applicability across multiple scenarios. Standardised 'generic publics' imagined within disaster response plans can obscure the diversity of individuals, households and communities impacted. EM professionals must balance the need for plans to be accessible to multiple responders and adaptable enough to accommodate the nuances of multiple kinds of incidents, whilst also accounting for and serving diverse needs of individuals, households and communities.

In our experience, isolated and limited activities at a practitioner level engage with disaster impacts on SOGIESC-diverse people. Afzal (2022) revealed that the organisational culture of the London Fire Brigade conceived diversity training as 'a tick box exercise' and that 'the current approach to learning, training and development was not promoting gender'. Whilst attempts to 'educate' staff about equality, diversity and inclusion focus on raising awareness about protected characteristics and responder biases, they have very little impact on behaviour change (Dobbin & Kalev, 2018). This assumes individuals are willing and have the agency to challenge systems and structures that traditionally do not consider marginalised experiences in every day or emergency circumstances and have examples of how to change their organisation.

Here, we consider what advances could be made by practitioners to anticipate, incorporate and respond to the needs of SOGIESC-diverse people in UK disaster management practice. At this point it seems pertinent to be transparent about ourselves as practitioners and our personal perspectives.

Louise

I, a white, cis-gender social science researcher and EM consultant, identify as a gay or bisexual woman. I am neither ashamed about any of the above statement nor have I had the misfortune, knowingly at least, of being discriminated for my sexuality, although in my current contract, in a country where being gay is illegal, I grew my hair out so that I might pass more easily as 'just' a woman, not a 'gay' woman. In my consultancy work, I frequently end up being the only woman working with older, white, ex-military men, so I already stand out as different. Writing this feels kind of awkward and has made me confront that I do not normally actively engage with or think about being gay in the context of work, but maybe I should. I feel self-conscious mixing a statement about my sexuality with a statement about my job—two sides of me that do not normally interact. I have never hidden it from anyone, but I have not mentioned it in work situations because it seemed irrelevant (and in some countries where I have worked it would have been an unwelcome or illegal admission). Perhaps this is because EM is about trying to document a way of responding for multiple hypothetical futures which have not happened yet.

An 'all-hazards' approach leads to concise documents, adaptable to multiple scenarios. However, this results in addressing standardised populations not marginalised ones. The more nuanced aspects of non-standardised individuals, populations and communities are left to be explored when the specific emergency unfolds. In the same way that all-hazards plans must be flexible to accommodate all-hazards, they must also accommodate all genders, sexualities and sex characteristics. The challenge is to do this more nuanced thinking in advance and not just in response to an emergency when these specific details may be more apparent. This might be different if the social dimensions of disaster were more actively engaged.

Matt

Reflecting on my journey and seeing it written down is daunting. I came out as a gay man in 2003 and started my EM career in 2006. At work, I navigate a complicated dance between my identity and my professional life. I am fortunate not to have faced direct discrimination but recognise the micro-aggressions and vulnerability I have occasionally felt, including pressure to conform in teams dominated by uniformed services. I have grown more confident being myself at work since 2016, but it took until 2023 to have the confidence to join others (including many who do not identify as LGBTQ+) in wearing a rainbow lanyard as a symbol of solidarity.

I have found myself increasingly frustrated at workplace diversity and inclusion initiatives focused on awareness rather than action. Having a long-term role in the recovery efforts from the Grenfell Tower fire disaster in London—which killed 72 people—I have seen first-hand the disproportionate impacts that tragedy can have on marginalised people. This has instilled in me a sense of responsibility for change. Whilst I cannot speak for the broad spectrum of SOGIESC-diverse experiences, I recognise I can use my own experience and access to advocate and amplify a broader set of voices from peripheral spaces.

Writing here has caused us both to reflect on how we might personally enact EM practice in ways that we have not previously considered. However, it poses a conundrum: must we be explicit about our sexuality/gender to make a positive change when there could be real or perceived repercussions? If we do not openly tell our stories, are we also performing some kind of disaster injustice (Lukasiewicz & Baldwin, 2020)? A SOGIESC-diverse person has neither the ability nor the responsibility to represent all SOGIESC-diverse people. Similarly, people with mainstream SOGIESC are not precluded from being active and purposeful allies. Beyond being more open about ourselves, what opportunities are there to address SOGIESC-diverse-related disaster (in)justices created within UK EM?[1] How can EM professionals become professional allies of SOGIESC-diverse people?

[1] We have specifically chosen to describe our profession as EM rather than Disaster and Emergency Management (DEM). In the UK there is no legal definition of disaster, only 'Emergency' (Civil Contingencies Act, 2004). Therefore, in a legal sense, the UK does not differentiate between emergency and disaster. We acknowledge, however, that there are a range of different ways of delineating between emergency and disaster. These are usually on the grounds of scale and impact, and to many people, scholars, practitioners and policy

About This Chapter

We speak directly to the contributions UK EM practitioners make towards a greater *disaster justice* (Lukasiewicz & Baldwin, 2020). We offer personal reflections as UK practitioners on the role of professional allyship and recommendations for further research and action. Our interest is organisations that provide day-to-day emergency response (the police, fire, ambulance and local government), rather than humanitarian organisations working at national or broader scales (footnote 1). We believe that our limited review of the research on SOGIESC and EM practice is the first review of its kind to be undertaken in the UK.

In the UK, the abbreviation LGBTQ (Lesbian, Gay, Bisexual, Transgender/transsexual and Queer) or similar variations are used to describe SOGIESC-diverse people. It is important to note that these concepts are not universal and represent a Western, and in this discussion, UK perspective (UN IOM, 2020). We use 'SOGIESC diverse' as an umbrella term for individuals whose sexual orientations, gender identities, gender expressions and/or sex characteristics fall outside the mainstream cultural categories in any location (unless citing or referring to a specific study or document, in which case we use that study's terminology).

We begin by reviewing the academic, practitioner and other grey literatures before highlighting three main routes through which EM practitioners can become SOGIESC-diverse allies. We focus on inclusion in three areas, developing materials, planning and decision-making, and training and exercising.

Literature Review

Research demonstrates stigmas relating to SOGIESC diversity can marginalise and compromise the security and well-being of LGBTI persons when coupled with 'rigid normative systems within the context of emergency response' (Rumbach & Knight, 2014). However, SOGIESC

makers, they are materially different. It is our opinion that in many respects the UK does not consider itself to have disasters. We have national emergencies, but not disasters. The second reason for using 'EM' is that there is no standardised collective job title for a person who plans for emergencies on a large scale. Health service colleagues might call themselves Emergency Preparedness, Resilience and Response (EPRR) professionals; others use the term resilience or emergency planner. In our experience, people working domestically in the field of EM in the UK do not call themselves disaster planners.

characteristics are rarely explicitly included in UK EM activities or documentation, and little academic research has engaged with the topic to date. Local Resilience Forums (LRFs) are the multi-agency groups that coordinate emergency management at a local level in the UK.[2] We reviewed publicly available information and documentation provided on all LRF websites in England (a full list is included in the appendix). We accessed each LRF website, using the link published by the Cabinet Office (2024). We used 'LGBT' and 'vulnerable' as key words and searched all webpages in the website and downloadable documents (e.g., response plans and household preparedness advice).

Our review highlighted that no LRFs mentioned LGBT (or similar) on their webpages or publicly accessible, downloadable documents. Vulnerability is mentioned infrequently and typically relates to the elderly or the young in relation to a loss of utilities. Missing, is a broader discussion about what makes a person vulnerable; the challenges they might face being open and expressing their needs to responders, their families or communities; and how this might relate to different kinds of emergencies. It is important to note that an absence of SOGIESC on LRF webpage material could indicate either that discussions around particular sensitivities and vulnerabilities of SOGIESC-diverse people are not happening or are addressed in non-public spaces. In our experience SOGIESC-diverse needs are not adequately addressed. They do not feature as a consideration in governance structures, guidance or professional discussions in UK EM. The national guidance for responders on identifying vulnerable people in a disaster fails to mention gender or sexuality at all (Cabinet Office, 2008).

This is unsurprising, given that only six countries mention LGBTQIA+ people within their official disaster preparedness, response or recovery policies: Australia, Canada, Ecuador, Fiji, Poland, Sweden (Blanchard, 2021). Globally, there is no single driver (e.g., UN-led programme of work or policy) pushing to understand the disaster impacts on gender and sexual minorities (Seglah & Blanchard, 2021), in part because the Sendai Framework fails to consider the needs of SOGIESC-diverse individuals (United Nations, 2015). The lack of policy and guidance relating to people with diverse SOGIESC characteristics affects 'the inclusion of these people in preparing for, responding to and recovering from disasters' (Seglah & Blanchard, 2021). The people most marginalised and

[2] Known as Local Resilience Partnerships in Scotland.

disempowered in everyday life are the least likely to be involved in or represented during emergency or disaster response; therefore, SOGIESC-diverse people should be more visible in EM (*ibid*). The needs, requirements and capacities of SOGIESC-diverse individuals are poorly recognised in plans and arrangements globally. This is also the case in the UK.

A review of academic literature identified the extent to which research has explored SOGIESC aspects of EM in the UK. We used the Lancaster University library search engine to search all their databases and library catalogue. We used 'LGB' and a wildcard to account for articles which included LGBTQ/LGBTQIA/LGBTQ+, combined with 'emergency management', 'disaster management' or 'emergency planning', and 'United Kingdom' or 'UK', or 'England', or 'Wales' or 'Scotland.' This search elicited very few results, which aligns with Blanchard's (2021) review of country-specific policy and research, in which the UK does not feature at all.

Findings from other countries and humanitarian contexts provide a counterpoint to our discussion of UK cases and data. Following Blanchard (2021), we found that researchers working on aspects of gender and sexuality in disasters typically concentrate on empirical cases and data from outside the UK. Research from Canada (Cianfarani, 2012) showed that gender and sexual minorities are largely absent from policy and academic work on disasters, and that they need to be protected immediately post-disaster, not months later (Public Safety Canada, 2016, p. 34). Dominey-Howes et al. (2014) called for greater attentiveness to the diverse needs of LGBTI populations, based on data from multiple case studies across Haiti, the United States of America (USA), India, Java and the Philippines.

In Australia, faith-based analyses showed LGBTI people are not treated equally (Dominey-Howes et al., 2016). Elsewhere in Australia, ongoing and historical engagement with institutions has been shown to affect people's willingness to engage with the same institutions in emergencies (Leonard et al., 2018). Parkinson et al. (2022) found that sexual minority women were subject to the compounding impacts of sexuality *and* gender-based discrimination as employees and in the provision of EM services. Women are poorly represented within EM leadership, meaning women's contributions to leadership and knowledge are overlooked in disaster reconstruction and recovery (Australian Institute for Disaster Resilience (AIDR), 2023b). In Australia and New Zealand, LGBTI communities were shown to be resistant and resilient, and emergency responders showed the potential for being attentive to diverse needs

(Dominey-Howes et al., 2018). According to Parkinson et al. (2023, p. 9) 'there is very little existing literature on gender and emergency guidelines in the Australian context' partly because it is at the intersection of several poorly researched topics.

Dwyer and Woolf (2018) explore LGBTIQ+ exclusion in humanitarian policy and practice, whilst others draw on data from non-Western contexts to argue for more sensitivity towards alternative and expansive concepts of gender in disasters and disaster risk reduction (DRR) (McSherry et al., 2015; Gaillard, Gorman-Murray, & Fordham, 2017a; Gaillard, Sanz, et al., 2017b). In Japan, failures of disaster policy and ethical considerations meant that LGBT people experienced bullying and discrimination in shelters after revealing their gender identities, as well as feeling isolated and that their recovery had been hindered (Yamashita et al., 2015). In the USA, Goldsmith et al. (2021, 2023) suggest recommendations for policy makers and responders including involving LGBTQ+ people in planning, wider access to insurance, comprehensive anti-discrimination policies, better training and education and specific inclusion of LTGBTQ+ in plans and documents.

In the UK, we found that COVID was a catalyst for research on SOGIESC-diverse issues. Hafford-Letchfield and Westwood's (2022) research on older LGBT+ people during the pandemic determined that not enough is being done by UK governments to recognise and address 'the needs and concerns of LGBT+ older people during emergencies' in general. They argue that governments should draw on the existing capacity, care networks and agency of the LGBT+ ageing community (Westwood et al., 2021; Toze et al., 2023).

Research into LGBTIQ+ identities during COVID lockdowns in the UK and Brazil highlighted the challenges faced by LGBTIQ+ people including 'mental health and isolation' and noted that 'Cis-heteronormative communications were a particular issue' (Haworth, 2021, p. 2). Strategies for responding to crisis in the UK and Brazil 'often overlooked' the needs and experiences of LGBTIQ+ people in relation to the spaces and family structures of disaster, leading to poor mental health and isolation from support networks and identity-affirming spaces (Haworth, 2021). The impact of lockdown was challenging for some SOGIESC-diverse people who were obliged to stay with unsupportive families, had limited access to support networks and had to navigate the maintenance and presentation of their work identities in a home space (Haworth et al., 2023).

Outside of COVID, these vulnerability factors (isolation, families of choice[3] vs. biological families, gender presentations) are also very important in EM situations. Whilst COVID clearly offered an opportunity to consider some SOGIESC-diverse aspects of emergencies in the UK, we hope that future research engages with other kinds of emergency, to open up more nuanced understandings of their impacts on SOGIESC-diverse people in planning and response.

Elsewhere in the UK, Davies et al. (2023) surveyed emergency response and preparedness staff in the National Health Service (NHS) in London to gauge their understanding of diversity, including but not limited to gender and sexual diversity. Whilst this offers an initial route, the report did not explore what these results might mean or how staff sexual orientation or gender identity impacts emergency preparedness and response in the NHS.

Grey literature from outside the UK offers suggestions for practitioners on including aspects of gender and sexuality in disaster planning and response in relation to issues highlighted in academic research. Academic articles tend to examine the problem from the perspectives of those affected by disaster, rather than those delivering the response. Practitioner-focused resources include a leaflet developed for American health centres on service provision to LGBT people during disaster (National LGBT Health Education Centre, n.d.); a Queer/Trans guide to hurricane season (Imagine Water Works, 2019); a cultural competence guide for emergency responders and volunteers from the Human Rights Campaign (no date); and a specific guide to including the LGBT community in disaster preparedness (National Association of County and City Health Officials—NACCHO, 2019). These resources provide information on how they were developed and might assist UK practitioners develop similar materials.

Discussion: Promoting SOGIESC-Diverse Inclusion and Allyship

We suggest three strategic areas to 'give a voice' (Gorman-Murray et al., 2018) to SOGIESC-diverse people in EM activities and enhance allyship from UK practitioners. Being specific about how to be sensitive to the challenges, impacts and opportunities presented by SOGIESC-diverse

[3] This is a term used to describe a group of people in a person's life who fulfil the role of the family, without being biologically or legally related.

people will help practitioners move beyond identifying what more can be done, to enacting organisational change. These strategic areas are inclusion in resource development, inclusion in planning spaces and inclusion in training and exercising. AIDR's Gender and Emergency Management Action Checklist would be an excellent place to start the process (2023a).

Co-produced Resources

The first route for inclusion is the co-production of resources underpinning emergency response. EM practitioners and policy makers should engage with SOGIESC-diverse people when making plans and producing materials (Goldsmith et al., 2021, 2023). For example, funding could be made available nationally with LRFs to support a coherent approach including SOGIESC-diverse needs in emergency preparedness activities such as risk assessment, planning, training and exercising.

Organisations with in-depth understandings of the needs, challenges and capacities of SOGIESC-diverse people should be appropriately compensated as active co-producers of these resources. The Greater London Authority (GLA) commissioned a report (2024) based on research with equalities groups, which[4] explored the EM needs of equality groups and those they represent. Researchers found that a perceived lack of incentives to engage and a perceived inability to influence decision makers hinders active involvement in EM structures. Any materials produced should be shared nationally, for example, by inclusion on the LRF websites.

Established LGBTQ+ organisations could be invited to partner in this activity at varying scales, including grassroots community organisations and leaders who are easily overlooked but provide direct support at local levels. This might assist with building more inclusive language in policy and practice. SOGIESC-diverse people have a long history and capacity for community organising, activism (Epstein, 1996) and self-reliance (Dominey-Howes et al., 2018), all critical in disasters. As well as learning from their experiences and insights of how disasters affect SOGIESC-diverse people, grassroots groups representing different aspects of the SOGIESC-diverse community could offer new approaches for use in the

[4] We are aware of this through the course of our work; however, little is currently published (https://www.london.gov.uk/programmes-strategies/fire-and-city-resilience/london-resilience-partnership/preparing-your-community-emergencies).

broader context of community resilience, community cohesion and emergent volunteerism.

The introduction of The Terrorism (Protection of Premises Act) 2025, (informally called Martyn's Law) to protect public safety represents an opportunity for EM organisations to work in new and focused ways with owners and patrons of queer spaces in considering venue safety and incident response arrangements. Named after Martyn Hett, a gay man who was killed in the 2017 Manchester Arena terrorist attack, the proposed law requires those responsible for certain venues to consider the threat from terrorism and implement appropriate and proportionate mitigation measures. Event organisers and those managing public spaces will need to be prepared, ready to respond and know what to do in the event of an attack. This will be enacted through enhanced security systems, staff training and clearer processes (Protect UK, 2023). London has already mapped major queer spaces across the capital (GLA, 2023), so has a starting point for engaging with places already sensitive to the needs of the queer community.

Importantly, 'Warm Spaces' (Warm Spaces, 2024) and 'Cool Spaces' (RCSC, 2023) have been established for people to shelter from extreme temperatures. As SOGIESC-diverse people are disproportionately homeless (Stonewall, 2018) they are more likely to be affected by extreme weather conditions. Engaging with groups representing SOGIESC-diverse perspectives could make these spaces safer for SOGIESC-diverse people by ensuring their family unit and gender presentation are accommodated in the design. For people who find themselves at the 'intersection' of being 'unhoused *and* part of the LGBTQIA+ community' an inclusive EM approach is required so that they receive immediate support that is linked to inclusive longer-term support that sensitively addresses their needs (Blanchard, 2024, p. 8).

Examples of resources which might be developed for UK ontologies of disaster include:

- SOGIESC-diverse considerations in processes and procedures (e.g., registration forms), and in spaces and emergency shelters
- template household emergency plans with SOGIESC-diverse considerations
- guidance on SOGIESC-diverse impacts and competencies aligned to UK risks and hazards
- guidance on understanding SOGIESC-diverse vulnerability in different emergencies

Inclusion at the Table

The second route to inclusion is to invite diversity into the spaces where the impacts and vulnerabilities of disaster are considered and planned for. The UK National Risk Register (NRR) is the government's assessment of the likelihood and potential impact of various risks that may directly affect the UK and its interests. It directly and indirectly frames practitioner understandings of who and what might be at risk by acting as a starting point for thinking about disaster impacts.

The register apparently considers the 'disproportionate impacts of the reasonable worst-case scenarios on vulnerable individuals and groups' but does not include specific details about which groups were considered, who participated or how the consultation informed the final NRR wording. Furthermore, the words 'gender' or 'diversity' are conspicuous in their absence (HM Government, 2025). It would be beneficial, not just for SOGIESC-diverse people, but for everyone, if these kinds of national documents considered, more explicitly, who is likely to be affected by scenarios in the NRR, and which kinds of scenarios make different groups of people vulnerable/more vulnerable. An intersectional perspective might also be useful here (Haworth et al., 2023).

One of the three core principles in the UK government's National Resilience Framework (HM Government, 2022, p. 2) asserts that 'resilience is a "whole of society" endeavour' and that 'we must be more transparent and empower everyone to make a contribution'. The framework notes a need to 'understand which groups are acutely vulnerable to local and national risks' (p. 17) but avoids drawing upon existing research on those disproportionately impacted by disasters. In parallel to our own observations, Alexander (2024) recently praised the framework for its practicality, but recommends the government revise it to better account for the needs of a broader spectrum of society including gender, minorities and disabilities.

One of the principles of UK emergency response is *Subsidiarity*, which means '[d]ecisions should be taken at the lowest appropriate level, with co-ordination at the highest necessary level. Local responders should be the building block of response for an emergency' (Cabinet Office, 2013). Structural inclusion should therefore extend to local resilience activities, where most activity and decision-making take place. Some decisions are nonetheless determined at a national level. Therefore national-level structures must also incorporate a consideration of the kinds of impacts that policy decisions could have on different individuals, groups and demographics. Including SOGIESC-diverse considerations in UK Resilience Forum agendas and working groups could be one such mechanism (Cabinet Office, 2023). There is an opportunity to be explicit about SOGIESC-diverse vulnerability to specific threats (malicious intent) and hazards (non-malicious), so that emergency planners, as well as SOGIESC-diverse people, can purposefully address these in local and personal arrangements.

LRFs coordinate multi-agency EM activities and provide collective approaches towards risk assessment, planning, training and exercising. They are structured with subgroups looking at specific risks or capabilities, and there are likely ways those existing subgroups could incorporate SOGIESC-diverse needs or new groups be formed. Another route would be for the Emergency Planning Society[5] to lead the way on establishing a group with an interest in SOGIESC diversity.

A suggested initial focus is for LRFs to extend invitations to locally active SOGIESC-diverse groups to have a 'seat at the table' in relevant subgroups which consider:

- How to incorporate SOGIESC diversity into community impact assessments, ensuring those undertaking assessments have cultural competency and awareness of relevant context.
- The design, layout and functionality of emergency shelters. For instance, explore using known and trusted queer community spaces to supplement existing provision or finding discrete ways to indicate that a location is inclusive.
- How people, families and relationships are described (e.g., the terms responders use) and accounted for during emergencies (e.g., in forms and data collection processes) and afforded equal status (Chu, 2017; UN IOM, 2020; Goldsmith et al., 2021).

[5] The Emergency Planning Society is a UK professional body for those working in resilience, emergency planning and disaster management.

- The provision of psycho-social support—to ensure this reflects and respects SOGIESC-diverse requirements.
- The specific capture of learning relating to SOGIESC diversity in emergencies, including sensitively collecting oral histories about SOGIESC-diverse people's experiences of emergencies—case studies in their own words.[6]

Inclusion in Training and Exercising

A third route for inclusion of SOGIESC diversity is within training and exercising activities. We have previously written character profiles to be used in live exercises to stimulate consideration of specific needs reflecting the range of issues responders might have to manage. In developing characters for Exercise Unified Response (EUR, a four-day urban search and rescue exercise held in London in 2016), I (Louise) would imagine the stories of the people around me on my morning commute and the considerations there could be if our train were involved in a major incident. Character details are provided to an 'actor' to play during the exercise, often sourced from within responding organisations or local colleges. There are some flaws with this approach.

First, the characters' stories in the exercise were limited by my own partial understandings, perspectives and knowledge of what the impact the scenario would have meant for my SOGIESC-diverse characters. More SOGIESC-diverse people could have been involved in the exercise planning and character development.

Second, the people tasked with roleplaying these characters often paid extremely superficial attention to the profiles provided. We have both observed actors half-heartedly provide details to emergency responders omitting key details (age, medical conditions, family relationships) from their character sheets. The actors likely did not think about or understand the consequences of this. Whilst the broader effectiveness of how emergency managers approach planning and conducting exercises in general is out of scope here,[7] we note that even the limited SOGIESC-diverse input into the planning for EUR barely made it to the surface during the exercise itself.

[6] The LGBT Foundation (2020) already does this for other stories, but not directly focusing on emergencies https://lgbt.foundation/oral-histories.

[7] Although something both authors would be keen to explore!

Finally, we need to think about vulnerability within disaster scenarios more broadly than just age and disability. In exercises involving medical casualties, people with pre-existing physical disabilities (e.g., amputees) are often engaged to participate. Moulage[8] simulates traumatic injuries, but to our knowledge the same attempt at realism is not sought for other kinds of vulnerability. Whilst the engagement of people with disabilities 'for realism' may indicate a systemic recognition of physical and medical disability, it reveals that social dimensions of vulnerability are less well recognised.

We therefore invite emergency planners to consider:

- Are we creating realistic and representative diversity in disaster training and exercising?
- Who do we involve during the creation of scenarios, characters and storylines?
- Who is tasked with providing that input during the exercise?
- How could this be evaluated later to identify opportunities for learning and process improvement?

Conclusion

There are many things that we, as UK disaster and EM practitioners, can be doing to become better SOGIESC-diverse allies. In closing, we have adapted Stonewall's (2017) five ways to get started as an ally specifically for the EM audience.

1. Familiarise yourself with SOGIESC diversity language—think about the terms you use and how this might impact people in emergency situations.
2. Discover the challenges facing the LGBT community today—understand why SOGIESC-diverse people might experience disproportionate impacts in emergencies.
3. Educate yourself—find out about specific SOGIESC-diverse experiences in emergencies, and the possible impact of your EM plans.
4. Build relationships with new partners—include local groups representing SOGIESC diversity in your planning meetings, training and exercising.

[8] The art of creating mock injuries for the purpose of training.

5. Use your voice, but step back when needed—take action to consider SOGIESC-diverse experiences in the planning and response to emergencies and make space to listen to those directly impacted.

Each person affected by a disaster will be impacted differently—'there is no single LGBTQIA+ experience' (Haworth et al., 2023)—the question is whether the formal structures for planning, preparing and decision-making during response can and will incorporate a more mature approach to exploring the SOGIESC-diverse facets of disasters, and what the role of practitioners who identify as SOGIESC diverse and allies could be to bring change.

Appendix

Table 18.1 The Local Resilience Forum websites reviewed by the authors in December 2023 and January 2024

	LRF Name	*Website*
Northwest	Cheshire	https://cheshireresilience.org.uk/policies/
	Cumbria	https://www.cumbriaprepared.org.uk/
	Greater Manchester	https://www.gmemergencyplanning.org.uk/
	Lancashire	https://www.lancashireprepared.org.uk/
	Merseyside	https://www.merseysideprepared.org.uk/
Northeast	Cleveland	https://www.clevelandemergencyplanning.info/
	Durham and Darlington	https://www.ddfire.gov.uk/county-durham-and-darlington-prepared
	Northumbria	https://www.northumberland.gov.uk/About/Partners/Northumbria-Local-Resilience-Group.aspx
Yorkshire and Humber	Humber	https://www.humberlrf.gov.uk/
	North Yorkshire	https://www.northyorks.gov.uk/north-yorkshire-local-resilience-forum/about-north-yorkshire-local-resilience-forum
	South Yorkshire	https://www.southyorks.police.uk/find-out/south-yorkshire-local-resilience-forum/
	West Yorkshire	https://www.westyorkshireprepared.org.uk/

(*continued*)

Table 18.1 (continued)

	LRF Name	*Website*
West Midlands	Staffordshire	https://www.staffordshireprepared.gov.uk/Home.aspx
	Warwickshire	https://www.warwickshire.gov.uk/community-safety/warwickshire-prepared
	West Mercia	https://www.westmercia.police.uk/police-forces/west-mercia-police/areas/west-mercia/about-us/about-us/west-mercia-local-resilience-forum2/
	West Midlands	https://www.dudley.gov.uk/council-community/emergencies/west-midlands-conurbation-local-resilience-forum/
East Midlands	Derbyshire Resilience Partnership	https://www.derbyshireprepared.org.uk/risks-derbyshire/community-risk-register/
	Leicestershire	https://www.llrprepared.org.uk
	Lincolnshire	https://www.lincolnshireresilienceforum.org/prepare/
	Northamptonshire	https://www3.northamptonshire.gov.uk/councilservices/fire-safety-and-emergencies/emergencies/emergency-prevention-and-advice/Pages/default.aspx
	Nottingham and Nottinghamshire	https://www.nottinghamshire.police.uk/police-forces/nottinghamshire-police/areas/about-us/about-us/additional-services/nottinghamshire-local-resilience-forum-lrf/
East of England	Bedfordshire and Luton	https://www.centralbedfordshire.gov.uk/info/169/bedfordshire_local_resilience_forum_blrf/1371/priority_services
	Cambridgeshire and Peterborough	Page not found—found by googling: https://www.cpprepared.org.uk
	Essex	http://www.essexprepared.co.uk/are-you-prepared
	Hertfordshire	https://www.hertfordshire.gov.uk/services/fire-and-rescue/are-you-ready-for-anything.aspx
	Norfolk	https://www.norfolkprepared.gov.uk
	Suffolk	https://suffolkprepared.co.uk/get-prepared/prepare-yourself/look-out-for-others/

(*continued*)

Table 18.1 (continued)

	LRF Name	*Website*
South West	Avon and Somerset	https://www.somersetprepared.org.uk/news/
	Dorset	https://www.dorsetprepared.org.uk
	Devon, Cornwall and Isles of Scilly	https://www.dcisprepared.org.uk/what-we-do/emergency-plans/
	Gloucestershire	https://www.glosprepared.co.uk/preparing-your-family/
	Wiltshire and Swindon	https://wiltshireandswindonprepared.org.uk/prepare/personal-resilience/yourself
South East	Hampshire and IoW	https://www.hants.gov.uk/community/emergencyplanning/prepareyourfamily
	Kent	https://www.kentprepared.org.uk/community-resilience
	Surrey	https://www.surreycc.gov.uk/community/emergency-planning-and-community-safety/emergency-planning/surreys-local-resilience-forum
	Sussex	https://www.sussex.police.uk/srf
	Thames Valley	https://www.thamesvalleylrf.org.uk/prepare/prepare-yourself/
London	London	https://www.london.gov.uk/programmes-strategies/fire-and-city-resilience/london-resilience-forum?ac-30121=30109 and also http://www.londonprepared.gov.uk

Conflict of Interest The authors have no conflict of interests.

References

Afzal, N. (2022). *Independent culture review of London fire brigade*. Retrieved February 27, 2024, from https://www.london-fire.gov.uk/media/8183/independent-culture-review-of-lfb-report.pdf

AIDR. (2023a). *Gender and emergency management action checklist*. Retrieved January 13, 2024, from https://www.aidr.org.au/media/10494/1b_gem-guidelines-checklist-december-2023.pdf

AIDR. (2023b). *Gender and emergency management guidelines*. Retrieved January 13, 2024, from https://knowledge.aidr.org.au/resources/national-gender-and-emergency-management-guidelines/

Alexander, D. (2024) *Is the UK National Resilience Framework missing something?* Retrieved February 27, 2024, from https://nationalpreparednesscommission.uk/publications/is-the-uk-national-resilience-framework-missing-something/

Blanchard, K. (2021). *LGBTQIA+ People and Disasters—Global policy tracker.* DRR Dynamics Ltd. Retrieved January 13, 2024, from https://bit.ly/DRRD_LGBT

Blanchard, K. (2024). *Intersectionality, marginalised groups and disasters: Unhoused people, sex workers and LGBTQIA+ populations.* Retrieved February 29, 2024, from https://irp.cdn-website.com/cde3424c/files/uploaded/Unhoused%20report-2.pdf

Cabinet Office. (2008). *Identifying people who are vulnerable in a crisis: guidance for emergency planners and responders.* Retrieved January 13, 2024, from https://www.gov.uk/government/publications/identifying-people-who-are-vulnerable-in-a-crisis-guidance-for-emergency-planners-and-responders

Cabinet Office. (2013). *Responding to emergencies: The UK central government response concept of operations.* Retrieved February 15, 2024, from https://assets.publishing.service.gov.uk/media/5a7a44b0ed915d1fb3cd6a5f/CONOPs_incl_revised_chapter_24_Apr-13.pdf

Cabinet Office. (2023). *Meeting notes for UK Resilience Forum.* Retrieved January 7, 2024, from https://www.gov.uk/government/publications/meeting-notes-for-uk-resilience-forum

Cabinet Office. (2024). *Guidance: Local resilience forums: contact details.* Retrieved February 15, 2024, from https://www.gov.uk/guidance/local-resilience-forums-contact-details

Chu, K. C. (2017). *Why queer people need chosen families.* Retrieved August 6, 2023, from http://www.vice.com/en_us/article/ywbkp7/why-queer-people-need-chosen-families

Cianfarani, M. (2012). Integrating diversity into disaster risk reduction (DRR): A literature review. *International Association of Emergency Managers Bulletin, 29*, 26–27.

Davies, F., Sellwood, C., & Subramaniam, S. (2023). Understanding the diversity of the NHS emergency preparedness, resilience and response (EPRR) community in London. *Journal of Contingencies and Crisis Management, 31*(4), 648–657. https://doi.org/10.1111/1468-5973.12466

Dobbin, F., & Kalev, A. (2018). Why doesn't diversity training work? The challenge for industry and academia. *Anthropology Now, 10*(2), 48–55. https://doi.org/10.1080/19428200.2018.1493182

Dominey-Howes, D., Gorman-Murray, A., & McKinnon, S. (2014). Queering disasters: On the need to account for LGBTI experiences in natural disaster contexts. *Gender, Place and Culture: A Journal of Feminist Geography, 21*(7), 905–918. https://doi.org/10.1080/0966369X.2013.802673

Dominey-Howes, D., Gorman-Murray, A., & McKinnon, S. (2016). Emergency management response and recovery plans in relation to gender and sexual minorities in New South Wales, Australia. *International Journal of Disaster Risk Reduction, 16*, 1–11. https://doi.org/10.1016/j.ijdrr.2016.02.004

Dominey-Howes, D. Gorman-Murray, A., & McKinnon, S. (2018). On the disaster experiences of sexual and gender (LGBTI) minorities: Insights to support inclusive Disaster Risk Reduction Policy and Practice. *Australian Journal of Emergency Management: Monograph*, 3, 60–68. Retrieved February 15, 2024, from https://knowledge.aidr.org.au/media/6031/diversity-in-disaster-monograph.pdf

Dwyer, E., & Woolf, L. (2018). Down by the river addressing the rights, needs and strengths of Fijian gender and sexual minorities in disaster risk reduction and humanitarian response. *Edge Effect,* 1–52. Retrieved February 15, 2024, from https://www.edgeeffect.org/wp-content/uploads/2019/03/Down-By-The-River-May2018.pdf

Easthope, L. (2018). *The recovery myth the plans and situated realities of post-disaster response* (1st ed.). Palgrave Macmillan.

Easthope, L. (2022). *When the dust settles.* Hodder and Stoughton.

Elstow, L. (2023). *Getting the measure of it: Radiation knowledge construction in Japan since 2011.* Ph.D. Thesis, Lancaster University. https://doi.org/10.17635/lancaster/thesis/2002

Epstein, S. (1996). *Impure science: AIDS, activism, and the politics of knowledge.* University of California Press.

Gaillard, J. C., Gorman-Murray, A., & Fordham, M. (2017a). Gender and sexual minorities in disaster. *Gender, Place and Culture: A Journal of Feminist Geography, 24*(1), 18–26. https://doi.org/10.1080/0966369X.2016.1263438

Gaillard, J. C., Sanz, K., Balgos, B. C., Dalisar, S. N. M., Gorman-Murray, A., Smith, F., & Toelipe, V. (2017b). Beyond men and women: A critical perspective on gender and disaster. *Disasters, 41*(3), 429–447. https://doi.org/10.1111/disa.12209

GLA. (2023). *Cultural infrastructure map.* Retrieved January 7, 2024, from https://www.london.gov.uk/programmes-strategies/arts-and-culture/space-culture/cultural-infrastructure-plan-and-toolbox/cultural-infrastructure-map

GLA. (2024). *GLA community resilience equalities engagement report.* Internal Report, Greater London Authority.

Goldsmith, L., Raditz, V., & Mendez, M. (2021). Queer and present danger: Understanding the disparate impacts of disasters on LGBTQ+ communities. *Disasters, 46*(4), 946–973. https://doi.org/10.1111/disa.12509

Goldsmith, L., Mendez, M., & Raditz, V. (2023). Disaster response must help protect LGBTQ+ communities. *Issues in Science and Technology.* Retrieved February 26, 2024, from https://issues.org/disaster-response-lgbtq-goldsmith-mendez-raditz/

Gorman-Murray, M. K. S., Dominey-Howes, D., Nash, C. J., & Bolton, R. (2018). Listening and learning: Giving voice to Trans experiences of disasters. *Gender, Place and Culture: A Journal of Feminist Geography, 25*(2), 166–187. https://doi.org/10.1080/0966369X.2017.1334632

Hafford-Letchfield, T. M., & Westwood, S. (2022). Unheard voices: A qualitative study of LGBT+ older people experiences during the first wave of the COVID-19 pandemic in the UK. *Health and Social Care in the Community, 30*(4), e1233–e1243. https://doi.org/10.1111/hsc.13531

Haworth, B. T. (2021). *Learning from LGBTIQ+ experiences of COVID-19 in the UK for future crises: Considerations for policymakers and practitioners.* Retrieved January 7, 2024, from https://www.alliancembs.manchester.ac.uk/media/ambs/content-assets/documents/news/the-manchester-briefing-on-*Covid-19*-b31-wb-5th-march-2021.pdf

Haworth, B. T., Barros Cassal, L. C., & de Paula Muniz, T. (2023). No-one knows how to care for LGBT community like LGBT do': LGBTQIA+ experiences of *COVID-19* in the United Kingdom and Brazil. *Disasters, 47*(3), 584–607. https://doi.org/10.1111/disa.12565

HM Government. (2022). *The UK Government Resilience Framework.* Retrieved February 15, 2024, from https://www.gov.uk/government/publications/the-uk-government-resilience-framework

HM Government. (2025). *National Risk Register 2025.* Retrieved April 22, 2025, from https://assets.publishing.service.gov.uk/media/67b5f85732b2aab18314bbe4/National_Risk_Register_2025.pdf published 16 Jan 2025

Human Rights Campaign. (n.d.). *Working with the lesbian, gay, bisexual and transgender community: A cultural competence guide for emergency responders and volunteers.* Retrieved February 15, 2024, from https://assets2.hrc.org/files/assets/resources/EmergencyResponders_-_LGBT_Competency.pdf

Imagine Water Works. (2019). *Queer/Trans guide to hurricane season.* Retrieved February 15, 2024, from https://www.imaginewaterworks.org/queer-trans-guide-to-hurricane-season/

Leonard, W., Duncan, A., Parkinson, D., & Archer, F. (2018). Findings from the first Victorian study of the experiences and needs of LGBTI communities in emergencies. *Australian Journal of Emergency Management, 3*, 31–34.

LGBT Foundation. (2020). *Hidden figures: The impact of the COVID-19 pandemic on LGBT communities in the UK* (3rd ed.). Retrieved August 6, 2023, from https://lgbt.foundation/coronavirus/hiddenfigures

Lukasiewicz, A., & Baldwin, C. (2020). *Natural hazards and disaster justice.* Palgrave Macmillan.

McSherry, A., Manalastas, E. J., Gaillard, J. C., & Dalisay, S. N. M. (2015). From Deviant to Bakla, strong to stronger: Mainstreaming gender and sexual minorities into Disaster Risk Reduction in the Philippines. *Forum for Development*

Studies, 42(1), 27–40. https://www.tandfonline.com/doi/pdf/10.1080/08039410.2014.952 330

NACCHO. (2019). *How to include the LGBT community in disaster preparedness.* Retrieved January 7, 2024, from https://www.hstoday.us/subject-matter-areas/emergency-preparedness/how-to-include-the-lgbt-community-in-disaster-preparedness/

National LGBT Health Education Centre. (n.d.). *Emergency preparedness and Lesbian, Gay, Bisexual and Transgender (LGBT) people: What health centres need to know.* Retrieved February 15, 2024, from https://lgbtqiahealtheducation.org/wp-content/uploads/Emergency-Preparedness-for-LGBT-People-Final.pdf

Parkinson, D., Duncan, A., Leonard, W., & Archer, F. (2022). Lesbian and bisexual women's experience of emergency management. *Gender Issues, 39*(1), 75–98. https://doi.org/10.1007/s12147-021-09276-5

Parkinson, D., Duncan, A., Joyce, K., Jeffrey, J., Archer, F., Weiss, C., Gorman-Murray, A., McKinnon, S., & Dominey-Howes, D. (2023). *Gender and Emergency Management (GEM) Guidelines—A literature review.* Retrieved January 13, 2024, from https://knowledge.aidr.org.au/media/5373/gem-literature-review.pdf

Protect UK. (2023). *"Martyn's Law"—What you need to know.* Retrieved August 6, 2023, from https://www.protectuk.police.uk/news-views/martyns-law-what-you-need-know

Public Safety Canada. (2016). *Seventh Annual National Roundtable on disaster risk reduction: Understanding disaster risks.* Retrieved January 13, 2024, from https://www.publicsafety.gc.ca/cnt/rsrcs/pblctns/pltfrm-dsstr-rdctn-2016/pltfrm-dsstr-rdctn-2016-en.pdf

RCSC. (2023). *Greater London Authority: Cool spaces.* Retrieved August 6, 2024, from https://www.kcsc.org.uk/noticeboard/greater-london-authority-cool-spaces

Rumbach, J., & Knight, K. (2014). Gender and sexual minorities in humanitarian emergencies. In L. Roeder (Ed.), *Issues of gender and sexual orientation in humanitarian emergencies. Humanitarian solutions in the 21st century* (pp. 33–74). Springer.

Seglah, H. A., & Blanchard, K. (2021). *LGBTQIA+ people and disasters.* Retrieved February 15, 2024, from https://irp.cdn-website.com/cde3424c/files/uploaded/LGBTQIA%2B%20report-3.pdf

Stonewall. (2017). Come out for LGBT: Becoming an active LGBT ally. *Stonewall.* 10 October. Retrieved August 6, 2023, from https://www.stonewall.org.uk/about-us/news/come-out-lgbt-becoming-active-lgbt-ally

Stonewall. (2018). *LGBT in Britain: Home and communities.* Retrieved February 27, 2024, from https://www.stonewall.org.uk/resources/lgbt-britain-home-and-communities-2018

Toze, M., Westwood, S., & Hafford-Letchfield, T. (2023). Social support and unmet needs among older trans and gender non-conforming people during the *COVID-19* 'lockdown' in the UK. *International Journal of Transgender Health, 24*(3), 305–319. https://doi.org/10.1080/26895269.2021.1977210

UN IOM (UN International Organisation of Migration). (2020). SOGIESC: Full glossary of terms. Retrieved August 6, 2023, from https://www.iom.int/sites/g/files/tmzbdl486/files/documents/IOM-SOGIESC-Glossary-of-Terms.pdf

United Nations. (2015). *The Sendai Framework for Disaster Risk Reduction 2015–2030.* Retrieved January 7, 2024, from www.preventionweb.net/files/43291_sendaiframeworkfordrren.pdf

Warm Spaces. (2024). *Warm Spaces.* Retrieved February 15, 2024, from https://warmspaces.org/about

Westwood, S., Hafford-Letchfield, T., & Toze, M. (2021). Physical and mental well-being, risk and protective factors among older lesbians /gay women in the United Kingdom during the initial *COVID-19* 2020 lockdown. *Journal of Women and Aging, 34*(4), 501–522. https://doi.org/10.1080/08952841.2021.1965456

Yamashita, A., Gomez, C., & Dombroski, K. (2015). Segregation, exclusion and LGBT people in disaster impacted areas: Experiences from the Higashinihon Dai-Shinsai (Great East-Japan Disaster). *Gender, Place and Culture: A Journal of Feminist Geography, 24*(1), 64–71.

PART V

No Longer an Optional Extra: Queer-Inclusive Disaster Policies and Services

CHAPTER 19

A Queer Paramedic's Journey into Emergency Management Research: Research Positionality Statement

Brigid Larkin

Introduction

This vignette on positionality provides a reflective exploration of the researcher's subjective stance and personal engagement within the context of their research study 'Australian LGBTIQA+ paramedics[1] experiences of the workplace when responding to disasters: a mixed methods study.' The study seeks to explore what could be described as the 'everyday disaster' (Davidson et al., 2020) and what queer people may experience when working as a paramedic in Australia, as well as what happens when large-scale disaster events throw regular operations up into the air; the ways that everything lands can affect different groups of people in different ways,

[1] Lesbian, gay, bisexual, transgender, intersex, queer, asexual, and other sexually or gender diverse people.

B. Larkin (✉)
Edith Cowan University, Joondalup, WA, Australia

D. Dominey-Howes et al. (eds.), *Queering Disasters, Climate Change and Humanitarian Crises*, Sustainable Development Goals Series, https://doi.org/10.1007/978-981-96-3857-4_19

especially queer people (Larkin, 2019). This vignette explores the intricate web of the researcher's identity, background, and experiences, acknowledging how these elements influence research processes. This introspective piece illuminates the researcher's biases, values, and perspectives, shedding light on the potential impact of their subjectivity on the study. The author is a queer PhD candidate at Edith Cowan University.

The Journey into Emergency Management Research

I grew up in a regional town where I completed all of my schooling before moving to a major city to study a BHSc in paramedicine at university. Growing up I was always aware of a general undercurrent of discomfort lurking in my subconscious which was associated with a feeling of being an outsider, even among friends and family. In my teenage years, I came to realise that a large part of this was related to my sexuality and knowing that I was unable to adhere to heteronormative standards in almost any way. I felt that the town I grew up in was dangerous for queer people—one gay male acquaintance had been kicked out of home as a teenager for being gay, others had been violently assaulted, and most queer people who had sufficient resources to move away did so. I came to an understanding that my sexual identity was something that had to be deliberately managed to maintain my safety. The conscious management of my identity continued during my undergraduate study, which prevented me from showing up authentically and forming meaningful relationships with my peers. This was largely in part because there was nothing in that environment that indicated any degree of safety for queer people, so it was safer not to even test the boundaries.

After completing my undergraduate degree, I got a job with an ambulance service in Australia and went through the standard career progression structure to become a fully qualified paramedic. I ended up working there for a period of 11 years, over which period there were a range of highs and lows. Like the university environment, when I started working there, I was unaware if it was a safe workplace for queer people and a lot of my emotional labour was dedicated to managing the disclosure and expression of my sexual identity.

As I developed a better understanding of what parts of me I felt safe to bring to work, I began to delve more into the organisation's attitude to queer people. At one time, I asked a supervisor to find out why the ambulance service was the only emergency service that did not participate in the pride parade. The response from management was, 'it's not in the image of the organisation to participate in an event like that.' To me, it felt like the organisation wasn't openly supportive of queer identities and therefore, I distanced myself from who I was as a means of self-protection. I questioned if a highly successful and varied career could be possible if I was an authentic expression of myself.

So I hid those parts of myself, rarely talking about my life with my partner or queer friends. Early in my career, I developed a reputation for being cold and distant, which meant I had few friends at work. But as a junior ambulance officer, I did generally have a reputation as a good clinician, so I decided I was willing to sacrifice my personal reputation because I was at least considered competent. Maybe that was the best I could hope for if making genuine connections with people was off the table. Later in my career, my peers were nominating me for national recognition for my leadership and advocacy efforts; however, it seemed that none of their nominations made it to the recognition body. Coincidentally, this also happened to highly respected and successful non-queer colleagues of mine who were vocal queer allies.

Managing my sexual identity at work was one thing, but managing my gender identity was on an entirely different level. The organisation appeared to demonstrate a surface-level acceptance of diverse sexualities, although there seemed to be a lack of deeper understanding of what true acceptance entails. If you had asked senior leadership whether they supported sexually and gender-diverse individuals, they would likely have affirmed their support, though without clear reflection on what that commitment actually means in practice. Some organisational practices, however, seemed to suggest gaps in their approach to inclusion. These included the leadership's hesitation to implement a formal sexual harassment policy, the use of non-disclosure agreements in cases of workplace harassment, the requirement for staff to obtain managerial approval to change gendered uniform allocations, and instances where individuals with potentially discriminatory views were promoted into leadership roles. One of those people argued with me for his right to use homophobic slurs at work because he apparently doesn't mean it 'like that.' Most egregious of all was the deliberate undermining of investigations into workplace sexual harassment and blatant refusal to investigate violence perpetrated against other staff.

I did, however, manage to have one victory with policy change. I created a report on the internal occupational health and safety system, raising the possibility of reporting the restrictive gender categories on internal systems to a government workplace safety regulator; if they refused to change it this time, I would go to the regulator with all of the evidence that I had raised this issue before. Symbolically, I did this on Trans Day of Visibility, and even though nobody followed up with me personally about the report, some positive changes were made to the system. Executive leadership later announced and touted this change as a significant contribution to inclusivity in the organisation. I received no acknowledgement for advocating for my communities, but as long as there was some practical improvement happening (even if it was not perfect), I could accept sitting with the invisibility knowing that I had made a difference for my gender diverse colleagues, some of whom had expressed their gratitude publicly on internal online platforms.

It took 10 years for me to finally begin the coming out processes for my gender identity. I feared that once I started down this path, it would mean the end of my career with the organisation. And it was. If I was going to be truly honest about who I was, then I also had to be honest that I was unlikely to feel safe in this organisation. It felt like I was standing under a towering wave of animosity for anyone who dared to step outside of what I considered to be highly strict and confining definitions of what and who a paramedic was meant to be. Simply acknowledging that I could not conform to those societal norms, triggered a wave of pressure, attempting to compel me back into a perceived expectation of what a paramedic should conform to. Being honest about who I was also meant being honest about what I would need to feel safe and welcome at work. And being honest about that meant that I would subject myself to possible denials of reasonable requests, where I could be told my concerns were irrelevant, my well-being was irrelevant, my safety was irrelevant. It's difficult to describe what entering those conversations where I was advocating for my communities was like. It can be tempting to label the outcome simply as 'nothing,' because no practical changes were made. But really, the outcome was the harm done to my communities that felt like they were forced to internalise deep-seated feelings of irrelevance, stemming from being ignored and rejected for so many years. The continued silence of my colleagues was the outcome.

After 10 years I was unable to bear the emotional fatigue of my self-fragmentation. The exhaustion of hiding drove me to ask for changes to be made to some of the harmful practices that had been built. I did not want to have to go through the humiliating process of asking my supervisor to approve my request for a uniform allocation change to a different sex, giving them the power to decide whether I was worthy enough to present myself authentically. After some changes in senior leadership one year, the organisation did enter itself in the pride parade in the state capital city. However, I did not want to have the incredibly ironic experience of registering for the pride parade without having the option to enter my personal details correctly because at the time, the only options were male and female. It wasn't uncommon for other cis heterosexual colleagues to dish out homophobic abuse at queer people, and I didn't want to have to witness this type of violence in my workplace. And I never wanted to have to write about this. I never wanted to experience the pain that drove me to this moment. I never wanted to have to give up on a career path that I had planned to follow for another 10, 20, or more years. But here I am.

The organisation's baseline attitude towards queer identities appeared to be one of caution and detachment, which seemed to become more entrenched during times of crisis. In my experience, a common response to inclusion-related issues in emergency services tends to be along the lines of 'We don't have time for that right now; we're focused on addressing immediate, pressing challenges.' This stance in Australian emergency services has also revealed itself in previous Australian research (Parkinson et al., 2018). The meagre resources (if any at all) that were dedicated to issues of inclusion were diverted to emergency management or business continuity, which is of course entirely reasonable. However, whenever this happens it seems to be the same groups of people losing out over and over again, getting left behind and still suffering the harms of work environment which feels toxic to them.

During the Covid-19 pandemic (a time when access to healthcare was of monumental importance), I requested that the ambulance service make a public commitment to the human rights of queer people—specifically the right to access healthcare, as the organisation has a number of primary health clinics and Australian doctors are able to refuse to treat queer patients due to conscientious objection.[2] While the public commitment would have had a significant impact on local queer communities and only

[2]As defined by the Australian Medical Association (2019): 'A conscientious objection occurs when a doctor, as a result of a conflict with his or her own personal beliefs or values, refuses to provide, or participate in, a legal, legitimate treatment or procedure which would be deemed medically appropriate in the circumstances under professional standards.'

involved the creation of a referral process and a press release, senior leadership refused repeatedly. I made this request multiple times and was given a different answer as to why it was refused every single time.

At times, it seemed that the response from leadership was a standard, policy-driven approach that did not fully address the core issues. On other occasions, the rationale for not addressing concerns was framed as being due to the perceived effort involved, or the belief that some policies were not evidence-based. It was deeply disheartening to wear my uniform and work for an organisation where I felt that the support and care I needed from colleagues, including medical professionals, was not always forthcoming. I resented putting that shirt on. And I resented that they refused to articulate why everyone else's human rights are not *my* human rights or my communities' human rights, or so it seemed to me. To go to work feeling that your employer views you as less than human is a uniquely humiliating experience.

As the Covid-19 pandemic continued, the mental and emotional load of working in a high-risk, poorly resourced environment with constantly changing advice and protocols took its toll, and my colleagues and I were worn down into varying degrees of burnout. There was an increasing sense that we were losing our collective camaraderie that had held us together through difficult times before. I often wore two small pride flag pins, the rainbow flag and the transgender pride flag, partly as an act of protest, but also as a symbol of my personal commitment to inclusive practice for patients and colleagues. During a busy pandemic shift, I had a manager ask me about the pride pins I was wearing that day as he had never seen the trans flag before. When I explained what it was, his reaction was to tell me this 'joke'—*What do you call a trans person when you set them on fire? An LGBT-BBQ.* I turned my back and walked away silently. I did not have the energy to confront him or anyone else who heard that conversation and stayed silent. I was already tired of everything, including seeing what I considered to be bigoted people getting promotion after promotion all around me, perpetuating toxic and incompetent leadership practices. I reported the conversation to Human Resources but refused to initiate any kind of formal investigation, which was bound to take months and possibly push me over the edge of mental and emotional exhaustion. I felt that I would not get the outcome I was looking for, and the whole thing possibly would disappear without meaningful action taken. I came to realise that when the organisation's resources were stretched so thinly and I was so emotionally depleted, I felt that I was far less likely to both ask for and receive the support I needed to feel any sense of safety at work as a queer person.

Later in the Covid-19 pandemic, and having had so many experiences of workplace sexism, homophobia, and transphobia grinding away at me for almost ten years, I decided to set up two employee networks—one women's network and one pride network. My hope was that a collective voice would be more persuasive in the push for inclusive change. I chose to focus on the women's network first, knowing there would be a swell of support from female colleagues who had reached their limit in dealing with workplace misogyny. The committee had several meetings with senior leadership during which we felt that their responses did not fully address the concerns surrounding workplace sexual violence, and we were left with unfulfilled promises for meaningful change. I struggled to arrange a meeting between the Pride Network committee and the executive, knowing that if they had been unwilling to acknowledge the gravity of issues like workplace sexual assault, they would likely be reluctant to engage with the struggles faced by queer employees. At times, it felt as though my queer colleagues and I were being treated as symbolic figures rather than as individuals with agency and voices of our own. Senior leadership often presented themselves as champions of positive change, but our experiences suggested a different reality, one where our needs and concerns seemed overlooked. The emotional weight of this situation made it difficult to engage in discussions, feeling that efforts for inclusion might be dismissed or unmet. I found myself in a difficult position—feeling both too outspoken and not enough, too vocal about the need for change yet not seen as having the power to drive it.

The final straw that led me to resign was an email sent to me accidentally by one of the executives. This email revealed that multiple people from different departments had been colluding together to sabotage an investigation into a colleague who had sexually assaulted a queer colleague who was a good friend of mine. I had been fighting so hard to advocate for my queer colleagues and make the service a better place for us to be, perhaps mistakenly believing that my biggest adversary was the ignorance of senior leadership. Somewhere in the back of my mind I must have thought that the battle over ignorance was winnable, even from a position in the lower ranks of the organisation. But this showed me the battle was not simply with ignorance, it was with overt hostility and the dehumanisation of queer people. I couldn't cope with seeing my queer colleague treated like that, or with the thought that if I was sexually assaulted at work, the organisation could sabotage any related investigation. I sent in my resignation letter feeling forced into the decision to do so, but also knowing that I fought as hard as I could to make any positive change.

I see the struggles of my queer colleagues and recognise that they are also my struggles; in our collective experiences is our collective discomfort and pain, preventing us from realising our authenticity, undermining our agency to realise our complete self-expression in all aspects of our life so we only allow fragments of ourselves into the workplace. My understanding of these processes is what drove me to undertake this research on the workplace experiences of queer paramedics during disasters so that this pain might find a voice and allow for an understanding of the practices and structures that allow it to manifest. However, even though the focus of this chapter has been mostly centred around workplace injustices, there are glimmers of strength and joy that shine through the tragedies, and it will be a priority of this study to also make room for and share these stories. As Martin (2022) puts it, 'we're only as educated as the narratives that are shared.'

Conflict of Interest The author has no conflict of interests.

References

Australian Medical Association. (2019). *Position statement, conscientious objection—2019*. Australian Medical Association Limited. Viewed 23 Jan 2024 from https://www.ama.com.au/position-statement/conscientious-objection-2019

Davidson, T., Milbrandt, T., & Park, O. (2020). Introduction. In T. Davidson & O. Park (Eds.), *Seasonal sociology*. University of Toronto Press.

Larkin, B. (2019). Pride and prejudice: LGBTIQ community responses to disaster events worldwide. *Australian Journal of Emergency Management, 34*, 60–66.

Martin, R. (2022). *Dr. Raquel Martin*. Dr. Raquel Martin, viewed 21 February 2024. https://www.raquelmartinphd.com/

Parkinson, D., Leonard, L., Duncan, A., & Jeffery, J. (2018). *Identifying the experiences and needs of LGBTI communities before, during and after emergencies in Victoria*. Gender and Disaster Pod and GLHV@ARCSHS, La Trobe University. https://www.genderanddisaster.com.au/wp-content/uploads/2018/06/Identifying-the-experiences-and-needs-of-LGBTI-emergencies-FINAL.pdf

CHAPTER 20

Identities in Disasters: opportunities and challenges for developing inclusive disaster risk governance networks for gender and sexual minorities in Dominica and neighbouring small Caribbean islands

Lowine Stella Hill

Introduction

Amidst calls for climate and disaster justice, the experiences of marginalised identities including gender and sexual minorities (GSM) are often overlooked (Rumbach & Knight, 2014; Gaillard et al., 2017; Haworth et al., 2022). GSM encompasses a range of non-heteronormative identities, extending beyond lesbian, gay, bisexual, trans, queer, and other gender and sexual diverse (LGBTQ+). It includes those who may not openly identify as LGBTQ+ due to many factors including stigma and discrimination (Agard-Jones, 2009; Kempadoo, 2009; Musquet, 2022).

L. S. Hill (✉)
School of Environment, Resources and Sustainability, University of Waterloo, Waterloo, ON, Canada
e-mail: lowine.hill@uwaterloo.ca

D. Dominey-Howes et al. (eds.), *Queering Disasters, Climate Change and Humanitarian Crises*, Sustainable Development Goals Series, https://doi.org/10.1007/978-981-96-3857-4_20

Emerging scholarship is beginning to address the needs and vulnerabilities of GSM in disasters and climate change contexts (Fordham, 2012; Gaillard et al., 2017), yet their voices remain largely excluded from disaster risk governance (DRG) processes (Dominey-Howes et al., 2014; Gorman-Murray et al., 2017; Yamashita et al., 2017; Haworth et al., 2022). The experiences and insights presented here are part of a study that sought to generate an intersectional, place-based DRG framework; develop insights on intersectional governance opportunities (or barriers) through the understanding of trust and social networks; and lessons that could be applied beyond Dominica to other similar small islands in the Wider Caribbean Region (WCR) (Hill et al., 2023; Hill, 2024). Few post-disaster GSM studies exist in the Caribbean, except for IGLHRC and SEROvie's (2011) work in Haiti after the 2010 earthquake and Dunn's (2016) study on marginalised masculinities in Caribbean disaster risk management (DRM) processes. This study is intended to add to the growing field of 'queering disasters,' drawing attention to a region little investigated and the lived experiences, needs, and capacities of GSM in Dominica.

The chapter provides background information on the study's geographic and cultural context, outlines data collection methods, and analyses findings on trust, networks, governance, and intersectionality (Hill et al., 2023). Insights from this analysis contribute to the development of more inclusive, justice-oriented DRG processes tailored to GSM. They help advance the broader dialogue on creating a 'fit-for-purpose' DRG rather than a 'one-size-fits-all' approach (Lukasiewicz & Baldwin, 2020).

Background

Dominica and the WCR provide a case study for examining the intersection of disasters, social vulnerabilities, and networks. The region is susceptible to numerous natural hazards and the impacts of escalating climate change. Recent events have occurred in Haiti, Dominica, Guadeloupe, Montserrat, and Saint Vincent and the Grenadines (SVG) among others (Barclay et al., 2019; Marcelin et al., 2016; Wilkinson, 2015; Wilkinson et al., 2016). Whilst most information here is from Dominica, the study also includes participants from Guadeloupe and Montserrat.

Dominica is a small independent island nation located on the northern part of the Windward Islands group in the WCR. It is characterised by mountainous terrain with around 70% of Dominica's land unsuitable for agriculture, due to erosion risks or waterlogging (Barclay et al., 2019).

Dominica faces multiple hazards, including hurricanes (the most recent being Hurricane Maria in 2017), earthquakes, landslides, volcanic eruptions, intense rainfall, slope instability, storm surges, and tsunamis (Wilkinson et al., 2016; Barclay et al., 2019). Guadeloupe and Montserrat are north of Dominica and face similar hazards. Guadeloupe is part of France and comprises six main islands and several uninhabited islets. Montserrat is a British Overseas Territory, half of which is an exclusion zone following the 1995 volcanic eruption (Wilkinson, 2015). Category-5 Hurricane Irma impacted both islands two weeks prior to the landfall of Category 5 Hurricane Maria across all three islands in 2017 (Météo-France, 2017).

Context

Hurricane Maria is the case study for the empirical results presented here. The Category 5-hurricane made landfall in September 2017 and devastated Dominica, forcing decision-makers to reassess the country's current DRM plans and the capacity of governance mechanisms to address present and future disasters (CoD, 2017). This initiated the drafting of Dominica's Climate Resilience and Recovery Plan 2020–2030 and the creation of the Climate Resilience Execution Agency for Dominica (CREAD), which draw directly on the Sendai Framework's inclusive, 'leave no one behind' approach to disaster risk reduction (DRR) (GCoD, 2020; IFRC, 2021). Whilst the framework promotes a rights-based approach to DRR and climate resilience, many plans default to a 'one-size-fits-all' model, often neglecting the identity-based needs of GSM. At the time of writing, only Barbados has formally recognised GSM within its development strategy.[1]

The lack of specific GSM inclusion in international, regional, and local DRR policies is reinforced and facilitated by cultural and institutional attitudes and practices. In fact, the WCR presents an interesting dichotomy. Social cohesion is a core component of close-knit island communities who come together to offer collective support, response, and recovery. Yet, socio-cultural, religious, and other factors and the nature of everyone knowing everyone's business reinforce marginalisation, particularly for GSM (Curtis, 2011; Agard-Jones, 2009; Musquet, 2022). Caribbean societies' cultural, political, and economic systems stem from histories of slavery, plantation economies, and colonisation, resulting in unique, localised identities that are cohesive, complex, and mixed but also fragmented (Curtis, 2011; Musquet, 2022; Hill et al., 2023). This fragmentation arises from the diverse demographics of the descendants of African slaves, the region's Indigenous Peoples, European slave masters and explorers, and other ethnic groups, all contributing to complex and often conflicting social dynamics (Mohammed, 2011).

These colonial legacies continue to influence Caribbean gender and sexuality politics. English-speaking islands like Jamaica[2] and SVG[3] maintain anti-buggery laws, criminalising consensual same-sex relations. However, these laws are being challenged and repealed in Antigua and Barbuda, Barbados, Saint Kitts and Nevis, and Trinidad and Tobago (Dickson et al., 2022). Despite progress, the influence of a religious 'politics of respectability' persists, enforcing heterosexual marriage as the norm along with conservative gender roles and expectations (Giraud, 1999; Thomas, 2004; Agard-Jones, 2009). In contrast, French territories like Guadeloupe and Martinique follow France's legal code, which provides protections and union rights to same-sex couples, although societal attitudes remain conservative (Agard-Jones, 2009). Finally, in Dominica, sections of the Sexual Offences Act (SOA) criminalising consensual same-sex activities were successfully challenged in the High Court, and on April 22, 2024, these sections were nullified, reaffirming human rights and dignity for GSM. This decision is expected to have significant legal and social implications for GSM in Dominica.

Theoretical Approach and Methods

Theoretical Principles

This study used a mixed-methods approach (Creswell, 2009) and an intersectional, place-based lens to analyse social networks within DRG processes in Dominica. Six principles inform this approach: (1) individuals are multidimensional and complex, (2) identities and vulnerability are not predefined, (3) spatial and temporal differences influence identity expression, (4) materiality of ecological systems influences intersectionality, (5) power relations are central to the emergence of social processes and epistemologies, and (6) positionality plays an important role in defining DRR (Hill et al., 2023). The research is organised into four themes: (1) navigating identities; (2) victimisation and vulnerability; (3) the importance of scale, place, and time; and (4) power defines access and agency.

Hill et al. (2023) emphasise intersectionality, which recognises the interconnectedness of race, class, gender, sexuality, and other social identities in shaping experiences of oppression and agency across different levels (Crenshaw, 1989, 1990). This perspective acknowledges that individuals are not homogeneous, and their experiences within DRG are influenced by multiple identity-related factors. Consequently, 'intersectional DRG' is defined as the place-based and intersecting dimensions of identities and power within decision-making processes that ensure that policies, strategies, and actions aimed at DRR centre the individual, collective vulnerabilities and social/relational inequalities.

Methods

Due to the status of GSM in Dominica, open discussions about sexual orientation and participation in DRG networks were not feasible. As a result, an anonymous online survey was developed using Qualtrics© (https://www.qualtrics.com), and shared through social media platforms and the WhatsApp groups of local formal and informal LGBTQ+ organisations. This method reaches marginalised individuals who are difficult to engage (Sterzing et al., 2018; Braun et al., 2021) and enables their participation without necessitating that they disclose their identity (McDermott et al., 2013). The survey, featuring both closed and open-ended questions, explored the structures, relationships, and challenges GSMs face within DRG networks. Although the survey focused on Dominica, responses were also received from Guadeloupe and Montserrat, with 38 participants in total, including 17 from Dominica (Table 20.1).[4]

Table 20.1 Respondents' demographic variables

Category	*Responses*	*Number of respondents*
Sex assigned at birth	Male	19
	Female	7
	Unknown	1
Location	Dominica	17
	Guadeloupe	5
	Monserrat	1
	Other	5
Sexual orientation	Lesbian/woman who has slept with women	3
	Gay/man who has slept with men	9
	Queer	1
	Bisexual	7
	No label/unknown	5

Table 20.1 provides an overview of the approximately 70% of respondents who provided answers to the demographic questions. The categories 'Lesbian/woman who have slept with women' and 'Gay/man who has slept with men' encompassed individuals who may not identify as LGBTQ+ but acknowledge that they do not conform to heterosexual norms.

Between April and December 2021, the project conducted interviews and in-person data collection with 120 community members ('Community member #') and 10 key institutional informants ('KI#') across Dominica. This data was used to understand mainstream attitudes (assumed to be primarily heterosexual and cisgender) towards GSM needs and to contextualise and illustrate the insights provided by GSM individuals. Pseudonyms ensured participant anonymity, and responses were provided in French, English, and Antillean Créole, and translated by the author. Although participants were not asked about their sexual orientation, sociometric surveys measured social structures, dynamics, and relationships within DRR networks. They also allowed for structural analysis to empirically establish the hierarchy and leadership of a social network (Wasserman & Faust, 1994). Community members were asked to identify the presence and nature of their relational ties with formal and informal DRR institutions and their information sources, and questions about how *gender and sexual identities* affected access to disaster information and overall community relationships.

Finally, institutional KI semi-structured interviews were conducted using open-ended questions on the (1) links to national and regional governance structures; (2) roles and responsibilities within national and regional disaster governance structures; (3) initiatives, projects, strategies, and general involvement with the studied communities; (4) effectiveness and outcomes of these activities, and (5) significance and visibility of GSMs and other sexual identities within DRG processes (see Hill (2024) for more information).

Results and Discussion

Four main themes emerged from the survey and interview data: (1) navigating identities; (2) victimisation and vulnerability; (3) the importance of place, scale, and time; and (4) how power defines access and agency.

Navigating Identities

Tajfel (1974) defines identity as 'an individual's self-concept which derives from the knowledge of their membership of a social group (or groups) together with the emotional significance attached to that membership.' Identities shape the groups and networks in which an individual is included or excluded over their lifetime. The study of identities is increasingly central to climate change adaptation (Adger et al., 2011; Barnett et al., 2021; Fresque-Baxter & Armitage, 2012) and, to a lesser extent, to DRR (Carr et al., 2015). In small islands, the study of identities within DRG processes reveals a key tension: while they foster social norms and community cohesion, they can also perpetuate systemic inequalities, undermining cohesion and trust post-disaster (Curtis, 2011; Templeton et al., 2020; Foley, 2023).

For GSM, navigating identity expression in disasters is complex. Although 66% of GSM respondents felt their identity did not affect their post-disaster experience (e.g. after Hurricane Maria), 72% concealed aspects of their gender or sexual identity to maintain community acceptance. Additionally, 34% reported negative experiences due to their identity, such as emotional (19%), financial (19%), psychological (15%), and verbal challenges (12%). This highlights the varied impacts of identity on GSM individuals in DRG, with some experiencing minimal effects while others face significant challenges. As Moise, Dominican GSM Respondent #1, recounted following Hurricane Maria:

> The needs of the LGBTQ community are often overlooked because of the stigma that goes with being LGBTQ in a generally homophobic society. In some cases, support and assistance are denied to people who belong to that community. Having members of the LGBTQ community working with disaster preparedness and planning can help after for their needs and curb the issue of discrimination in post disaster situations.

Over 70% of community interviewees indicated that a provider's sexuality does not impact the credibility or willingness to share disaster information. As Joel, a Community Respondent # FFKal022, said, the relevance of information surpasses the provider's identities:

> Personally, [...] it's the information they're giving you. It has nothing to do with the person. Whether they are lesbian, they're gay ... they're giving me information. If it's something I can use, it's okay with me. I don't have a problem.

This suggests that GSM individuals are trusted and supported as vital community members, notwithstanding their gender and sexual identities. However, this inclusivity contrasts with instances of marginalisation, particularly for those with intersecting identities facing compounding oppressions (Curtis, 2011, p. 84). Further, this situation reflects broader patterns of marginalisation, such as faith-based aid organisations excluding aid to GSMs in Haiti after the 2010 earthquake (IGLHRC and SEROvie, 2011).

When GSM respondents were asked how their identities influenced their interactions during or after disasters, some noted that their gender had a more significant impact than their sexuality, reflecting the complex interplay of identities in disaster. Erica, GSM Respondent #10, states:

> we are human like everyone and [are] able to play a vital role within the community … disaster affects everybody and not just the LGBTQ community. Everyone should be involved to be honest. Regardless the gender or the ethnicity.

These experiences highlight the need for disaster policies to adopt nuanced gender and sexuality frameworks and acknowledge the role of identity dynamics within communities. Current policies and decision-making processes tend to homogenise the vulnerability and the experience of marginalised communities by oversimplifying the complex, intersecting factors that shape their identities, capacities, and options for action (Haworth et al., 2022). This has been particularly visible in Dominica's resilience building policies (i.e. GCoD, 2020). Dominica showcases a complex multidimensionality where individuals are allowed to assert their unique identities, attributes, and experiences, but only within certain (arguably flexible) community norms and values. This suggests that there is a certain level of social acceptance or tolerance for individuality and diversity within community norms and values that might be subject to interpretation, negotiation, and change.

Victimisation and Vulnerability

Understanding the dynamics of GSM victimisation and vulnerability within DRG processes is crucial. Victimisation refers to the multifaceted individual, community, and systemic factors that contribute to the marginalisation and harm of GSM in disasters (Trundle et al., 2023). Vulnerability includes the characteristics and circumstances of a person, community, system, or asset that make it susceptible to the damaging effects of environmental and social change (Adger, 2006; Blaikie et al., 2004), affecting their well-being, resource access, and resilience (Armitage & Plummer, 2010; Dominey-Howes, 2018).

Victimisation often results in restricted access to services, compromised safety, and limited social support. GSM individuals rely heavily on chosen families and community networks for support, which disasters disrupt (IGLHRC and SEROvie, 2011). Policy-driven disaster networks were established in Dominica post-Hurricane Maria (GCoD, 2020). Nonetheless, about 72% of GSM respondents reported they received no aid from these networks; instead, they relied on family and friends (29%), professional networks (19%), and limited government post-disaster relief (23%). In contrast, approximately 80% of community respondents (108 of 120) received some form of assistance from these local and national disaster networks.

This gap suggests potential barriers for GSM individuals in accessing mainstream relief due to fears of discrimination, limited outreach, or lack of information (Biholar, 2014; Haworth et al., 2022). Additionally, over 20% of GSM respondents identified kinship networks as their primary disaster-information source (Fig. 20.1), indicating a preference for personal, informal networks over formal channels due to limited trust in local media, which may be seen as politicised (Marvin, KI #006). This indicates potential mistrust or information dissemination issues impacting GSM's access to formal disaster relief.

Figure 20.1 presents the community's sources of disaster information (light grey) and GSM sources (dark grey). The category 'Word of mouth' includes information from kinship groups/ties, i.e. family, friends, and acquaintances. It appears that information from kinship networks is more important for gender and sexual minorities than for other community members.

Kinship networks provide support and belonging following disasters (Faas & Jones, 2017; Foley et al., 2023). On the question of who should be involved in DRG processes, James, GSM respondent #8, stated:

> During a disaster everyone should be involved. If everyone is included in the planning process, it mitigates the risk of person from a certain group being left out, gay people will always look out for their kind during a disaster.

Further, Jocelyn, GSM respondent #5, stated:

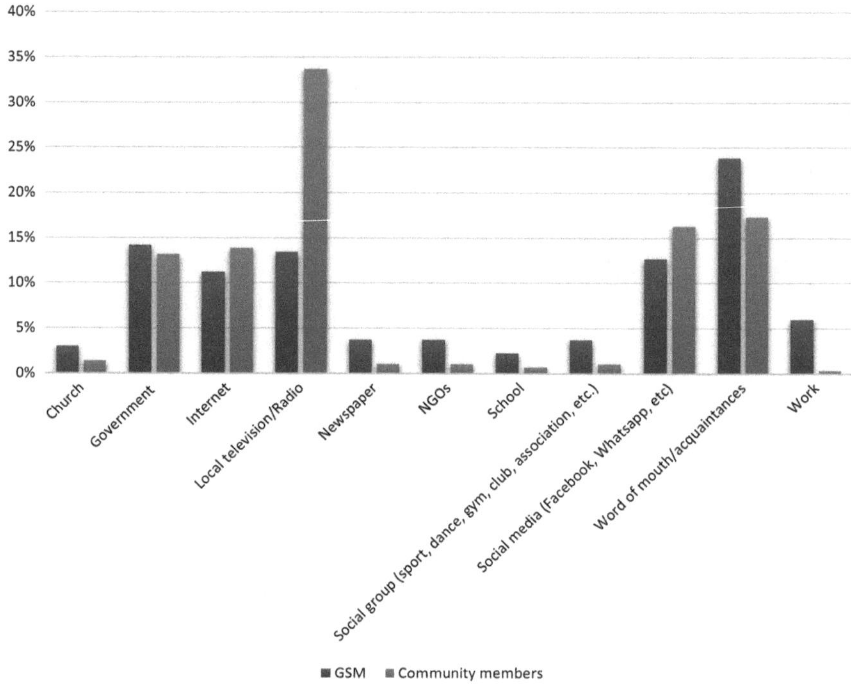

Fig. 20.1 Community respondents' and GSM's participants sources of disaster information

> If these groups are not being consulted and are continuously ignored by policy makers, then the vision is for the groups themselves to create parallel societies that offers a more effective and united front to address DRR (as compared to those in power).

These insights highlight four dimensions for addressing the vulnerability and victimisation of GSM individuals within DRG processes. First, they call for deliberate efforts towards inclusive DRG processes that recognise and address the unique needs of GSM individuals and a plurality of voices (e.g. Gorman-Murray et al., 2018; Pincha & Krishna, 2008). This pluralisation of voices involves creating spaces where people can express their identity-specific vulnerabilities. It also involves drafting concrete and measurable outcomes within an intersectional place-based governance framework that promote accountability and incorporate a diversity of thoughts and ways of doing.

Second, these statements highlight the strong sense of community and solidarity within GSM groups, reflecting shared resilience and support mechanisms during disasters (IGLHRC and SEROvie, 2011; Haworth et al., 2022). Third, they acknowledge the increased vulnerability of GSM individuals, who, for many reasons, may disproportionately rely on informal community networks (IGLHRC and SEROvie, 2011; Biholar, 2014; Dunn, 2016; Gaillard et al., 2017). Last, they emphasise the importance of proactive GSM representation within DRG processes. As Judith, Guadeloupe GSM respondent #20, states:

> Plus de personnes avec une ethnie différente ou une sexualité différente devraient occuper de grands postes [dans les systèmes de gestion des catastrophes] [même s'il peuvent être] eux aussi victimes de discrimination (Translation, More people of different ethnicity or sexuality should hold senior positions [within DRG processes—emphasis added] [even though they too may be] discriminated against).

This perspective adds complexity, especially in places like Guadeloupe, where high-level decision-making occurs far away in France. Moreover, recognising the potentially harmful focus on victimisation and rights violations is crucial. Yet, it is important to acknowledge that an exclusive emphasis on these aspects might inadvertently reinforce a narrative of continuous victimhood and helplessness. This perspective calls for alternative DRG framings that limit victimisation, address identity-specific vulnerabilities, and support agency beyond physical vulnerabilities. This alternative framing highlights systemic oppressions entrenched in culture and customs (Jacobs, 2019).

The Importance of Place, Scale, and Time

Valentine (2007) argues that the expression of identities is contingent upon specific spatial and temporal contexts, a notion that is particularly pertinent to DRG as it allows identities to be situational within broader social and institutional frameworks (Hill et al., 2023). This perspective highlights the significance of scale and place for the contextualisation of identities—the scale at which decisions are made versus the one at which risk is experienced, and the outcomes of decisions made at both individual and institutional levels.

A lack of trust within networks can hinder the effectiveness of governance networks across jurisdictional scales (Klijn et al., 2010). Policymakers in Dominica acknowledge the need for inclusivity (i.e. IFRC, 2021; GCoD, 2020); however, the institutional frameworks required to do so are limited. At national and local levels, the recognition that communities are relatively small prompts caution against singling out identities, as noted by KI#006. At regional and international levels, discussions regarding the explicit inclusion of GSM and other marginalised identities are considered a sovereign issue that can only be mentioned upon national stakeholders' request (KIs #007 and #008). This stance complicates GSM rights advocacy at all levels, with sovereignty sometimes invoked to avoid protective measures for GSM individuals. Moreover, it reflects a reactive rather than proactive approach to GSM inclusion, potentially exacerbating disaster vulnerabilities by placing the onus on domestic individuals or groups to champion their rights and recognition.

Overlooking identities and their associated marginalisation, coupled with the restricted participation of these identities within DRG processes, can reduce the effectiveness of DRR efforts for the whole population, at both local and institutional levels (Blackburn, 2014; Djoudi et al., 2016; Jacobs, 2019; Haworth et al., 2022). As highlighted by KI#002:

> In building resilience within the population and excluding another population who may significantly be impacted by everything that you're doing, [you might reach your project targets and goals] [...], even if it's not explicitly stated I should probably focus on them [excluded vulnerable population] because in focusing on them, I then strengthen the resilience of everybody else. (Erica, KI #002)

Another GSM respondent in Dominica echoed this sentiment regarding inclusivity and decision-making, stating:

> If decisions are made on behalf of people most-affected by the decision (i.e. there is a lack of consultations or solicitation of inputs from the groups most-affected), then the decision-makers fail to include and integrate valuable information in the policies making process. In fact, the policies might be at risk of becoming more harmful than helpful in the long-run. (Leroy, GSM respondent #12)

The misalignment between disaster policies at national and international levels and the local needs of GSM is evidenced by the informal, identity-specific networks established by GSM. These networks address historical exclusion within DRG by providing GSM a platform to voice concerns, share experiences, and advocate for their unique needs (Turner et al., 2008; Grove, 2013; Curtis, 2011, p. 84). They enable tailored disaster response strategies that bridge resource and information gaps specific to GSM vulnerabilities.

Similarly, the concept of place significantly influences identity-specific governance arrangements. According to Hill et al. (2023), 'place' offers an entry point to examine identities and their problem-solving capabilities within governance networks. Trust is fundamental in these networks for effective learning and information exchange (Klijn et al., 2010), impacting DRR and environmental adaptation. For instance, the cultural concept of 'Koudmen'[6] is a 'voluntary community self-help system, grounded in people and partnership' that often transcend individual identities (Joseph, 2023). In disasters, communities practise 'Koudmen' by addressing needs as they arise, regardless of social identities (Mulot, 2000; Pittman & Armitage, 2017). As Jonathan, a community respondent (FFMer004) shared, 'There were two ladies that had [issues during Maria] and I just showed them how to put [back their house together]. It is my duty to help whenever I can.' Whilst it's uncertain if similar support would be extended to a gay male couple, this case involving a lesbian couple suggests that community belonging and the 'Koudmen' ethos can foster cohesion and trust beyond social identities.

Power Defines Access and Agency

Biases and power dynamics significantly impact disaster relief and recovery, determining who gains access to support and resources (Grove, 2013). Disaster justice highlights that power differentials are rooted in deep-seated social, economic, and governance inequalities, which shape individuals' disaster coping strategies and access to support networks (Blackburn, 2014; Gahman & Thongs, 2020; Rivera, Rivera, 2022). Here, limited power does *not* imply powerlessness but calls for an examination of governance and justice issues.

An intersectional framework recognises that individuals embody multiple identities. In Dominica and Guadeloupe, GSM prioritise personal freedom over strict identity labels, questioning some Western-centric frameworks (see Agard-Jones, 2009; Kempadoo, 2009; Musquet, 2022). This research illustrates that many GSM navigate marginalisation by forming parallel networks through their intersecting identities, embedding themselves in wider community structures and mitigating exclusion. This does not eliminate discrimination but shows GSM individuals exercising agency within their networks.

Cultures of kinship and resourcefulness are integral to some Caribbean Islands, shaping DRG networks and outcomes. These networks can foster solidarity but may also reinforce norms that marginalise minority or dissident voices (Foley et al., 2023; Hill et al., 2023). Recognising these dynamics and valuing individual and group agency are essential to developing equitable DRG strategies that reflect diverse identities. Such strategies should also enable individuals to voice their needs, building trust and cohesion within small island communities. As stated by Joanne, GSM respondent #15:

> Many spaces that create safe mental and physical conditions are never considered so gender identity familiar and qualified individuals should also be involved in the strategic planning and delegation as well as resource distribution of natural disasters.

Positionality—a key factor in power and agency—remains underexplored within environmental disciplines, despite its significance in shaping DRR agendas and decisions (Hill et al., 2023). While some community members recognise the reliability of disaster information from GSM individuals, openly disclosing one's sexuality is frowned upon, especially for men. This fear of association can lead to social exclusion, contrasting with the support the lesbian couple received, reflecting entrenched social norms where association with male GSM individuals is viewed more negatively. As stated by Arlington, a community respondent (FFDav014):

> For example, people can just see you talking to me and think you are my girlfriend. So, it is a similar thing. Sometimes, people can say things if they see me talking to [a gay person] and if I receive assistance, others could say that I am involved with that person. So, I rather starve than to interact with someone who is gay. Maybe if the person comes to me and I do not know their sexuality. But if I am aware of it, I will not interact with the individual.

These attitudes perpetuate discrimination and exclusion and may deter GSM individuals from seeking help or participating in disaster efforts. In fact, during the in-person interviews one of the fisherfolks initially stated that a person's sexuality did not play any role in determining a person's trustworthiness. However, they later changed their answer following disparaging comments from an older community member. This highlights how peer pressure and social hierarchies shape societal dynamics and decision-making processes (Bowleg, 2008; Bennett et al., 2020). Dominant community narratives often overshadow progressive views, with power asymmetries affecting resource allocation and trust within communities (Vallet et al., 2020) that can act as barriers to systemic transformation.

Conclusions and Way Forward

This analysis provides insights into the experiences of GSM within DRG processes in Dominica and several other neighbouring small Caribbean Islands, highlighting the complexities of navigating identities, victimisation, and vulnerability, as well as the impact of place, scale, and power dynamics on access and agency. Dominica and the WCR exemplify both unity and exclusion, as individuals navigate community norms and conflicts, revealing that a one-size-fits-all approach cannot address identity expression within DRG networks. Instead, a place-based, intersectional perspective is essential to understand the multidimensionality and dynamics associated with identity within these networks.

The findings emphasise the need for an approach that values and leverages the diverse experiences of GSM, allowing for a richer understanding of how intersecting identities and marginalisation affect DRG processes. Small communities, like those in Dominica, pose specific barriers, as identity-specific needs often remain unvoiced due to risks of further marginalisation. Consequently, GSM individuals may rely on other identities to maintain social cohesion and trust within their communities. The findings also challenge the static, uniform, and normative understandings of gender, sexuality, and vulnerability that dominate disaster policy and practice. As a key informant noted, it is important to work beyond the narrow definition of gender and sexual identities to address broader societal concerns. Although the legal status of GSM in the Caribbean poses challenges, this research highlights the importance of investigating how disasters exacerbate or mitigate pre-existing social inequalities.

It is necessary to acknowledge the research limitations due to the exclusion of direct intersectional questions during interviews. This was a methodological choice reflecting the region's socio-political sensitivities around sexuality. This constraint highlights the challenges GSM individuals face in openly discussing their identities, which may have affected the comprehensiveness of the findings. This constraint highlights the intricate balance between ethical research and the need for a nuanced exploration of identities within DRG, emphasising the ongoing struggle for visibility and recognition faced by GSM communities in the Caribbean. Despite these limitations, this study takes a crucial first step in recognising GSM contributions to DRG in Dominica and the WCR. While implementing inclusive policies may be challenging due to young, fragmented yet centralised bureaucracies (Cashman, 2017; Hinds, 2019), regional collaboration can promote shared learning and inclusive DRG strategies. Promoting acceptance and establishing feedback mechanisms are essential steps towards integrating GSM voices, reducing stigma, and fostering more responsive governance across the region.

Conflict of Interest This research project was funded by Canada's International Development Research Centre (IDRC) through the IDRC Doctoral Research Awards. The author's research has been further supported with a SSHRC Joseph-Armand Bombardier Canada Graduate Scholarship Doctoral Award. The author has no other conflicts of interest.

References

Adger, W. N. (2006). Vulnerability. *Global Environmental Change, 16*(3), 268–281. https://doi.org/10.1016/j.gloenvcha.2006.02.006

Adger, W. N., Barnett, J., Chapin, F. S., & Ellemor, H. (2011). This must be the place: Underrepresentation of identity and meaning in climate change decision-making. *Global Environmental Politics, 11*(2), 1–25. https://doi.org/10.1162/GLEP_a_00051

Agard-Jones, V. (2009). Le Jeu de Qui? Sexual politics at play in the French Caribbean. *Caribbean Review of Gender Studies, 3*, 19. https://sta.uwi.edu/crgs/november2009/abstracts/Agard-Jones.pdf

Armitage, D., & Plummer, R. (Eds.). (2010). *Adaptive capacity and environmental governance*. Springer-Verlag. https://doi.org/10.1007/978-3-642-12194-4

Barclay, J., Wilkinson, E., White, C. S., Shelton, C., Forster, J., Few, R., Lorenzoni, I., Woolhouse, G., Jowitt, C., Stone, H., & Honychurch, L. (2019). Historical trajectories of disaster risk in Dominica. *International Journal of Disaster Risk Science, 10*(2), 49–165. https://doi.org/10.1007/s13753-019-0215-z

Barnett, J., Graham, S., Quinn, T., Adger, W. N., & Butler, C. (2021). Three ways social identity shapes climate change adaptation. *Environmental Research Letters, 16*(12), 124029. https://doi.org/10.1088/1748-9326/ac36f7

Bennett, K., Neef, A., & Varea, R. (2020). Embodying resilience: Narrating gendered experiences of disasters in Fiji. In A. Neef & N. Pauli (Eds.), *Community, environment and disaster risk management* (pp. 87–112). Emerald Publishing Limited. https://doi.org/10.1108/S2040-726220200000022004

Biholar, R. (2014) *Masculinities and the practice of Dominica's national gender policy*. IDRC Research Report 106430-001. International Development Research Centre: Ottawa, ON Canada. https://sta.uwi.edu/igds/ppgj/documents/IDL-53628.pdf

Blackburn, S. (2014). The politics of scale and disaster risk governance: Barriers to decentralisation in Portland, Jamaica. *Geoforum, 52*, 101–112. https://doi.org/10.1016/j.geoforum.2013.12.013

Blaikie, P., Cannon, T., Davis, I., & Wisner, B. (2004). *At risk: Natural hazards, people's vulnerability and disasters*. Routledge.

Bowleg, L. (2008). When Black + Lesbian + Woman ≠ Black Lesbian woman: The methodological challenges of qualitative and quantitative intersectionality research. *Sex Roles, 59*(5), 312–325. https://doi.org/10.1007/s11199-008-9400-z

Braun, V., Clarke, V., Boulton, E., Davey, L., & McEvoy, C. (2021). The online survey as a qualitative research tool. *International Journal of Social Research Methodology, 24*(6), 641–654. https://doi.org/10.1080/13645579.2020.1805550

Carr, E. R., Abrahams, D., & De la P. A. T., Suarez, P. and Koelle, B. (2015). Vulnerability assessments, identity and spatial scale challenges in disaster-risk reduction: Original research. *Jamba: Journal of Disaster Risk Studies, 7*(1), 1–17. https://doi.org/10.4102/jamba.v7i1.201

Cashman, A. (2017). Why isn't IWRM working in the Caribbean? *Water Policy, 19*(4), 587–600. https://doi.org/10.2166/wp.2017.100

Commonwealth of Dominica (CoD). (2017). *Post-disaster needs assessment: Hurricane Maria—September 18, 2017*. Commonwealth of Dominica, Roseau.

Crenshaw, K. (1989). Demarginalizing the intersection of race and sex: A Black feminist critique of antidiscrimination doctrine. In *Feminist theory and antiracist politics. University of Chicago Legal Forum*, 1 (31).

Crenshaw, K. (1990). Mapping the margins: Intersectionality, identity politics, and violence against women of colour. *Stanford Law Review, 43*(6), 1241–1300.

Creswell, J. W. (2009). *Research design: Qualitative, quantitative, and mixed methods approaches* (3rd ed.). Sage Publications.

Curtis, T. (Ed.). (2011). *Islands as crossroads: Sustaining cultural diversity in small island developing states.* UNESCO.

Dickson, A., Loft, P., Robinson T., Walker N. (2022). *LGBT+ rights and issues in the Caribbean.* Commons Library Research Briefing. Retrieved from https://researchbriefings.files.parliament.uk/documents/CBP-9436/CBP-9436.pdf

Djoudi, H., Locatelli, B., Vaast, C., Asher, K., Brockhaus, M., & Basnett Sijapati, B. (2016). Beyond dichotomies: Gender and intersecting inequalities in climate change studies. *Ambio, 45*(3), 248–262. https://doi.org/10.1007/s13280-016-0825-2

Dominey-Howes, D. (2018). Hazards and disasters in the Anthropocene: Some critical reflections for the future. *Geoscience Letters, 5*(1), 7. https://doi.org/10.1186/s40562-018-0107-x

Dominey-Howes, D., Gorman-Murray, A., & McKinnon, S. (2014). Queering disasters: On the need to account for LGBTI experiences in natural disaster contexts. *Gender, Place and Culture, 21*(7), 905–918. https://doi.org/10.1080/0966369X.2013.802673

Dunn, L. (2016). Integrating men and masculinities in Caribbean disaster risk management. In E. Enarson & B. Pease (Eds.), *Men, Masculinities and Disaster* (pp. 209–218). Routledge. https://doi.org/10.4324/9781315678122-18

Faas, A. J., & Jones, E. C. (2017). Social network analysis focused on individuals facing hazards and disasters. In *Social network analysis of disaster response, recovery, and adaptation* (pp. 209–218). Elsevier. https://doi.org/10.1016/B978-0-12-805196-2.00002-9

Foley, A., Brinklow, L., Corbett, J., Kelman, I., Klöck, C., Moncada, S., Mycoo, M., Nunn, P., Pugh, J., Robinson, S., Tandrayen-Ragoobur, V., & Walshe, R. (2023). Understanding "Islandness". *Annals of the American Association of Geographers, 0*, 1–18. https://doi.org/10.1080/24694452.2023.2193249

Fordham, M. (2012). Gender, sexuality and disaster. In B. Wisner, J. C. Gaillard, & I. Kelman (Eds.), *Routledge handbook of hazards and disaster risk reduction* (pp. 424–435). Taylor and Francis.

Fresque-Baxter, J. A., & Armitage, D. (2012). Place identity and climate change adaptation: A synthesis and framework for understanding: Place identity and climate change adaptation. *Wiley Interdisciplinary Reviews: Climate Change, 3*(3), 251–266. https://doi.org/10.1002/wcc.164

Gahman, L., & Thongs, G. (2020). Development justice, a proposal: Reckoning with disaster, catastrophe, and climate change in the Caribbean. *Transactions of the Institute of British Geographers.* https://doi.org/10.1111/tran.12369

Gaillard, J. C., Gorman-Murray, A., & Fordham, M. (2017). Gender and sexual minorities in disaster. *Gender, Place and Culture, 24*(1), 8–26. https://doi.org/10.1080/0966369X.2016.1263438

Giraud, M. (1999). Une construction coloniale de la sexualité [A propos du multipartenariat hétérosexuel caribéen]: A propos du multipartenariat hétérosexuel caribéen. *Actes de la recherche en sciences sociales, 128*(1), 46–55. https://doi.org/10.3406/arss.1999.3293

Gorman-Murray, A., Morris, S., Keppel, J., McKinnon, S., & Dominey-Howes, D. (2017). Problems and possibilities on the margins: LGBT experiences in the 2011 Queensland floods. *Gender, Place and Culture, 24*(1), 37–51. https://doi.org/10.1080/0966369X.2015.1136806

Gorman-Murray, A., McKinnon, S., Dominey-Howes, D., Nash, C. J., & Bolton, R. (2018). Listening and learning: Giving voice to trans experiences of disasters. *Gender, Place and Culture, 25*(2), 166–187. https://doi.org/10.1080/0966369X.2017.1334632

Government of the Commonwealth of Dominica (GCoD). (2020). *Dominica Climate Resilience and Recovery Plan 2020–2030.*

Grove, K. (2013). Hidden transcripts of resilience: Power and politics in Jamaican disaster management. *Resilience, 1*(3), 193–209. https://doi.org/10.1080/21693293.2013.825463

Haworth, B. T., McKinnon, S., & Eriksen, C. (2022). Advancing disaster geographies: From marginalisation to inclusion of gender and sexual minorities. *Geography Compass, 16*(11), e12664. https://doi.org/10.1111/gec3.12664

Hill, L. (2024). *Understanding island-wide intersectional disaster risk governance: The place of social networks in the Commonwealth of Dominica.* UWSpace. http://hdl.handle.net/10012/20301

Hill, L. S., Armitage, D., Collins, A. M., & Pittman, J. (2023). Principles for the consideration of intersectionality in place-based disaster risk governance in islands. *Sustainable Development, sd.2684.* https://doi.org/10.1002/sd.2684

Hinds, K. (2019). Caribbean political culture, governance, and participation. In K. Hinds (Ed.), *Civil society organisations, governance and the Caribbean community* (pp. 55–77). Springer International Publishing. https://doi.org/10.1007/978-3-030-04396-4_3

International Gay and Lesbian Human Rights Commission and SEROVie. (2011). *The Impact of the Earthquake, and Relief and Recovery Programs on Haitian LGBT People.* Retrieved from https://outrightinternational.org/sites/default/files/504-1.pdf

International Red Cross and Red Crescent (IFRC). (2021). Integrating CCA and DRR laws and policies towards a climate-resilient development: lessons from The Commonwealth of Dominica, (authored by Tommaso Natoli). Geneva.

Jacobs, F. (2019). Black feminism and radical planning: New directions for disaster planning research. *Planning Theory, 18*(1), 24–39. https://doi.org/10.1177/1473095218763221

Joseph, R. (2023). The spirit of Koudmen: The genesis of identity, community and cooperation in Saint Lucian Society. In V. Cuffy & J. Carr (Eds.), *Creole cultures, Vol. 1: Safeguarding creole intangible cultural heritage* (pp. 83–102).

Springer International Publishing. https://doi.org/10.1007/978-3-031-24275-5_5

Kempadoo, K. (2009). Caribbean sexuality—Mapping the field. *Caribbean Review of Gender Studies*, 3. https://sta.uwi.edu/crgs/november2009/journals/Kempadoo.pdf

Klijn, E.-H., Edelenbos, J., & Steijn, B. (2010). Trust in governance networks: Its impacts on outcomes. *Administration and Society, 42*(2), 193–221. https://doi.org/10.1177/0095399710362716

Lukasiewicz, A., & Baldwin, C. (Eds.). (2020). *Natural hazards and disaster justice: Challenges for Australia and its neighbours*. Palgrave Macmillan. https://doi.org/10.1007/978-981-15-0466-2

Marcelin, L. H., Cela, T., & Shultz, J. M. (2016). Haiti and the politics of governance and community responses to Hurricane Matthew. *Disaster Health, 3*(4), 151–161. https://doi.org/10.1080/21665044.2016.1263539

McDermott, E., Roen, K., & Piela, A. (2013). Hard-to-reach youth online: Methodological advances in self-harm research. *Sexuality Research and Social Policy, 10*(2), 125–134. https://doi.org/10.1007/s13178-012-0108-z

Météo-France. (2017). *Compte-Rendu du Passage de l'ouragan IRMA sur les ÎLES des Antilles Francaises*. Direction interrégionale Antilles-Guyane. https://atlas.amicale-des-ouragans.org/dossiers/irma2017/irma2017_cr_meteofrance.pdf.

Mohammed, P. (2011). *Of Poteau mitans, bedis, vèvè and things: Caribbean island identities and cultural production*. In Curtis, T. (Ed.). Islands as crossroads: sustaining cultural diversity in small island developing states. UNESCO

Mulot, S. (2000). "Je suis la mère, je suis le père!" : l'énigme matrifocale. Relations familiales et rapports de sexe en Guadeloupe. Anthropologie sociale et ethnologie. Ecole des Hautes Etudes en Sciences Sociales (EHESS),. Français. fftel-00266923v2. Available from https://tel.archives-ouvertes.fr/tel-00266923v2/document

Musquet, C. (2022). *Être homosexuel(le) aux Antilles: 15 témoignages*. Caraïbéditions.

Pincha, C., & Krishna, H. (2008). *Aravanis: Voiceless victims of the tsunami*. Humanitarian Practice Network. https://odihpn.org/publication/aravanis-voiceless-victims-of-the-tsunami/

Pittman, J., & Armitage, D. (2017). How does network governance affect social-ecological fit across the land–sea interface? An empirical assessment from the Lesser Antilles. *Ecology and Society, 22*(4). https://doi.org/10.5751/ES-09593-220405

Rivera, D. Z. (2022). Disaster colonialism: A commentary on disasters beyond singular events to structural violence. *International Journal of Urban and Regional Research, 46*(1), 126–135. https://doi.org/10.1111/1468-2427.12950

Rumbach, J., & Knight, K. (2014). Gender and sexual minorities in humanitarian emergencies. In L. W. Roeder (Ed.), *Issues of gender and sexual orientation in humanitarian emergencies: Risks and risk reduction* (pp. 33–74). Springer International Publishing. https://doi.org/10.1007/978-3-319-05882-5_3

Sterzing, P. R., Gartner, R. E., & McGeough, B. L. (2018). Conducting anonymous, incentivized, online surveys with sexual and gender minority adolescents: Lessons learned from a national polyvictimization study. *Journal of Interpersonal Violence, 33*(5), 740–761. https://doi.org/10.1177/0886260517744845

Tajfel, H. (1974). Social identity and intergroup behaviour. *Social Science Information, 13*(2), 65–93. https://doi.org/10.1177/053901847401300204

Templeton, A., Guven, S. T., Hoerst, C., Vestergren, S., Davidson, L., Ballentyne, S., Madsen, H., & Choudhury, S. (2020). Inequalities and identity processes in crises: Recommendations for facilitating safe response to the COVID-19 pandemic. *British Journal of Social Psychology, 59*(3), 674–685.

Thomas, D. A. (2004). *Modern blackness: Nationalism, globalization, and the politics of culture in Jamaica*. Duke University Press.

Trundle, G., Jones, K. A., Ropar, D., & Egan, V. (2023). Prevalence of victimisation in autistic individuals: A systematic review and meta-analysis. *Trauma, Violence, and Abuse, 24*(4), 2282–2296. https://doi.org/10.1177/15248380221093689

Turner, N., Gregory, R., Brooks, C., Failing, L., & Satterfield, T. (2008). From invisibility to transparency: Identifying the implications. *Ecology and Society, 13*(2), 7. https://doi.org/10.5751/ES-02405-130207

Valentine, G. (2007). Theorizing and researching intersectionality: A challenge for feminist geography. *The Professional Geographer, 59*(1), 10–21. https://doi.org/10.1111/j.1467-9272.2007.00587.x

Vallet, A., Locatelli, B., Barnaud, C., Makowski, D., Quispe Conde, Y., & Levrel, H. (2020). Power asymmetries in social networks of ecosystem services governance. *Environmental Science and Policy, 114*, 329–340. https://doi.org/10.1016/j.envsci.2020.08.020

Wasserman, S., & Faust, K. (1994). *Social network analysis: Methods and applications*. Cambridge University Press. https://doi.org/10.1017/CBO9780511815478

Wilkinson, E. (2015). Beyond the volcanic crisis: Co-governance of risk in Montserrat. *Journal of Applied Volcanology, 4*(1), 3. https://doi.org/10.1186/s13617-014-0021-7

Wilkinson, E., Lovell, E., Carby, B., Barclay, J., & Robertson, R. E. A. (2016). The dilemmas of risk-sensitive development on a small volcanic island. *Resources*, 5(2), Article 2. https://doi.org/10.3390/resources5020021

Yamashita, A., Gomez, C., & Dombroski, K. (2017). Segregation, exclusion and LGBT people in disaster impacted areas: Experiences from the *Higashinihon Dai—Shinsai* (Great East-Japan Disaster). *Gender, Place and Culture, 24*(1), 64–71. https://doi.org/10.1080/0966369X.2016.1276887

CHAPTER 21

No One Left Behind?: The Impact of the Climate Emergency on Gender and Sexual Minorities and Their Marginalisation in the Sustainable Development Goals (SDGs)

Andrew Dalton

Introductory Notes on Terminology

This chapter uses the term 'gender and sexual minorities' (SGM), which maintains inclusivity and recognises the variety of lived experiences and identities of those who may not use LGBTQI+[1] to culturally describe

[1] The acronym LGBTQI+ stands for lesbian, gay, bisexual, transgender, queer/questioning, and intersex. The plus (+) encompasses additional identities such as asexual, which are assumed but not named in the shortened acronym.

A. Dalton (✉)
ReportOUT, Newcastle-upon-Tyne, UK
e-mail: drewdalton@reportout.org

D. Dominey-Howes et al. (eds.), *Queering Disasters, Climate Change and Humanitarian Crises*, Sustainable Development Goals Series, https://doi.org/10.1007/978-981-96-3857-4_21

Fig. 21.1 The Guna Yala Islands. (Image credit: Author)

themselves or to locate themselves politically. It also uses the acronym LGBTQI+ when citing sources that use or infer this term.

The term 'Climate Emergency' recognises that the terminology of 'climate change' risks dulling the significance of the changes and the interventions needed for the scale of the issues faced. *Climate change* suggests a naturally occurring phenomenon which can (potentially) be ignored. The use of the term *Climate Emergency* reflects a changing language used by reporters, activists, scientists, and the public (Climate Emergency Declaration, 2024). Due to the recognition of the urgency of the Climate Emergency itself, this chapter uses the term to describe the process and effects of climate change, as well as the urgency needed to respond. 'Climate change' is only used where cited from other sources (Fig. 21.1).

What Is the Climate Emergency?

The Intergovernmental Panel on Climate Change (IPCC) publishes global, peer-reviewed reports on the role of human activities in contributing to the Climate Emergency. In 2013, its Sixth Assessment Report stated unequivocally that changes in climate are human induced and are real (IPCC, 2023). Burning fossil fuels since pre-industrial levels changed the atmosphere's composition (the greenhouse effect), increasing the average global temperature between 1.1 and 1.2 °C (UNEP, 2023). Between 2011 and 2020, temperatures were the warmest on record, with each of the previous four decades becoming warmer than any decade since 1850 (UNEP, 2023).

Due to environmental interconnectedness, Climate Emergency consequences include intense droughts, water scarcity, rising sea levels, flooding, melting polar ice, catastrophic storms, and declining biodiversity (McCormick, 2022; United Nations, 2023a). This has been felt significantly in some parts of the world where visible effects of the Climate Emergency can be seen through Extreme Weather Events (EWEs) such as drought, wildfires, tornadoes, and hurricanes. In 2018, the IPCC launched a report on the impact of global warming should planetary temperatures exceed the 1.5 °C 'guard rail' suggested by the IPCC (which is anticipated by 2030) (IPCC, 2018). The IPCC (2018) report also outlined a series of catastrophic environmental disasters affecting life on Earth should temperatures exceed this level and stated that without global support and with rapid changes caused by human carbon emissions, the Earth will experience increasingly harmful EWEs. Furthermore, the same report forecasts a temperature rise of 2.5–10 °C over the next century, leading to a range of catastrophic effects.

International declarations and government policies aim to maintain global temperature rises to no more than 1.5 °C, albeit unevenly. Most agencies agree that this would help avoid the worst of the climate impacts and maintain a liveable climate; however, forecasts already point to a 3 °C temperature rise by 2100. Consequently, approximately 2349 jurisdictions in 40+ countries have declared a 'Climate Emergency' via a global declaration, which includes over 1 billion citizens (Climate Emergency Declaration, 2024).

As narratives around the Climate Emergency have grown in number and frequency, they have increasingly been underpinned by a rhetoric of 'it affects all of us' as a great equaliser. Whilst they are largely true and adaptation will be required everywhere, actions need to be targeted at the most vulnerable people with the least resources. The effects of the Climate Emergency are being felt unevenly by minority and marginal groups, including refugees, migrants, women, people living in extreme poverty (United Nations, 2023b), and gender and sexual minorities (Dalton, 2023).

Absence of Gender and Sexual Minorities Within the United Nations Sustainable Development Goals

The United Nations launched the Sustainable Development Goals in 2015 as part of Agenda 2030—an attempt to build on the successes and failures of the eight 2000–2015 Millennium Development Goals (MDGs). The MDGs were criticised for being written in an office in the Global North, lacking consultation with other nation states and actors, and being 'development' for those countries who 'need developing' (Fehling et al., 2018). These lessons were incorporated in the drafting of the Sustainable Development Goals (SDGs) as part of Agenda 2030. The SDGs were developed in consultation with over 60+ nation states, agreed upon by 193 nation states, encompassing the Global North and the Global South. The SDGs comprise 17 goals, with a stated aim to 'leave no one behind' and to 'reach the furthest behind first' (UN SDGs, 2023). All goals are interconnected and have a human rights framework at their heart. They recognise efforts to end poverty and other deprivations as key. These efforts are coupled with strategies to reduce inequalities; stimulate economic growth; grow sustainable cities with safe homes; and tackle climate change (OHCHR, 2023).

A shared grievance between indigenous peoples and gender and sexual minorities is how the SDGs were developed and the lack of consultation with representatives of both populations. Indigenous peoples were marginalised in the design of the Goals and unnamed in the SDG literature despite forming 5% of the world's population (IISD, 2019; Amnesty International, 2023). Similarly, gender and sexual minorities (see also LGBTQI+ people) were overlooked, or ignored, in the SDG development and implementation (Dalton & Smith, 2023), despite comprising 10%+ of the world's population (Ipsos, 2021). Simultaneously, many indigenous groups have concepts of a third gender, non-normative sexual relations, or non-normative gender expressions (see later discussion).

Additionally, 'in the implementation of the Sustainable Development Goals, LGBTQI+ individuals have been left behind in several different ways' (Martin & Smith, 2023, p. 7), including invisibilisation, or the active process of editing out gender and sexual minorities from the SDGs. The inclusion of both populations in the development of the SDGs was problematic for some governments and nation states. For example, in discussions about the inclusion of gender and sexual minorities during the initial SDG development phase, a powerful voting bloc of nation states vetoed Agenda 2030. Policymakers were concerned that the inclusion of LGBTQI+ people and their communities might imply support for gender and sexual minorities from otherwise hostile and deeply homophobic and transphobic countries (Izugbara et al., 2022; Brown, 2023).

The international development sector has been critiqued as deeply heteronormative (Jolly, 2011, 2021; Dalton, 2023). Government programmes, non-governmental organisations (NGOs), and international non-governmental organisations (INGOs) and their donors often insensitively overlook gender and sexual minorities in their programme design, delivery, and implementation (Dorey, 2018). Simultaneously, programmes reflect the cisnormative binary model of biological sex and gender in Goal 5, which speaks of 'women and girls' and 'men and boys' (Martin & Smith, 2023).

By side-lining sexual rights as part of international development, Elson (2018) argues we have 'pushed behind' some communities due to international developments' original founding on utilitarian ethics, placing an excessive focus on economic growth at the expense of factors such as liberation. Mills (2015), in his meta-analysis of research related to gender, sexuality, and development, found that 'mechanisms of exclusion' have repeatedly left gender and sexual minorities behind in development policies and practices. Mills (2015) concluded that the social exclusion and discrimination that gender and sexual minorities have long experienced will remain a fundamental threat, and so they will be left behind once again.

Similarly, the humanitarian aid and international peace and security sectors, and their associated bodies, have also been accused of heteronormativity and cisnormativity in their programme design (Hagen, 2016). The exclusion of gender and sexual minorities from the SDGs and the larger development sector serves as a backdrop for the wider exclusion of gender and sexual minorities at a nation-state level. For example, at this level humanitarian responses to disasters have ensured that gender and sexual minorities are overlooked, and wider social stigma and marginalisation of these groups are reinforced afterwards. Therefore, it is important to consider the 'real time' impact of the Climate Emergency on the lives of gender and sexual minorities, which is highlighted in the fate of the Guna Yala (and others), to which we turn.

The Impacts of the Exclusion of Gender and Sexual Minorities in Disaster Planning and Response

The Guna Yala (Fig. 21.1) is an archipelago of approximately 365 islands (of which 49 are inhabited) in northeast Panama and home to approximately 50,000 indigenous Guna people. The local economy focuses on subsistence agriculture, fishing, handicraft production, and tourism. The Guna left the Panama mainland hundreds of years ago and developed a deep-rooted history of trade and self-determination, helping them to resist Panamanian efforts at colonial (re)acculturation.

The Guna are characterised by a strong sense of gender equity. Women are property owners and decision-makers, with weddings focused on a woman undertaking a ceremonial abduction of a man, and husbands moving in with their new brides. Within the Guna, gender fluidity is common. Some men can become a third gender 'Omeggid' (meaning 'like a woman'), where they work and act like other women in the community. The Guna exhibit a strong acceptance of homosexuality and same-sex relationships (BBC, 2018).

It is predicted that significant rises in sea level will lead to flooding of sections of the Guna Yala archipelago. In a few decades these sections will be under seawater, leaving many of the Guna homeless and separated from their traditional lands (Science News, 2023). The Panamanian government has responded to the effects of the Climate Emergency by committing to build new shelters on the mainland for the Guna. However, this may come at a significant risk to the Guna's own unique culture. It may lead to changes in their social attitudes to gender and sexual diversity, including the adoption of external homophobic and transphobic beliefs and practices and alter the place and impact LGBTQI+ identity politics on their lives (BBC, 2018). The changing situation of the Guna people and the growing impacts of the Climate Emergency highlight the sensitive nexus between changing weather patterns due to climate change and the lives of gender and sexual diverse communities, not only in the Guna Yala archipelago but globally.

The Aravanis of Sri Lanka are another group who have been rendered invisible in state disaster planning, development, and humanitarian relief. The Aravanis are gender fluid and often portrayed as transgender or in some cases as intersex. They are subject to social stigma, systemic discrimination, and arbitrary arrest by the police, and their main income is from begging and dancing for events and weddings. When the 2004 Indian

Ocean Tsunami hit Sri Lanka, and the subsequent disaster relief programmes rolled out, data on the Aravanis, several of whom had perished, were not collected in disaster relief assessments conducted by the government, meaning they were denied bedding, shelter, and food (Pincha, 2008). A significant number of the Aravanis had lost their small cash savings in the tsunami and, importantly, lost their tools of employment, such as make-up, dancing costumes, and jewellery, with no recourse to governmental funding (Pincha, 2008).

Compounding these social injustices, all Sri Lankans needed to be registered as municipal residents to access government food aid, which involved prolonged contact and sharing an address with the police (if they had one). This blocked not only the Aravanis from seeking support but also other gender and sexual minorities, as homosexuality was criminalised in Sri Lanka at that time (The New Humanitarian, 2020). This unique mix of invisibilisation, marginalisation, and lack of response by state and international actors is not only present in disasters such as the 2004 tsunami but also present in other Climate Emergency–related disasters.

Globally, gender and sexual minorities are over-represented in unhoused statistics (where they are collected). Nation-state reports tend to show that younger members of gender and sexual minorities are more likely than many other societal demographics to live homeless, with gender and sexual minorities estimated to comprise between 20 and 40% of the global homeless population (Fraser et al., 2019). Causes of homelessness range from parental rejection, state homo/bi/transphobia, stigma, and forced eviction to precariousness of asylum or refugee status, under-employment, and unemployment (Fraser et al., 2019). In short, many gender and sexual minorities face deepening levels of poverty and marginalisation due to societal stigma and discrimination.

This precariousness means that many gender and sexually diverse homeless people are subjected to the immediate, on-the-ground effects of EWEs because of the impact of the Climate Emergency. This is true for the self-titled 'Gully Queens' of Kingston, Jamaica, a group of mainly younger gender and sexual minorities who have gathered in the storm gullies for protection in the city. The gullies are six to nine feet wide, and they lie below street level, draining rainwater, and the city's refuse out to sea (Dunn, 2016; Reuters, 2017). Already engaged in sex work, facing regular attacks from the public, and made homeless due to wider societal prejudice, the Gully Queens are at the mercy of the effects of the Climate Emergency as Jamaica is experiencing a rise in climate-induced flash-flooding (World Bank, 2023). The gullies, which have become a home, can flood in less than a minute, risking drowning the Gully Queens. These tangible links between institutional homophobia and transphobia, societal stigma, and the effects of the Climate Emergency on the lives of gender and sexual minorities have never been more pronounced.

Furthermore, the effects of the Climate Emergency are also a threat-multiplier in that they exacerbate existing social problems for gender and sexual minorities (ReportOUT, 2023). Devex (2020) notes that in times of lawlessness after disasters and the increased pressure on more scarce resources, there is an increase in homo/bi/transphobia which 'rise to the surface.' Perpetrators of homo/bi/transphobia use pre-existing stigma to gain resources or to enact violence or harm, with the perpetrators acting without impunity or consequence. In 2019, the devastation of Cyclone Idai in the city of Beira in Mozambique compounded existing issues of poverty, inadequate housing, and women's vulnerability to gender-based violence. Due to the devastation of Cyclone Idai, many poorly constructed shelters in which minority women lived were destroyed, leading to reports about the heightened visibility and therefore vulnerability, of lesbian and bisexual women to sexual violence and so-called corrective rape (The Ecologist, 2021). In the wake of a Climate Emergency disaster, the limitations of Sustainable Development Goal 5 (Gender Equality) become magnified. This goal is to 'achieve gender equality and empower all women and girls,' yet is problematic within its own defined binary version of gender (ReportOUT, 2021). This is further magnified in terms of humanitarian relief and development responses, which overlook or ignore gender-expansive populations.

When Super Typhoon Yolanda impacted the Philippines in 2013, responses to it saw gendered lines drawn in relief efforts (Nichols, 2019). Whilst actors and institutions pledged to respond using gendered frames, recognising the uneven impact facing women in relief efforts, the recovery efforts were based around a cisgender man-woman binary (Nichols, 2019), essentialising all women as one group rather than a group of people with differential needs (Junio, 2018). Consequently, many transgender and gender-diverse Filipino populations were excluded from livelihood initiatives, unable to access housing benefits and sanitation facilities (UNDP, ODI and GALAND, 2020). Within post-disaster contexts, a pressing need for many people becomes livelihood and employment. These become significantly more difficult to achieve due to deep societal prejudice and discrimination, which was noted in the after-effects of

Yolanda amongst queer Filipinos turned away from employment (Gaillard et al., 2016).

Even within nation states that have existing legislative protections for gender and sexual minorities against homo/bi/transphobia, their vulnerability to the effects of the changing climate is increased. In an example closely tied to the effects of the Climate Emergency, during the 2011 floods in Queensland, Australia, 43% of sexual and gender respondents felt afraid of the actual evacuation centres during the disaster and recovery process, fearing homophobia and transphobia (Dominey-Howes et al., 2014). Similarly, when Hurricane Katrina devastated New Orleans in 2005, the flooding left significant populations homeless, blocked sanitation services, and forced populations to gather in temporary camps and football stadiums. When seeking safety, transgender people were reported to face discrimination and were denied access to emergency shelters, alongside reported arrests of transgender individuals who used a bathroom reflecting their own gender in a temporary emergency shelter (D'Ooge, 2008; Goldsmith & Bell, 2022).

During this time, stigmatising practices of marginalisation surfaced as some religious leaders blamed gender and sexual minorities as the cause of the disaster. Subsequently, faith-based organisations could use someone's lack of official identification documents that matched their name or gender as justification for denying them aid (Knight and Sollom, 2012). These issues were compounded at the state level by what Bourdieu (1977) called symbolic violence (Bourdieu, 1977) as gender and sexual minorities relationships and family structures were not recognised in disaster and relief benefits. At that time, existing legislation only recognised married couples, which, in the United States (USA), did not include same-sex couples or non-heteronormative families (same-sex marriage was not introduced in the USA until 2015). The impact of this denial of resources was felt particularly amongst gender and sexual minorities who had children (Goldsmith & Bell, 2022).

There is also a lack of data on the specific health risks and issues faced by gender and sexual minorities in the face of Climate Emergency–induced disasters (Brody, 2021). Whilst there are shared vulnerabilities across all populations, the social and economic exclusion faced by minority populations has negative impacts on their resilience and capacity to minimise the risks and impacts of disasters (Outright International, 2020). Prior to a disaster, many gender and sexual minorities may not have access to the same levels of resources, capital, information, social networks, and assets as the general population, which affects their resilience after a climate disaster (Devex, 2020).

Intersections of health and well-being make responses to Climate Emergency events particularly problematic for some gender and sexual minorities, such as those living with HIV and AIDS. Societal stigmatisation around HIV is already significant. However, access to healthcare and health-seeking behaviours after climate disasters can be further curtailed or restricted, heightening vulnerability amongst people living with HIV. During Hurricane Maria in the North-eastern Caribbean in 2017, which mainly affected Puerto Rico and Dominica, many pharmacies were destroyed or lost electricity due to fallen power lines or damaged generators. Consequently, they were unable to get Anti-Retroviral Treatments (ART) or hormone replacement therapy prescription refills to those who needed them (Melin et al., 2018). Similarly, during Hurricane Katrina, many gender and sexual minorities who were living with HIV were reluctant to disclose their HIV status to emergency relief doctors who were not familiar to them (Haskell, 2021).

As the Climate Emergency presents more EWEs, additional and intersectional challenges will continue to be faced by gender and sexual minorities. Currently, the Sustainable Development Goals form the guiding mechanisms for humanitarian relief efforts and official responses to these events. However, due to their heteronormative and cisnormative framing, these are not sufficient to protect gender and sexual minorities as vulnerable populations regarding Goal 13 (i.e. Climate Action).

The Climate Emergency as a 'Beyond the Back of the Mind' Issue

I have argued elsewhere that the Climate Emergency is not only, as Giddens (2009) states, a 'back of the mind issue' but gender and sexual minorities have gone 'beyond the back of the mind' (Dalton, 2023). The 'Giddens paradox' argues that many electorates of many nation states cannot grasp the significance of the Climate Emergency as it appears to be too abstract, not dramatic enough, and 'distant' from their lives. As a 'back of the mind' issue, many populations will not see the results of the Climate Emergency until the extreme end of its global effects, which by then may be too late. In this political landscape, the longer the issue is debated and fought over (e.g. the long-lasting debate about environmental reparations for developing countries), the more pollution is happening, which is exacerbating the issue even further (Dalton, 2023). Finally, 'free riders' have ample opportunity to tackle the Climate Emergency, but instead wait for others to do so and so little is achieved as a united front (Dalton, 2023). The compounding impact of all three aspects of the Giddens paradox ensures that the Climate Emergency may appear to be distant, unaffected by populations, and future-facing only, despite the impact and increasing range of climate-related disasters.

For gender and sexual minorities across the globe, the Giddens paradox (2009) is also underpinned by three additional forces. Firstly, the 'no one is left behind' slogan as set out as part of Agenda 2030 and the Sustainable Development Goals is problematic in that the heteronormative and cisnormative aspects of the global goals automatically create exclusion. Secondly, gender and sexual minorities have become a 'beyond the back of the mind' (Dalton, 2023) issue in wider Climate Emergency discussions and thereby have been overlooked as a vulnerable group by politicians, many global bodies and development institutions. Some of these barriers to engagement emerge from the invisibility of gender and sexual minorities in agenda-setting. But some also arise from the state criminalisation (or overlooking) of queer bodies, alongside the heteronormative and cisnormative nature of the development and humanitarian aid industries.

Finally, this exclusion also manifests in ensuring that gender and sexual minorities have been effectively side-lined from wider Climate Emergency debates and narratives, as well as a wide disconnect between the climate justice movement and the LGBTQI+ movement (Dalton, 2023). There is significant queer representation in the climate justice moment (Weder & Swastika, 2021), with gender and sexual diverse minorities supporting climate politics in different ways. Yet, despite this overlap, sexual rights and climate change groups do not connect. Hitherto, these exacerbating factors have contributed to gender and sexual minorities' lack of engagement in Climate Emergency narratives, actions, and relief efforts.

How Are Gender and Sexual Minorities Responding to the Climate Emergency?

There is a growing movement of gender and sexual minorities acknowledging and responding to the impacts of the Climate Emergency on their own lives. From the academy, 'Queer Ecology' has emerged to dismantle heteronormative ideals of nature and to redefine the Climate Emergency narrative as no longer a hetero-natural vision for the future. It also recognises the need for more specialised, diverse, and localised support (Mattson, 2020). Mattson (2020) argues that through facing the impact and effects of both the Climate Emergency and state homophobia and transphobia, many gender and sexual minorities have adopted a positive 'queer futurism.' This technique looks forward to the future, even when it is painful to examine the past and the present, to offer a queer future of hopeful possibilities (Mattson, 2020). This queering of the Climate Emergency and its temporal nature has taken a relatively mundane topic such as the weather and used it to invoke a current crisis in queer lives, to reframe the known as something both familiar and strange (Chakrabarty, 2009).

Giddens (2009) hoped for the development of an 'ensuring state' and a strong combination of global actors to tackle the Climate Emergency, and whilst this is present in United Nations Conference of the Parties (COP) meetings, many gender and sexual minorities have started their own ways to ensure voices are heard and amplified. One is to 'queer the SDGs' (Daly, 2020) with global LGBTQI+ organisations such as ReportOUT (ReportOUT, 2021) and nation-state LGBTQI2S+ organisations such as Egale (Egale, 2023) in Canada, engaging with the SDGs and queering them to find inroads for engagement. Whilst the very exclusion of gender and sexual minorities seems apparent from the SDGs, the universalist taglines 'leave no one behind' and 'reach the furthest behind first' (UNSDGs, 2023) have provided ample ground for SDG engagement amongst some activists and LGBTQI+ social movements. Significantly, the statement behind all the goals that they are applicable to 'any other status' has allowed organisations to engage with the Voluntary National Review (VNR) process and the various mechanisms whereby research and evidence of homophobic and transphobic exclusion can be included in government and United Nations reports.

Reframing policy and planning so that they include SGMs has also started to emerge. The Asia Pacific Forum for Women, Law, and Development (APWLD, 2015) set a strong example of women-led action to inform Climate Emergency policy planning. This programme supported the collective work of indigenous women and queer groups in nine communities in Asia-Pacific—the Philippines, Nepal, India, Indonesia, Thailand, Papua New Guinea, Vietnam, and Bangladesh. The programme supported queer women (alongside heterosexual and cisgender women) to document their own experiences, responses, and needs to ensure that those most affected by the Climate Emergency can shape policies at the local, national, and international levels. As a result of this work, a statement and set of recommendations aimed at Association of Southeast Asian Nations (ASEAN) policymakers was produced.

Others have also pushed for further demographic data categories in international development to include gender and sexual minorities in development so that they can be recognised and given representation at a higher level (Galati, 2015). The approach of 'what is measured, matters' (Stuart and Woodroffe, 2016, p. 75) has led to the collection of data on gender and sexual minorities not only in some development evaluations, but also in Climate Emergency issues. In 2023, ReportOUT launched the 'Green in the Rainbow' study analysing data from LGBTQI+ organisations, their knowledge of the impacts of a changing climate on gender and sexual minorities, and their organisational Climate Emergency preparedness (ReportOUT, 2023). Other grassroots organisations, such as Queer X Climate, have developed a membership pledge of queer organisations and individuals to limit and account for their carbon emissions (Queer X Climate, 2022). Furthermore, Out for Sustainability (2023) has produced a toolkit for emergency equity preparedness. Others have offered additional frontline support, such as Organización Latina Trans en Texas (OLTT), a transgender Latinx immigrant-led organisation that has provided shelter, food, and resources to unhoused transgender individuals during Hurricane Harvey in 2017.

As a response to Goal 13 (Climate Action), in 2021 GiveOut launched the world's first 'LGBTQI Climate Fund' for individuals and businesses, stating that queer people are 'not sat at the table and are being excluded from government decision-making' (GiveOut, 2021; Reuters, 2021). Examples of success through this fund have included the Tonga Leitis Association (TLA) in the Kingdom of Tonga, who face extreme climate-induced cyclones, yet fear seeking safety in emergency shelters due to them being run by hostile religious organisations. The fund has enabled sensitivity training from the TLA to those who run the shelters, as well as amplifying the voices of the LGBTQ+ community in nation-state climate consultations (GiveOUT, 2023).

It is vital that gender and sexual minorities claim their space in the SDGs, their implementation, and within their corresponding development industries. Additionally, linkages between global SGMs and the effects and responses to the Climate Emergency need to be foregrounded in debates. A queering of ecological, environmental issues and responses to the Climate Emergency has begun to emerge, though it is in its infancy.

No One Left Behind?

Will no one be left behind by 2030? For SGMs, this appears to be unlikely. The heteronormative and cisnormative nature of the design of the SDGs, plus the invisibilisation of SGMs in SDG planning, has ensured that more vulnerable populations of SGMs are marginalised in both international development and its mechanisms of responding to climate-induced disasters. The Climate Emergency is a threat multiplier for these groups, bringing existing and new threats to the surface. These threats interact with and increase the negative impacts of the absence of any consideration of the effects of the Climate Emergency and post-disaster responses on SGMs. Existing homo/bi/transphobic inequality, stigma, and discrimination, alongside, in some nation states, the criminalisation of queer bodies, has meant that climate-induced disasters expose, magnify, and even deepen existing negative and prejudicial attitudes.

As SGMs live with and within nature, they face the systematic invisibilisation of their identities not only at strategic levels of policy, but also in humanitarian and development methods adopted to respond to climate changes in terms of sharing and allocating resources. What is clear is that climate impacts further elevate existing fault lines of homophobia and transphobia, exposing new and significant vulnerabilities for SGMs. Without the protection and elevation of gender and sexual minorities from global actors, nation states, and organisations, these fault lines will continue to lead to further injustices as the worsening effects of the Climate Emergency play out.

Conflict of Interest The author has no conflict of interest.

References

Amnesty International. (2023). *Indigenous peoples overview* (Online) https://www.amnesty.org/en/what-we-do/indigenous-peoples/

APWLD. (2015). *Statement from LGBTI and indigenous women affected by climate change.* (Online) https://apwld.org/statement-from-lgbti-and-indigenous-women-affected-by-climate-change/

BBC. (2018). *Guna Yala: The islands where women make the rules.* (Online) https://www.bbc.com/travel/article/20180813-guna-yala-the-islands-where-women-make-the-rules

Bourdieu, P. (1977). *Outline of a theory of practice.* Cambridge University Press.

Brody, A. (2021). *Mapping the linkages between climate change, health, gender and SOGIESC for the Asia-Pacific Region.* (Online) https://www.sparkblue.org/system/files/2021-02/Desk%20review_climate%20change_gender_health_Feb2021.pdf

Brown, S. (2023). Visibility or impact? International efforts to defend LGBTQI+ Rights in Africa. *Journal of Human Rights Practice, 15*, 506–522.

Chakrabarty, D. (2009). The climate of history: Four theses. *Critical Inquiry, 35*(2), 197–222.

Climate Emergency Declaration. (2024). *Climate Emergency Declarations in 2,349 jurisdictions and local governments cover 1 Billion Citizens.* (Online) https://climateemergencydeclaration.org/climate-emergency-declarations-cover-15-million-citizens/.

D'Ooge, C. (2008). Queer Katrina: Gender and sexual orientation matters in the aftermath of the disaster. In B. Willinger (Ed.), *Katrina and the women of New Orleans.* Tulane University.

Dalton, D. (2023). Have we left behind the rainbow warriors? The climate emergency and its impact on global queer people and their communities. In D. Dalton & A. Smith (Eds.), *Gender, sexuality, and the UN's SDGs* (pp. 17–40). Palgrave Macmillan.

Dalton, D., & Smith, A. (Eds.). (2023). *Gender, sexuality, and the UN's SDGs.* Palgrave Macmillan.

Daly, F. (2020). *LGBT+ inclusion and the UN sustainable development goals.* (Online) https://static1.squarespace.com/static/5bba53a8ab1a62771504d1dd/t/5f6b4f3a978b0513584d2280/1600868211690/SDG-LGBT+inclusion.pdf

Devex. (2020). *Responses to climate change must be LGBTQ-inclusive, experts say.* (Online) https://www.devex.com/news/responses-to-climate-disasters-must-be-lgbtq-inclusive-experts-say-98062

Dominey-Howes, D., Gorman-Murray, A., & McKinnon, S. (2014). Queering disasters: On the need to account for LGBTI experiences in natural disaster contexts. *Gender, Place and Culture, 21*(7), 905–918.

Dorey, K. (2018). *Leave no one behind. LGBT inclusion in development.* (Online) https://www.bond.org.uk/news/2018/02/leave-no-one-behind-lgbt%20inclusion-in-development

Dunn, L. (2016). Integrating men and masculinities in Caribbean disaster risk management. In E. Enarson & B. Pease (Eds.), *Men, masculinities and disaster.* Routledge.

Egale. (2023). *Queering the sustainable development goals in Canada.* (Online) https://egale.ca/awareness/queering-sustainable-development-goals-canada/

Elson, D. (2018). *Push no one behind*. CDP Background Paper Number 43. United Nations, Department of Economic and Social Affairs.

Fehling, M., Nelson, B. D., & Venkatapuram, S. (2018). Limitations of the millennium development goals: A literature review. *Global Public Health, 8*(10), 1109–1122.

Fraser, B., et al. (2019). LGBTIQ+ homelessness: A review of the literature. *International Journal of Environmental Research and Public Health, 16*(15), 26–77.

Gaillard, J. C., et al. (2016). Beyond men and women: A critical perspective on gender and disaster. *Disasters, 41*(3), 429–447.

Galati, A. J. (2015). Onward to 2030: Sexual and reproductive health and rights in the context of the sustainable development goals. *Guttmacher Policy Review, 18*(4), 77–84.

Giddens, A. (2009). *The politics of climate change*. Polity Press.

GiveOut. (2021). *Why climate change is an LGBT Issue*. (Online) https://giveout.org/why-climate-change-is-a-lgbt-issue

GiveOUT. (2023). *LGBTQI climate fund*. (Online) https://giveout.org/lgbtqi-climate-fund

Goldsmith, L., & Bell, M. (2022). Queering environmental justice: Unequal environmental health burden on the LGBTQ+ Community. *American Journal of Public Health, 112*(1), 79–87.

Hagen, J. (2016). Queering women, peace, and security. *International Affairs, 92*(2), 313–332.

Haskell, B. (2021). *Sexuality and natural disaster: Challenges of LGBT communities facing hurricane Katrina*. (Online) https://papers.ssrn.com/sol3/papers.cfm?abstract_id=2513650

International Institute for Sustainable Development (IISD). (2019). *No sustainable development without indigenous peoples*. (Online) https://sdg.iisd.org/commentary/guest-articles/no-sustainable-development-without-indigenous-peoples/

IPCC. (2018). Summary for policymakers. In *Global Warming of 1.5°C. An IPCC Special Report on the impacts of global warming of 1.5°C above pre-industrial levels and related global greenhouse gas emission pathways, in the context of strengthening the global response to the threat of climate change, sustainable development, and efforts to eradicate poverty*. (Online) https://www.ipcc.ch/sr15/chapter/spm/

IPCC. (2023). *Climate change 2013: The physical science basis*. (Online) https://www.ipcc.ch/assessment-report/ar6/

Ipsos. (2021). *LGBT+ Pride 2021 global survey*. (Online) https://www.ipsos.com/en/ipsos-lgbt-pride-2021-global-survey

Izugbara, C., et al. (2022). "The SDGs are not God": Policymakers and the queering of the sustainable development goals in Africa. *Development Policy Review, 40*(2), 1–16.

Jolly, S. (2011). Why is development work so straight? Heteronormativity in the international development industry. *Development in Practice, 21*(1), 18–28.

Jolly, S. (2021). Is development work still so straight? Heteronormativity in the development sector over a decade on. *Development in Practice, 33*(4), 463–474.

Junio, A. (2018). *KFAW. The 27th Foreign Correspondent Report: Sexual minorities and disasters in the Philippines: Where are they?* (Online) http://www.kfaw.or.jp/correspondents/docs/27-2_Philippines_E.pdf

Knight, K., & Sollom, R.. (2012). *Making disaster risk reduction and relief programmes LGBTI-inclusive: Examples from Nepal* (Online) https://odihpn.org/publication/making-disaster-risk-reduction-and-relief-programmes-lgbti-inclusive-examples-from-nepal/.

Marshall, G. (2015). *Don't even think about it: Why our brains are wired to ignore climate change.* Bloomsbury.

Martin, C., & Smith, A. (2023). Introduction. In D. Dalton & A. Smith (Eds.), *Gender, sexuality, and the UN's SDGs.* Palgrave Macmillan.

Mattson, L. (2020). *Elemental climate disaster texts and queer ecological temporality.* Graduate Theses and Dissertations. (Online) https://scholarcommons.usf.edu/etd/8252.

McCormick, J. (2022). *Introduction to global studies* (2nd ed.). Bloomsbury.

Melin, K., Maldonado, W. T., & López-Candales, A. (2018). Lessons learned from hurricane Maria: Pharmacists' perspective. *Ann Pharmacother, 52*(5), 493–494.

Mills, E. (2015). "Leave no one behind": Gender, sexuality, and the sustainable development goals. *Sexuality, Poverty and Law Evidence Report,* 154. Institute of Development Studies.

Nichols, E. (2019). *What the impact of Stonewall has taught me about environmental rights.* (Online) https://www.greenpeace.org/international/story/22806/what-the-impact-of-stonewall-has-taught-me-about-environmental-rights/

OHCHR. (2023). *OHCHR and the 2030 Agenda for sustainable development.* (Online) https://www.ohchr.org/en/sdgs

Organización Latina Trans en Texas (OLTT). (2023). *Homepage.* (Online) https://www.latinatranstexas.org/

Outright International. (2020). *Vulnerability amplified: The impact of the COVID-19 pandemic on LGBTIQ people.* (Online) https://outrightinternational.org/content/vulnerability-amplified-impact-*Covid--19*-pandemic-lgbtiq-people

Pincha, C. (2008). *Indian Ocean Tsunami through the gender lens: Insights from Tamil Nadu, India.* Earthworm Books.

Queer X Climate. (2022). *Pledge.* (Online) https://www.queersxclimate.org/pledge

ReportOUT. (2021). *Queering the sustainable development goals*. (Online) https://www.reportout.org/post/queering-the-sustainable-development-goals

ReportOUT. (2023). *A crisis of queer invisibility: Climate change as a risk multiplier for LGBTQ people*. (Online) https://www.reportout.org/green-in-the-rainbow

Reuters. (2017). *Jamaican LGBTQ youths escape persecution in city storm drains*. (Online) https://www.reuters.com/article/us-jamaica-lgbt-homeless-idUSKBN1685AY

Reuters. (2021). *UK charity launches world's first LGBT+ Climate fund*. (Online) https://www.reuters.com/article/us-climate-change-lgbt-trfn-idUSKBN2B32YD

Science News. (2023). *How an indigenous community in Panama is escaping rising seas*. (Online) https://www.sciencenews.org/article/indigenous-guna-panama-escape-rising-seas

Stuart, E., & Woodroffe, J. (2016). Leaving no-one behind: Can the sustainable development goals succeed where the millennium development goals lacked? *Gender and Development, 24*(1), 69–81.

The Ecologist. (2021). *Queer African voices call for climate justice*. (Online) https://theecologist.org/2021/jul/08/queer-african-voices-call-climate-justice

The New Humanitarian. (2020). *LGBTQ+ People left out by exclusionary COVID-19 aid practices*. (Online) https://www.thenewhumanitarian.org/news-feature/2020/06/24/LGBTQ-gender-coronavirus-discrimination-aid

UN SDGs. (2023). *Homepage*. (Online) https://sdgs.un.org/

UNDP, ODI and GALAND. (2020). *Making it work: Lesbian, bisexual and transgender women's economic empowerment in the Philippines*. (Online) https://www.undp.org/philippines/publications/making-it-work-lesbian-bisexual-and-transgender-womens-economic-empowerment-philippines

UNEP. (2023). *The climate emergency*. (Online) https://www.unep.org/climate-emergency

United Nations. (2023a). *What is climate change?* (Online) https://www.un.org/en/climatechange/what-is-climate-change

United Nations. (2023b). *Climate adaption*. (Online) https://www.un.org/en/climatechange/climate-adaptation

Weder, F., & Swastika, S. (2021). Advocacy for sustainability communication: Unseen potential of queer communicators in environmental, climate change and sustainability science. *Sustainability, 13*(24), 138–171.

World Bank. (2023). *Country: Jamaica*. (Online) https://climateknowledgeportal.worldbank.org/country/jamaica

CHAPTER 22

Hurricane Katrina and a Curious Closet for Queers: A Short Story

Lisa Overton

Introduction

Queering Academic Writing

Human stories are critical in widening our understanding of disasters, particularly their impacts on groups whose stories are not usually heard and those that have been pioneered by marginal scholars, particularly within black feminist thought (hooks, 1991; Mutch & Marlowe, 2013). In this experimental vignette, the traditional form of sharing stories in qualitative research is turned on its head by being framed as a short story. This short story introduces two main protagonists and three sub-characters to explore themes of gender, sexuality and life course.

The story is inspired by research conducted in 2012 that relied on life-history interviews of queer-identified people reflecting on their experiences growing up after Katrina. For some participants, this meant looking

L. Overton (✉)
Middlesex University, London, UK
e-mail: L.A.Overton@mdx.ac.uk

D. Dominey-Howes et al. (eds.), *Queering Disasters, Climate Change and Humanitarian Crises*, Sustainable Development Goals Series, https://doi.org/10.1007/978-981-96-3857-4_22

back on their teenage selves. The story presented here is a fictive interpretation of themes that ran through the interviews, themes that would otherwise remain hidden (Halperin, 1995).

Regulation of sexuality has consequences for all and can be particularly harmful to those who do not fit heterosexual norms (Nussbaum, 1999), such as LGBTQ+[1] identifying young people. On the other hand, it is important to recognise and celebrate 'queer joy' which, whilst bittersweet, can empower and inspire (Oxfam, 2024). In the story presented in this vignette, sisters Reina and Carmen are at a critical juncture in their life course, which includes discovering their sexualities while living through Hurricane Katrina. As an experimental and creative piece in an overall 'academic' book, it has been a refreshing experience to have a licence to play across genre and form, yet I still feel it necessary to caveat that because of this, the vignette may feel untidy—this is deliberate and, I hope, acts as a space for questions. Untidiness, after all, like queering, unsettles what we think we know and here I seek to make playful troublesome so that we might see thing afresh (Phillips & Kara, 2021, p. xi).

The story told here is about an overarching theme in my research: that sometimes it is everyday fear that is most frightening and everyday fear for a young queer person may differ. A disaster can amplify, mutate and emphasise such fears and simultaneously open up space for connection and positive change—for discovery and for joy.

The Importance of Not Doing What You Are Told

Chapter I: Evacuate

The highway was thick with cars and fumes, tangled traffic going on for miles. This is what an evacuation looks like, Reina thought, feeling the clammy heat against her skin, the scratch of her denim shorts, the fuzz of the car seat. Half-packed bags and cases were piled around her. The air inside the car felt foggy with the breath of bodies—her own, her sister Carmen's and her parents clogging the space. All the windows were open, but there was no air flow? Everyone in the city seemed to be evacuating at the same time. Everyone with a vehicle.

[1] Lesbian, gay, bi-sexual, transgender, queer and all other gender identities and sexual orientations.

'I can't believe uncle Arth wouldn't budge,' Reina's mum said to her dad, Reg, who was driving.

'He needed to stay to protect the house. I respect that.'

'At least he made aunt Jilly go with the kids,' Lucia said. Reina remembered uncle Arth's face. He looked worried and said he needed to stay because 'things are gonna get bad.'

Reina jolted forward—again. How they hadn't rear-ended the car in front she couldn't fathom.

Reg was exhausted in the stop-start-crawl they had been in for hours and they weren't even out of the city yet. 'Unbuckle and tell me what you see up there, Reina,' Reg asked.

'Cars, just cars for forever, all stacked up.' She curled back inside and hoisted up her denim shorts whilst trying not to wake Carmen, hitting her leg against the sharp edges of her sports bag. She pulled it onto her lap so she could stretch her legs out.

'We should have left earlier, Reg, I told you,' Lucia said.

'Shut up, woman! What do you want me to do about it, huh?' Reg tried to control the anger in his voice, strangling the wheel of the car and conscious that Reina was watching, conscious that Carmen might wake up. 'It's what we decided. We just have to wait.'

Every time they jolted, Reina looked up to see Reg's head lolling down, followed by her mum screaming at him to wake up.

The next jolt sent Reina flying forward, her bag clattering onto the centre console into the middle bit of the car. She caught it just in time before it slid all the way through to the stick shift.

Reg cursed. 'What is that bag?'

'Nothing,' Reina mumbled as she tried to vanish it away, stuffing it back into the footwell.

'Did you—you bought it with you, didn't you? When I told you we don't have the room!'

'It's not that big of a deal, dad. You're just grumpy and you won't admit you're too tired to pull over!'

'This is an emergency, Reina, don't you understand that? Some people can't even take themselves. You're a selfish little girl.'

'I'm almost seventeen!'

'If you have filled this car with anything but the essentials, you are in serious trouble,' Lucia joined in.

Reina imagined the long line of desperate families waiting for the coaches and had hoped no one would notice the funny lumpy shape of black sports bag she had packed it full of DVD box sets but only one change of jeans.

They all fell silent, Carmen still asleep. The strange hum of traffic blanketed them, and Reg locked his eyes on the road, they were moving more consistently now, the mood in the car lifting at the thought they might finally get some miles behind them.

Then crash. They'd hit the car in front.

This time Reina didn't fly forward but Carmen was dislodged down on top of Reina's sports bag in the footwell. Carmen was about to cry out, but Reina flashed her a glaring look, so instead she bit her lip and her big grey eyes welled up and her round little face went all red.

Reina felt even more ashamed as she mouthed 'sorry baba' and held her little sister in her arms.

'I'm alright,' Carmen said. She knew that she needed to be brave because that's what Reina was doing, but she wished that Reina would stop calling her baba, but she kept quiet. This was what they did in their family. Carmen also knew to keep quiet about the box sets of DVDs with the women on the covers that looked like Sex and the City but weren't—they dated each other instead of men and was called the L Word like it was meant to be some big secret.

After the cursing and swearing at Lucia for not paying attention and at Reina for distracting him, Reg got out of the car to check the damage.

'What were you doing, man, you went right in the back of me!' shouted a huge, hulking man.

'There's no damage. Look, I was barely going forward,' Reg placated, bowing his head and his shoulders in deference.

'You shouldn't be driving; you look like you fell asleep.'

'Hey, you slammed your breaks on!'

Fear had taken hold of Reg now and that made him angry; he was ready to fight and both men squared up to each other as their voices grew louder and harsher.

They were saved by the shrill sound of wailing coming from something inside the hulking man's car. A boy, no more than six years old, his fearful sobs directed not at the looming hurricane but at his own father and this strange other tiny man.

Carmen started to cry.

'Not now Carmen; be a big baby girl now like your sister,' Lucia scolded.

‘Mum! Look what you’re shouting has done now,’ Reina almost sobbed herself as she took Carmen into her arms again.

‘Carmen, baby girl, I’m sorry, mama’s just tired, that’s all, it’ll all be ok.’

Carmen didn’t say anything. She wanted to though. She wanted to say that she is almost thirteen years old and that she isn’t a baby girl or a big baby girl for that matter.

Others had gotten out of their cars now to see what was going on, but there wasn’t much to see. Still, there was panic rising in the crowd, their energies tense, unpredictable, like they were ready to tear her dad and the hulking man into shreds just to get a little bit closer to safety. The storm was rolling in and there was nothing anyone could do except evacuate or hanker down.

Reina looked down at Carmen in her arms.

‘I’m sorry Carmen, I know you’re being brave for us; mom and dad are just scared but they don’t know how to tell us that because they want to protect us.’

‘You don’t tell me anything either.’ Even Reina doesn’t take her seriously; she is too busy these days hiding her secret life from their parents. Carmen knows though. She’s seen the messages on Reina’s phone to Freya, seen them hold hands and kiss like they are boyfriend and girlfriend. She knows this is something dangerous, to be concealed, and so she does the same. She and Jilly haven’t kissed each other though, not yet, but she wants to.

‘I want to protect you too baba.’ Reina saw the same thing in Carmen that she recognised in herself; she noticed it every time Carmen and her best friend Jilly touched hands, hugged each other cheek to cheek; they both loved in a way that felt right but would bring their parents shame. She wished they could understand so she wouldn’t have to hide. So she could be a better sister to Carmen, but she was afraid for her and hiding was safer.

Reg got back in the car and the crowd had dispersed. Maybe it was the wake-up call he needed because even though it took several more hours, they made it to the motel without any more problems.

Chapter II: Motel

The family had booked two adjoining rooms. For now, Reina and Carmen could share until their older sister, Kat, and the baby arrived, maybe with her husband, Josh, maybe not. A cacophony of voices from the room next door—both elated and terrified told Reina that Kat and the baby had arrived, but Josh wasn't with them. From what Reina could gather listening through the wall, there was disagreement about who was going where and Josh wasn't madly keen on bunking up with the entire family.

'What if he hasn't made it out?' Kat sobbed. Josh's cell was out of range and their house was by the canal. The baby started crying and Kat totally lost it.

I need to get Carmen out of here before mum goes off and dad loses his temper, Reina thought. She had to protect Carmen before the whole family erupted in its own chaotic storm. She hated that she had to choose.

'What are you doing, your thumbs will fall off!' she teased Carmen who was lying on one of the beds texting furiously. 'Let's go explore.'

'I just want to see if Jilly's alright.'

'She'll message soon; the networks are totally overloaded.'

'Don't you want to set up your DVD player?' Carmen smirked a bit.

'Come on. It'll be like a little adventure and anyway, we'll be back home in a couple of days.' Reina tried to sound sure, cheery.

Carmen's phone chimed.

'It's Jilly!' her face turned grave.

'Jesus, those crazy folk have decided to stay haven't they? Idiots!'

'Hey! Don't call them that. They've stayed every storm for like a hundred generations and it's always been ok.'

'Ok, ok. Text her back then let's get out of here,' Reina said, reminding herself she needed to be chill, for Carmen's sake.

Carmen couldn't pinpoint exactly when her feelings for Jilly started to change. Boys were gross, obviously. Girls could be friends or enemies to each other, but what she felt for Jilly was fizzy. She got belly flips so crazy at the thought of Jilly that she almost started laughing.

'There's like a cinema room or something down here,' Reina said, scanning the motel guide she'd taken with them from their room. 'It says it's on the lower ground floor.'

Carmen peered over. 'I don't think that's for us.'

'So? Haven't you noticed no one cares where we go right now?' Reina said as she pushed the door ahead of them into a gloomy corridor. 'Let's go!'

It was dark and creepy looking and Carmen wasn't entirely sure she wanted to go down there. What if they got lost or some weird man was down there?

'Fine, don't come, bye!' Reina teased as she set off ahead, the lights coming on automatically as she strode on.

'Wait!' Carmen called; she'd rather go with Reina than try to find her way back to their room.

They could hear the sound of some kind of movie soundtrack coming from the movie room when they got to the end of the hall. Someone was already in there.

'Oh, hi, I'm Spencer.'

'I'm Reina. This is Carmen. We just evacuated from New Orleans,' Reina said it like it was a badge of honour like she was trying to be all big about it which made Carmen embarrassed.

'They let me down here whenever I want because we come here all the time. Every time there's a storm. My stepdad works for this company here that monitors the storms.'

'Oh cool, so you like to get to live in a hotel all the time?' Reina said.

'Not all the time and it's pretty boring actually. Plus, my stepdad is...what's the nice way of saying an arsehole?'

Reina laughed and sat down on the sofa next to Spencer. 'I think it's just arsehole. Carmen, come sit down, she's being shy.'

'No, I'm not! You're being embarrassing.'

Reina laughed again and shrugged at Spencer.

Spencer's hair was kind of wild. Neither short nor long and her face was angular, with super intense blue eyes.

'So, what are you watching down here?'

'*The Godfather*, it's a classic. I'm going to be a filmmaker.'

'Oh really? What kind of films?'

Spencer paused and wondered, should they? Should they try this out on this stranger? 'It's kind of underground, like a bit out of the mainstream, I'm not sure I should say,' Spencer said, signalling to Carmen.

'Hey! I'm not a baby. I know a lot more than anyone thinks, actually.'

'Oh really?' Reina teased.

Carmen snapped. 'Yes. I do, actually. Like that you are totally flirting with her right now and she's flirting back.'

Reina went bright red, a sinking feeling in her stomach that she'd been caught.

'I'm gonna make films about changing the world for people like us.'

'People like us?' Carmen was curious now.

'Yeah. Like, you don't fit in, do you? Like your sister?'

'She fits in,' Carmen snorted with a jealous streak.

'But we're both hiding,' Reina said almost in a whisper.

'I'm hiding too,' Spencer said, still unsure to tell them the whole truth but unwilling to break the euphoric feeling of having her gender affirmed even though it was different to the one her parents and everyone in her real life knew her as. 'You wanna watch *The Godfather* with me? I can start it from the beginning and I've got popcorn.'

'Sure,' Carmen smiled and hopped over the back of the sofa, landing in between them.

Reina looked at Carmen, felt the ease that came over her and realised she needed to let her little sister be in charge and make decisions, for the first time really seeing her. In a space where no one belonged, three girls could finally be themselves for a while.

Notes from the Research Diary: Themes and Linked Literature

Gender Roles

On the one hand, we know a lot about gender roles post-disaster through the wealth of research about women in disaster (Enarson & Morrow, 1998; Buvinic et al., 1999; Delaney & Shrader, 2000; Enarson & Merles, 2004; Peek & Fothergill, 2008; Bradshaw, 2015). However, on the other hand, we know very little about more diverse gendered experiences during disasters, including gender and sexual identities and processes which means the needs and interests of those outside heteronormativity may be overlooked (Bradshaw & Fordham, 2013a; Overton, 2017). A small body of research has suggested a myriad of 'missing' groups in disaster research from sex workers (Valdés, 2009, p. 19), to lesbians (D'Ooge, 2008) to disabled women (Ishrad et al., 2012) to elderly women (Mitchell & Karr, 2014), to men and masculinities (Pearson, 2000; Enarson & Pease, 2016) to young women (Rashid & Michaud, 2000; Doppler, 2009; Overton, 2014, 2017). Within larger research projects, some have mentioned that young women have different feelings of fear of the future, health, marriage, family and intimate relationships than adult women and therefore will possess different capabilities and vulnerabilities (Anderson & Woodrow, 1998; Bradshaw & Linneker, 2009, p. 82; Saad, 2009, pp. 93 and148; Saito, 2012).

Queer-identifying teenage girls and young women are among those on the margins. Adult decisions about young people's lives bring about worry around fulfilling their expected gendered roles (Overton, 2017). Teenage girls, when they are presented in research as children, are mostly constructed as genderless (Fothergill & Peek, 2004; The Coalition for Adolescent Girls, 2012; Plan International, 2013). However, the findings in my research showed that girls experience disaster differently from adults, including adult women, and that this is amplified further for queer-identifying girls (Overton, 2017). The storied part of this paper illustrates how gender and sexuality are core identities for the two girls—Reina and Carmen—demonstrating a very different experience of evacuation and displacement to the adults.

Age and Stage in Life Course

Age and stage in life course have only been recognised within disaster risk reduction (DRR) since the 2010s. This shift in thinking was led by the International Development literature translating to the inclusion of girls in DRR (Levine et al., 2009; Coalition for Adolescent Girls, 2012). Adolescent girls and disasters then became the focus of Plan International's, 2013 'Because I am a Girl' annual campaign, further raising their profile as a separate group from adult women and children. However, whilst girls entered discourse (see Tanner, 2012; Plan, 2013), gender and disaster scholars were critical of how transformative this inclusion would be for gender norms (Bradshaw, 2013). Rather than 'empowering' girls, like other gender concerns that are now included in gender-responsive DRR and development initiatives (Bradshaw & Fordham 2013a, 2013b; Bradshaw, 2015; Chant, 2016), girls inclusion may also reinforce preconceived notions about women and girls as caregivers and service providers to their families rather than actually empower girls by positively transforming their lives (Bradshaw, 2013).

Gender, Sexuality and Youth

Sexuality is the main underlying theme threaded through the story. Reina's DVD collection of *The L Word* is a symbol of the importance to keep her queerness close to her over 'necessary' items. To Reina, these were necessary and reflected findings by Kroll-Smith and Gunter (1998) about how subjective disaster experiences are in real life. As a queer person who is not 'out' to her family, the DVD collection aided her in her sense of belonging. The theme of hiding is also linked to sexualities and teenagers' fears about family acceptance (D'Augelli et al., 2009, 2020). That Reina, Carmen and Spencer feel like they have to hide their sexuality from their 'real' lives is an important theme in the findings of my research (Overton, 2017). Parents finding out about their queer sexuality is a particular worry because of the fear that parents won't understand or will disown them, sometimes due to how parents talk about queer folk in a homophobic way and how they construct relationships as man/woman only. This is one of the many reasons why 'safe spaces' outside of family unity are important. For example, in the story, meeting Spencer, who understood their sexualities, enabled Reina and Carmen to find a safe space, regardless of the hurricane and parental challenges.

Conclusion

Disasters disrupt daily life, and this means those affected can find themselves in new and unfamiliar spaces, especially young people who have little control over what happens during an evacuation, for example. Existing stresses and tensions may amplify and make things more fraught—as in the story. Reg hasn't been made into an angry man because of Hurricane Katrina, but it certainly exacerbated it, and Carmen, being the older sister and feeling a sense of duty, wanted to protect her sister—something she usually does in 'normal' times. The journey to the hotel changes the environment for the two sisters who can then bond over their shared queer identity that they usually keep hidden from one another by making a friend 'like them.' We need, in disaster research, to consider a wider range of relationships, especially when taking a queering disaster approach. This will help us to better understand capacities and vulnerabilities to provide appropriate support to diverse groups of people and understand how power relations operate within families, both of origin and chosen as well as friendships and communities that are often vital to queer people. People themselves construct disasters differently than official legislators (Kroll-Smith & Gunter, 1998), but the research on how disasters are perceived by those who live through them has been limited and this short story provides a window into why it is important to listen. How will we know what people need, especially more marginalised groups, if we do not ask? If we do not listen?

Conflict of Interest The research drawn on in this vignette was funded by The Social Policy Research Centre at Middlesex University as a PhD scholarship, and ethics granted from Middlesex University. The author has no other conflicts of interest to declare.

References

Anderson, M. B., & Woodrow, P. J. (1998). *Rising from the ashes: Development strategies in times of disaster*. Westview Press.

Bradshaw, S. (2015). Engendering development and disasters. *Disasters, 39*(s1), 54–75.

Bradshaw, S., & Fordham, M. (2013a). *Gender, development and disasters*. Edward Elgar Publishing.

Bradshaw, S., & Fordham, M. (2013b). *Women and girls in disasters*. Department for International Development (DFID).

Bradshaw, S., & Linneker, B. (2009). Gender perspectives on disaster reconstruction in Nicaragua: Reconstructing roles and relations? In E. Enarson & P. Chakrabarti (Eds.), *Women, gender and disaster: Global issues and initiatives*. SAGE.

Buvinic, M., Vega, G., Bertrand, M., Urban, A. M., & Truitt Nakata, G. (1999). *Hurricane Mitch: Women's needs and contributions*. Inter-American Development Bank.

Chant, S. (2016). Galvanising girl for development? Critiquing the shift from "smart" to "smarter economics". *Progress in Development Studies, 16*(4), 314–328.

Coalition for Adolescent Girls. (2012). *Missing the emergency: Shifting the paradigm for relief to adolescent girls*. The Coalition for Adolescent Girls. Retrieved December 3, 2024, from http://www.coalitionforadolescentgirls.org/wp-content/uploads/2012/05/2012-05-23-Missing-the-Emergency-FINAL1.pdf

D'Augelli, A. R., Grossman, A. H., & Starks, M. T. (2009). Gender atypicality and sexual orientation development among lesbian, gay, and bisexual youth: Prevalence, sex differences, and parental responses. In R. M. Mathy & J. Drescher (Eds.), *Childhood gender nonconformity and the development of adult homosexuality* (1st ed., pp. 121–144). Routledge.

D'Augelli, A. R., Grossman, A. H., & Starks, M. T. (2020). Gender atypicality and sexual orientation development among lesbian, gay, and bisexual youth: Prevalence, sex differences, and parental responses. In *Childhood Gender nonconformity and the development of adult homosexuality* (pp. 121–144). Routledge.

D'Ooge, C. (2008). Queer Katrina: Gender and sexual orientation matters in the aftermath of the disaster. In B. Willinger (Ed.), *Katrina and the women of New Orleans*. Newcomb College Centre for Research on Women.

Delaney, P., & Shrader, E. (2000). Gender and post-disaster reconstruction: The case of Hurricane Mitch in Honduras and Nicaragua. *Decision review draft*. LCSPG/LAC Gender Team, The World Bank.

Doppler, J. (2009). *Gender and the Tsunami: Vulnerability and coping of Sinhalese widows and widowers on the South-West Coast of Sri Lanka*. Master's dissertation for Wien University.

Enarson, E., & Merles, L. (2004). International perspectives on gender and disaster: Differences and possibilities. *International Journal of Sociology and Social Policy, 24*(10/11), 49–93.

Enarson, E., & Morrow, B. (Eds.). (1998). *The gendered terrain of disaster: Through women's eyes*. Prager.

Enarson, E., & Pease, B. (Eds.). (2016). *Men, masculinities and disaster*. Routledge.

Fothergill, A., & Peek, L. A. (2004). Poverty and disasters in the United States: A review of recent sociological findings. *Natural Hazards, 32*, 89–110.

Halperin, D. M. (1995). *Saint Foucault: Towards a gay hagiography*. Oxford University Press.

hooks, b. (1991). Theory as liberation practice. *Yale Journal of Law and Feminism, 4*, 1–12.

Ishrad, H., Mumtaz, Z., & Levay, A. (2012). Long-term gendered consequences of permanent disabilities caused by the 2005 Pakistan earthquake. *Disasters, 36*(3), 452–464.

Kroll-Smith, S., & Gunter, V. J. (1998). Legislators, interpreters, and disasters: The importance of how as well as what is a disaster. In E. L. Quarantelli (Ed.), *What is a Disaster? A dozen perspectives on the question* (pp. 161–177). Routledge.

Levine, R., Lloyd, C. B., Greene, M., & Grown, C. (2009). Girls count: A global investment and action agenda. *Girls Count Report*. Coalition for Adolescent Girls, Retrieved December 3, 2024, from http://www.coalitionforadolescentgirls.org/wp-content/uploads/2012/03/Girls_Count_2009.pdf

Mitchell, D., & Karr, V. (2014). *Crises, conflict and disability: Ensuring equality*. Routledge.

Mutch, C., & Marlowe, J. (2013). Lessons from disaster: The power and place of story. *Disaster Prevention and Management, 22*(5), 385–394.

Nussbaum, M. C. (1999). *Sex and social justice*. Oxford University Press.

Overton, L. R. A. (2014). From vulnerability to resilience: An exploration of gender performance art and how it has enabled young women's empowerment in post-hurricane New Orleans. *Procedia Economics and Finance, 18*, 214–221.

Overton, L. (2017) *'Girls Interrupted': Young women 'growing up' in post-Katrina New Orleans: An exploration of the intersections of genders, sexualities and youth.* Doctoral dissertation, Middlesex University.

Oxfam. (2024). Queer Joy: What it is and why we need more of it. Retrieved November 21, 2024, from https://www.oxfam.org/en/queer-joy

Pearson, R. (2000). Which men, why now? Reflections on men and development. *IDS Bulletin, 31*(2), 42–48.

Peek, L., & Fothergill, A. (2008). Displacement, gender, and the challenges of parenting after Hurricane Katrina. *NWSA journal, 20*(3), 69–105.

Phillips, R., & Kara, H. (2021). *Creative writing for social research: A practical guide.* Policy Press.

Plan International. (2013). *The state of the world's girls 2013.* In *Double Jeopardy: Adolescent girls and disasters.* Plan International Report. Retrieved December 3, 2024, from The State of the World's Girls 2013: Adolescent Girls and Disasters | Plan International

Rashid, S. F., & Michaud, S. (2000). Female adolescents and their sexuality: Notions of honour, shame, purity and pollution during the floods. *Disasters, 24*(1), 54–70.

Saad, S. G. (2009). Environmental management and disaster mitigation: Middle Eastern gender perspective. In E. Enarson & P. Dhar Chakrabarti (Eds.), *Women, gender and disaster: Global issues and initiatives.* SAGE.

Saito, F. (2012). Women and the 2011 East Japan disaster. *Gender and Development, 20*(2), 265–279.

Tanner, T. (2012). Youth, gender and climate change: Moving from impacts to agency. In M. Perisic, M. Komarecki, & A. Minujin (Eds.), *Adolescent girls-cornerstone of society: Building evidence and policies for inclusive societies.* The New School supported by UNICEF, pp. 1550163. Retrieved December 3, 2024, from P60151-AdolescentGirls_print.pdf

Valdés, H. (2009). A gender perspective on disaster risk reduction. In E. Enarson & P. Chakrabarti (Eds.), *Women, gender and disaster: Global issues and initiatives.* SAGE.

Index[1]

[1] Note: Page numbers followed by 'n' refer to notes.

D. Dominey-Howes et al. (eds.), *Queering Disasters, Climate Change and Humanitarian Crises*, Sustainable Development Goals Series, https://doi.org/10.1007/978-981-96-3857-4

E

Kolibri 360 GmbH
Ferdinand-Jühlke-Straße 7
99095 Erfurt, Deutschland
produktsicherheit@kolibri360.de